*Springer Series in Molecular Biology*

---

*Series Editor: Alexander Rich*

Springer Series in Molecular Biology

*Series Editor:* Alexander Rich

---

Yeast Genetics
*Fundamental and Applied Aspects*
J.F.T. Spencer, Dorothy M. Spencer, A.R.W. Smith, eds.

Myxobacteria
*Development and Cell Interactions*
Eugene Rosenberg, ed.

DNA Methylation
*Biochemistry and Biological Significance*
Aharon Razin, Howard Cedar, Arthur D. Riggs, eds.

# Myxobacteria

## *Development and Cell Interactions*

Edited by
Eugene Rosenberg

With Contributions by

R. P. Burchard, M. Dworkin, M. Inouye, S. Inouye, D. Kaiser, H. Kleinig, D. Ralt, H. Reichenbach, E. Rosenberg, L. J. Shimkets, M. Varon, D. White, T. Yee, A. Zahavi, D. R. Zusman

With 68 Figures

Springer-Verlag
New York Berlin Heidelberg Tokyo

Eugene Rosenberg
Department of Microbiology
George S. Wise Faculty of Life Sciences
Tel-Aviv University
Ramat-Aviv
Israel

*Series Editor:*
Alexander Rich
Department of Biology
Massachusetts Institute of Technology
Cambridge, Massachusetts 02139 U.S.A.

Library of Congress Cataloging in Publication Data
Main entry under title:
Myxobacteria, development and cell interactions.
(Springer series in molecular biology)
Bibliography: p.
1. Myxobacterales. 2. Cell interaction.
I. Rosenberg, Eugene. II. Series.
QR81.M98 1984 589.9′8 84-1428


Softcover reprint of the hardcover 1st edition 1984



Typeset by University Graphics, Inc., Atlantic Highlands, New Jersey.

9 8 7 6 5 4 3 2 1

ISBN-13:978-1-4613-8282-9 e-ISBN-13:978-1-4613-8280-5
DOI: 10.1007/978-1-4613-8280-5

# Series Preface

During the past few decades we have witnessed an era of remarkable growth in the field of molecular biology. In 1950 very little was known of the chemical constitution of biological systems, the manner in which information was transmitted from one organism to another, or the extent to which the chemical basis of life is unified. The picture today is dramatically different. We have an almost bewildering variety of information detailing many different aspects of life at the molecular level. These great advances have brought with them some breath-taking insights into the molecular mechanisms used by nature for replicating, distributing and modifying biological information. We have learned a great deal about the chemical and physical nature of the macromolecular nucleic acids and proteins, and the manner in which carbohydrates, lipids and smaller molecules work together to provide the molecular setting of living systems. It might be said that these few decades have replaced a near vacuum of information with a very large surplus.

It is in the context of this flood of information that this series of monographs on molecular biology has been organized. The idea is to bring together in one place, between the covers of one book, a concise assessment of the state of the subject in a well-defined field. This will enable the reader to get a sense of historical perspective—what is known about the field today—and a description of the frontiers of research where our knowledge is increasing steadily. These monographs are designed to educate, perhaps to entertain, certainly to provide perspective on the growth and development of a field of science which has now come to occupy a central place in all biological studies.

The information in this series has value in several perspectives. It provides for a growth in our fundamental understanding of nature and the manner in which living processes utilize chemical materials to carry out a variety of activities. This information is also used in more applied areas. It promises to have a significant impact in the biomedical field where an understanding of disease processes at the molecular level may be the capstone which ultimately holds together the arch of clincial research and medical therapy. More recently in the field of biotechnology, there is another type of growth in which this science can be used with many practical consequences and benefit in a variety of fields ranging from agriculture and chemical manufacture to the production of scarce biological compounds for a variety of applications.

This field of science is young in years, but it has already become a mature discipline. These monographs are meant to clarify segments of this field for the readers.

Cambridge, Massachusetts

Alexander Rich
*Series Editor*

# Preface

This book celebrates ten golden years of myxobacterial research. Since the summer of 1974, a small group of microbiologists have met annually, two consecutive years in the United States and one in Europe, to discuss their research on myxobacteria. During these ten years there has been a dramatic change in myxobacteriology, both with regard to the types of problems examined and the approaches employed. Until the early 1970s most studies on myxobacteria were descriptive in nature and were concerned with dispersed growing cultures. During the last ten years the emphasis has shifted to the developmental cycle on solid media and the experimental approach to modern techniques in bacterial genetics and molecular biology. The result is that *Myxococcus xanthus* has become the *Escherichia coli* of Developmental Biology. Thus, current myxobacterial research has become an exciting subject not only for microbiologists, but developmental biologists, in general.

This is the first book devoted entirely to the myxobacteria. Each chapter has been written by an expert in the field. Except for an occasional suggestion by the Editor, the contents of the chapters were left to the descretion of the individual authors. The aim of the book is to bring concepts and methodology in myxobacterial research to a broad audience of investigators and teachers of microbiology and developmental biology. Toward this end, most chapters begin with an introduction in which the topic is discussed within a general biological framework. This is followed by a presentation of the "state-of-the-art" of that topic with specific regard to the myxobacteria. The final chapter contains an overview of myxobacterial research. Appendices are included which contain some of the most common methods for studying myxobacteria.

Eugene Rosenberg

# Preface

This book celebrates ten golden years of myxobacterial research. Since the summer of 1974, a small group of microbiologists have met annually, two consecutive years in the United States and one in Europe, to discuss their research on myxobacteria. During these ten years there has been a dramatic change in myxobacteriology, both with regard to the types of problems examined and the approaches employed. Until the early 1970s most studies on myxobacteria were descriptive in nature and were concerned with [illegible] cultures. During the last ten years the emphasis has shifted to the developmental cycle on solid media and the examination of specific [illegible] [illegible] [illegible] [illegible] ogists, but developmental biologists in general.

This is the first book devoted entirely to the myxobacteria. Each chapter has been written by an expert in the field. Except for an occasional suggestion by the Editor, the contents of the chapters were left to the discretion of the individual authors. The aim of the book is to bring concepts and methodology in myxobacterial research to a broad audience of investigators and teachers of microbiology and developmental biology. Toward this end, most chapters begin with an introduction in which the topic is discussed within a general biological framework. This is followed by a presentation of the "state-of-the-art" of that topic with specific regard to the myxobacteria. The final chapter contains an overview of myxobacterial research. Appendices are included which contain some of the most common methods for studying myxobacteria.

Eugene Rosenberg

# Contents

## 4. Nutrition, Metabolism, and the Initiation of Development
*Lawrence J. Shimkets*



## 5. Antibiotics and Lytic Enzymes
*Eugene Rosenberg and Mazal Varon*



## 6. Pigments of Myxobacteria
*Hans Reichenbach and Hans Kleinig*



## 7. Gliding Motility and Taxes
*Robert P. Burchard*



## 8. Genetics of Myxobacteria
*Dale Kaiser*

# Contributors

ROBERT P. BURCHARD Department of Biological Sciences, University of Maryland Baltimore County, Catonsville, Maryland 21228 U.S.A.

MARTIN DWORKIN Department of Microbiology, University of Minnesota, Minneapolis, Minnesota 55455 U.S.A.

MASAYORI INOUYE Department of Biochemistry, State University of New York at Stony Brook, Stony Brook, New York 11794 U.S.A.

SUMIKO INOUYE Department of Biochemistry, State University of New York at Stony Brook, Stony Brook, New York 11794 U.S.A.

DALE KAISER Department of Biochemistry, Stanford University, Stanford, California 94305 U.S.A.

HANS KLEINIG Institut für Biologie II der Universität, Schänzlestr. 9-11, D-7800 Freiburg, Federal Republic of Germany

DINA RALT Institute for Nature Conservation Research, The George S. Wise Faculty of Life Sciences, Tel Aviv University, Ramat Aviv, Israel

HANS REICHENBACH GBF, Gesellschaft für Biotechnologische Forschung, Department of Microbiology, Mascheroder Weg 1, D-3300 Braunschweig, Federal Republic of Germany

EUGENE ROSENBERG Department of Microbiology, The George S. Wise Faculty of Life Sciences, Tel Aviv University, Ramat Aviv, Israel

LAWRENCE J. SHIMKETS Department of Microbiology, University of Georgia, Athens, Georgia 30602 U.S.A.

MAZAL VARON Department of Microbiology, The George S. Wise Faculty of Life Sciences, Tel Aviv University, Ramat Aviv, Israel

DAVID WHITE Department of Biology, Program in Microbiology, Indiana University, Bloomington, Indiana 47405 U.S.A.

THOMAS YEE Department of Biochemistry, State University of New York at Stony Brook, Stony Brook, New York 11794 U.S.A.

AMOTZ ZAHAVI Institute for Nature Conservation Research, The George S. Wise Faculty of Life Sciences, Tel Aviv University, Ramat Aviv, Israel

DAVID R. ZUSMAN Department of Microbiology and Immunology, University of California, Berkeley, California 94720 U.S.A.

# 1

# Myxobacteria: A Most Peculiar Group of Social Prokaryotes

Hans Reichenbach*

We are approaching the centenary of the discovery of the myxobacteria as a group of organisms with an unusual and, at the time of their discovery, quite incredible life cycle (Thaxter 1892). This may thus be a good occasion to summarize our knowledge, patchy as it is, about these fascinating bacteria. In the last few years a growing number of microbiologists, molecular biologists, and microbial geneticists have become attracted by the myxobacteria, because these peculiar prokaryotes provide a unique system for studying the molecular basis of development and social interactions. Most of the morphogenetic studies and all of the genetics have been carried out on one particular organism, *Myxococcus xanthus,* so that this book will deal mainly with that species. In this chapter, however, I shall discuss the myxobacteria in a more general framework.

The myxobacteria are typical Gram-negative bacteria. In comparison to other Gram-negative bacteria, they show three outstanding peculiarities: (1) they move by gliding; (2) they have a spectacular morphogenetic potential; and (3) they are "social" organisms. As will be seen, these properties are not independent from one another and may find their explanation in the ecology of the myxobacteria.

## Gliding Motility

Gliding motility (i.e., movement along interfaces by creeping or sliding) is a characteristic that the myxobacteria share with a considerable number of other

*GBF, Gesellschaft für Biotechnologische Forschung, Department of Microbiology, Mascheroder Weg 1, D-3300 Braunschweig, Federal Republic of Germany.

Gram-negative bacteria: *Cytophaga*-like bacteria (CLB), *Lysobacter,* many cyanobacteria, several filamentous nonphototrophic bacteria such as *Vitreoscilla, Beggiatoa, Simonsiella, Leucothrix, Thiothrix, Saprospira, Herpetosiphon,* and filamentous phototrophs from the chlorobacteria branch *(Chloroflexus).* These gliders are definitely not all related to one another. The mechanism of gliding, which is discussed in Chapters 7, 8, and 11, is still largely enigmatic. Gliding is always connected with the extrusion of slimy material (Fig. 1.2) consisting, apparently, of polysaccharides (Sutherland and Thomson 1975; Sutherland 1979). Because of this gliding motility, the colonies of myxobacteria tend to spread, particularly on poor growth substrates. On

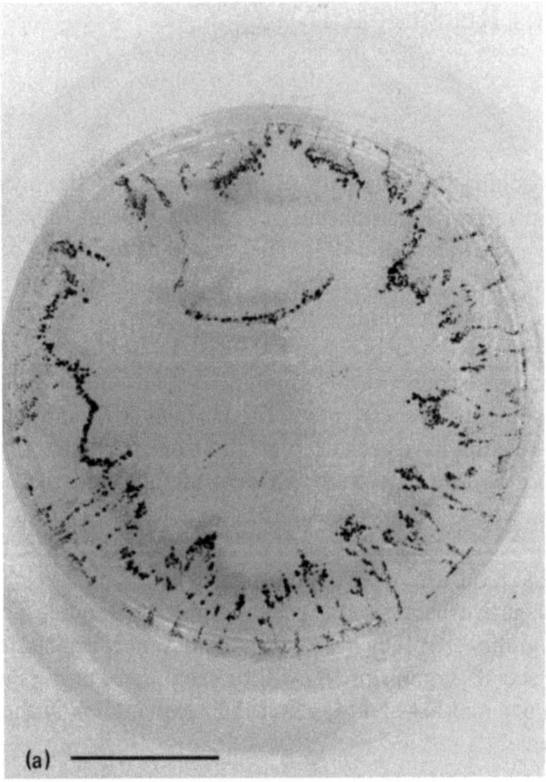


**Figure 1.1.** Swarm colonies of myxobacteria. (a) *Cystobacter fuscus:* Petri dish with fruiting swarm (bar = 25 mm). (b) *Myxococcus fulvus* on agar (bar = 220 μm). (c) *Cystobacter violaceus:* swarm with network of veins stained by the dark violet pigment characteristic for the organism (bar = 20 mm). (d) *Sorangium cellulosum* on yeast agar; on this substrate the organism penetrates into the agar and forms a swarm with radiating veins and feathery lobes (bar = 2.5 mm). (3) *Nannocystis exedens* corroding the agar (bar = 9.2 mm). (f) *Polyangium* sp. with radiating paths etched into the agar (bar = 5.6 mm).

agar plates, we usually observe large, thin, and delicate structures for which the term "swarm" seems quite appropriate (Figs. 1.1–1.3). In contrast to the swimming of flagellated bacteria, gliding is also possible in relatively dry environments.

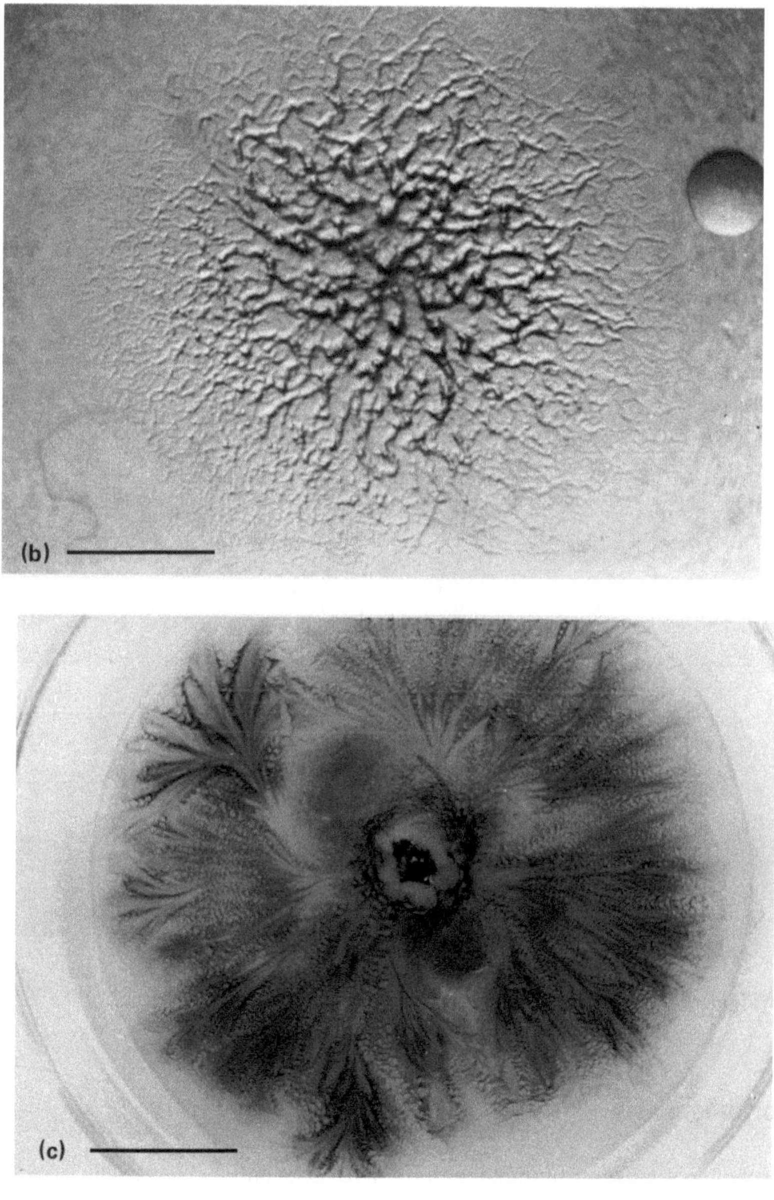


**Figure 1.1.** (Continued.)

**Figure 1.1.** (Continued.)

## Morphogenesis

Morphogenetic events in myxobacteria take place on two levels. The most spectacular single feature displayed by these bacteria is their ability to build fruiting bodies in a cooperative action involving hundreds of thousands of individual

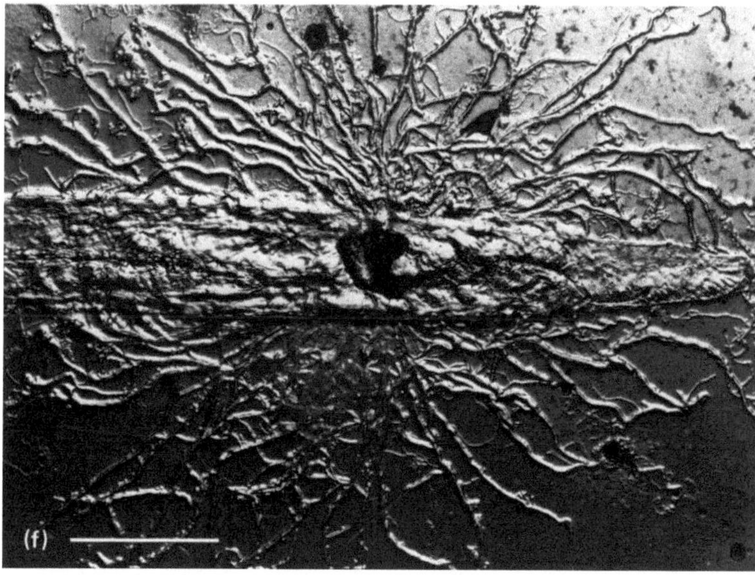


**Figure 1.1.** (Continued.)

cells (cooperative morphogenesis; Fig. 1.4; see also Figs. 1.8–1.14). Inside the maturing fruiting body the vegetative rod cells change their shapes and structures more or less extensively (cellular morphogenesis; Figs. 1.5–1.7), becoming resting cells, that is, the so-called myxospores. These two developmental processes have received considerable attention as model systems for the study of regulatory mechanisms in prokaryotic morphogenesis.

## Myxospore Formation

The resting cells of the myxobacteria, the myxospores, are fundamentally different from the bacterial endospores in their structure, physiologic properties, and the way in which they are formed. Rather, the myxospore is comparable to the microcyst of *Sporocytophaga myxococcoides,* which organism, contrary to earlier taxonomic schemes, is not related to the myxobacteria, and the cysts of *Azotobacter*. Indeed myxospores were called microcysts for some time. The degree of morphologic differentiation between vegetative cell and myxospore varies considerably with different myxobacteria. In all cases the entire vegetative cell gradually converts itself into one myxospore, often changing its shape quite noticeably in the process (Figs. 1.6 and 1.7). The most dramatic shape change is found among the members of the suborder Cystobacterineae. Here the myxospores are always much shorter and fatter than the vegetative

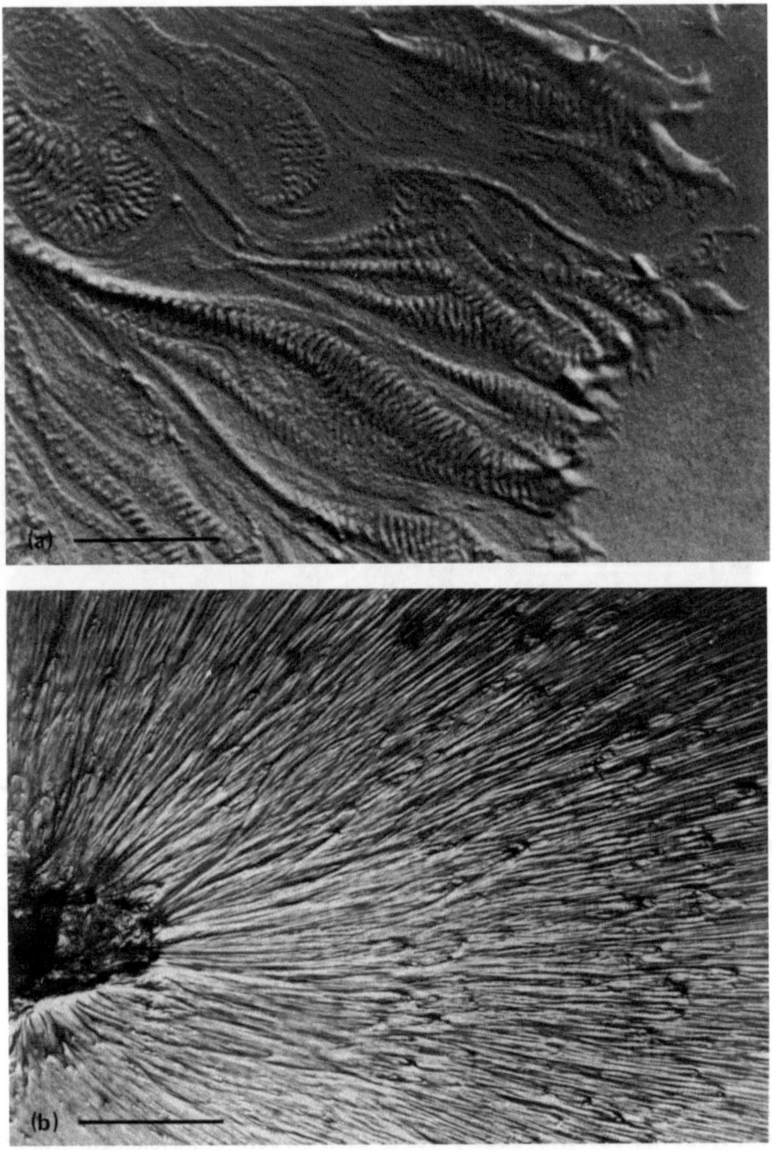


**Figure 1.2.** Myxobacterial swarms: edge and structural details. (a) *Stigmatella erecta:* swarm edge with the typical flamelike protrusions and trails of oscillating waves. Wild dissecting microscope (bar = 110 μm). (b) *Archangium serpens:* swarm with finely plicated slime sheet (bar = 2.5 mm). (c) *Stigmatella aurantiaca:* swarm edge (bar = 290 μm). (d) *Myxococcus fulvus:* in chamber culture on a thin agar layer, slime trails become visible (bar = 45 μm). (e) *Nannocystis exedens* in crude culture; swarm edge consists of meandering massive ridge of cells (bar = 275 μm). (f) *Polyangium* sp. in crude culture; marginal ridge disintegrated into small portions that continue to migrate away from the swarm center thereby etching paths into the agar surface (bar = 65 μm).

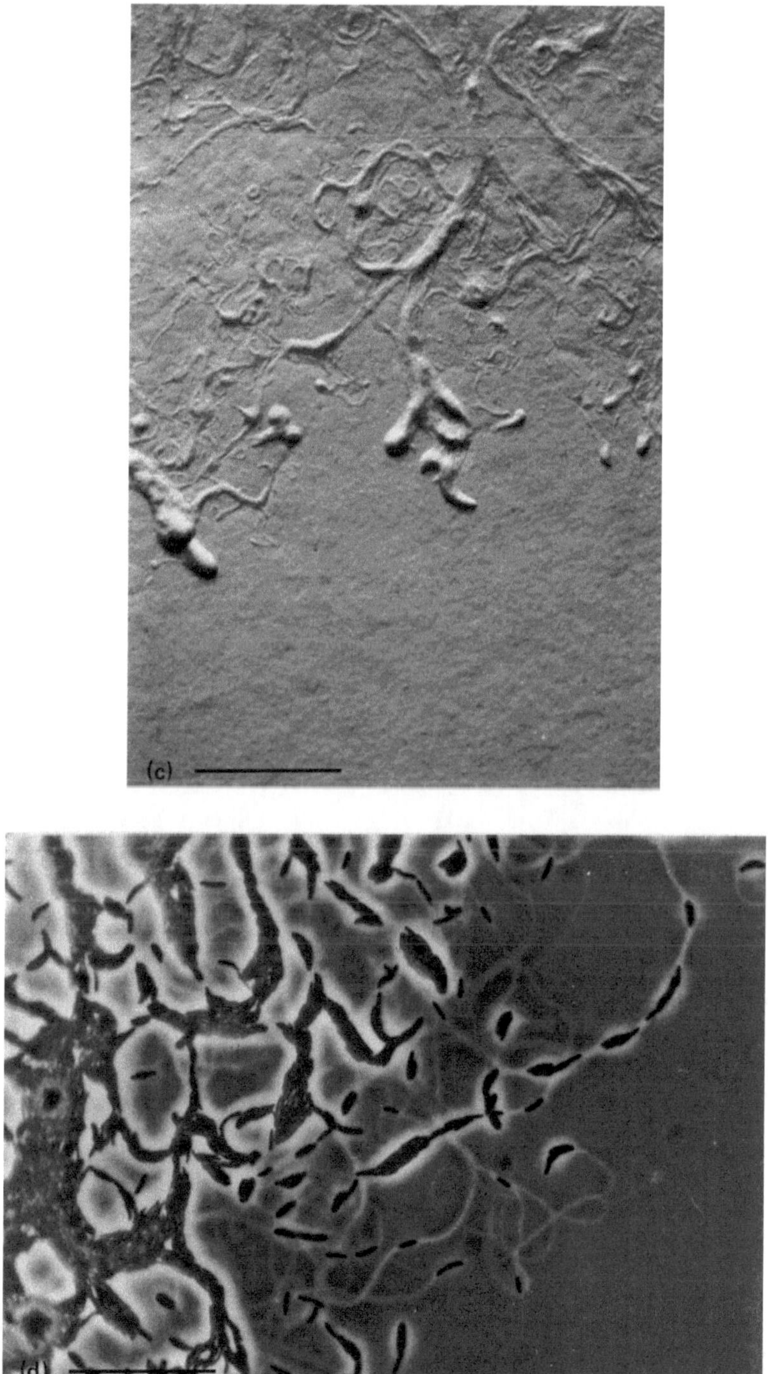


Figure 1.2. (Continued.)

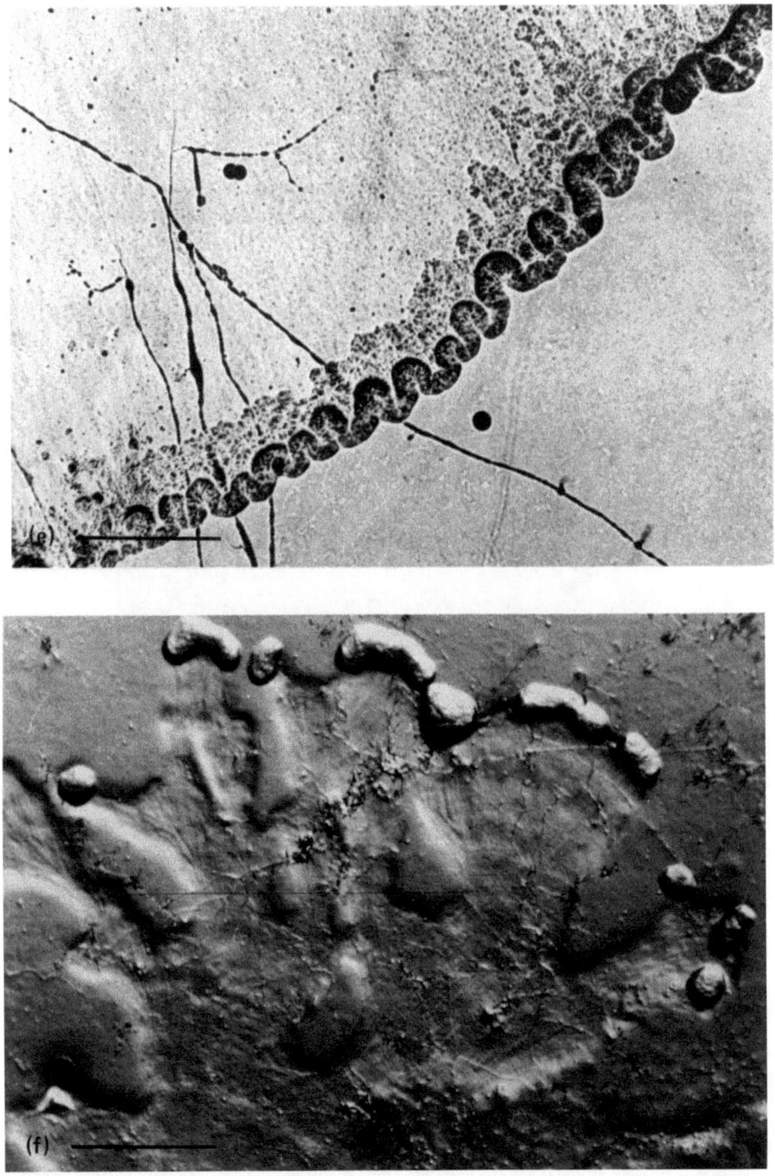


**Figure 1.2.** (Continued.)

cell; often they are spheric or nearly so. They are surrounded by a more or less heavy capsule and when mature, appear bright in phase contrast. The myxospores of the Sorangineae are not too different in shape and size from vegetative cells; they do not seem to be surrounded by a capsule, but sometimes they are bright in phase contrast.

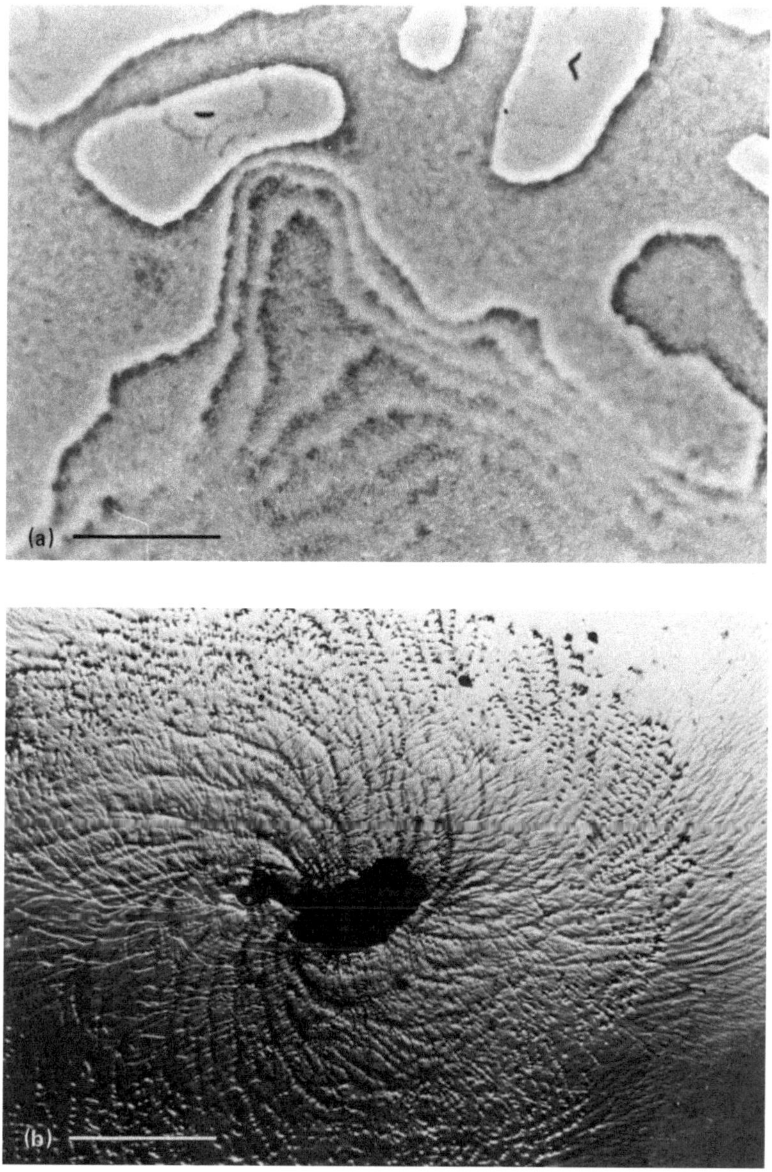


**Figure 1.3.** Myxobacterial swarms: structural details. (a) *Myxococcus xanthus* in chamber culture; on top of the marginal, still perforated monolayer of cells, series of additional layers developing (bar = 45 μm). (b) *Archangium serpens:* system of ridges studded with tiny fruiting bodies spirals out of the inoculation center (bar = 5.2 mm). (c) *Cystobacter fuscus:* at swarm edge large veins fan out (bar = 70 μm). (d) *Nannocystis exedens:* deep corrosion of the agar plate is typical for this organism (bar = 1.3 mm).

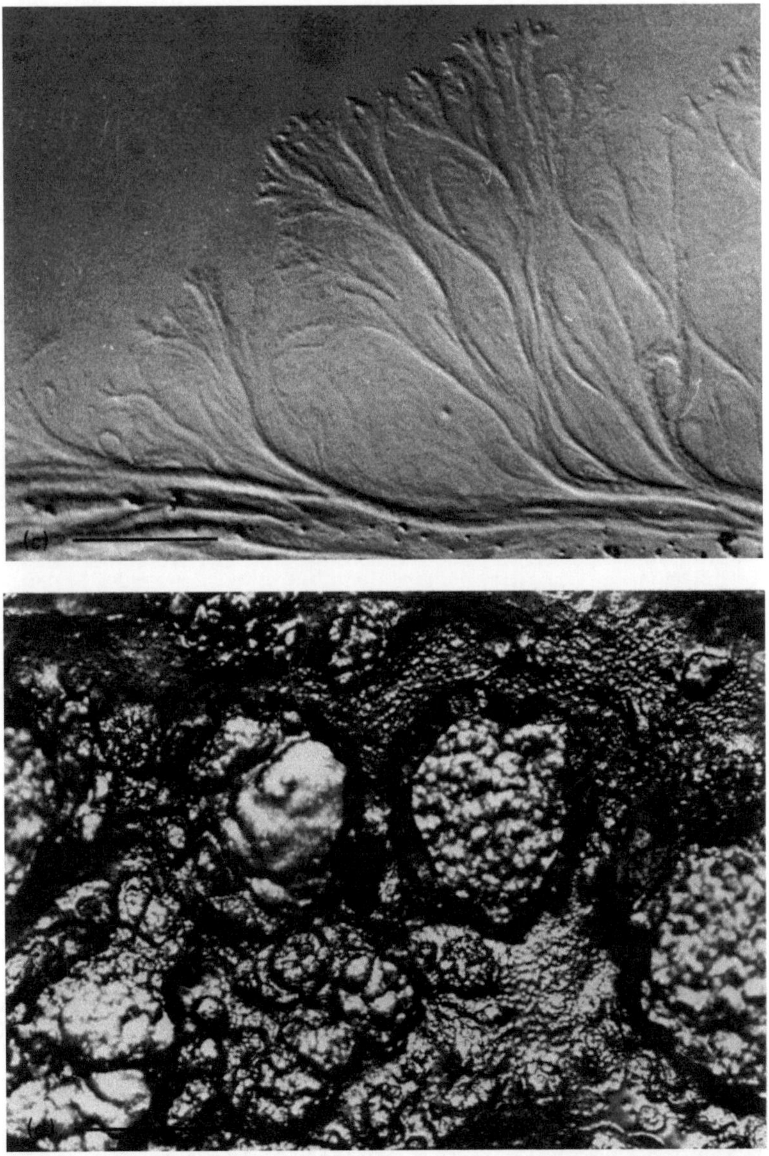

**Figure 1.3.** (Continued.)

An important property of myxospores is their desiccation resistance; vegetative cells always die as soon as they become dry. Resistance to desiccation is an ecologically meaningful property, because myxobacteria often live in environments that periodically become dry during either rainless or freezing seasons. Although it is so far not possible to preserve vegetative myxobacteria reli-

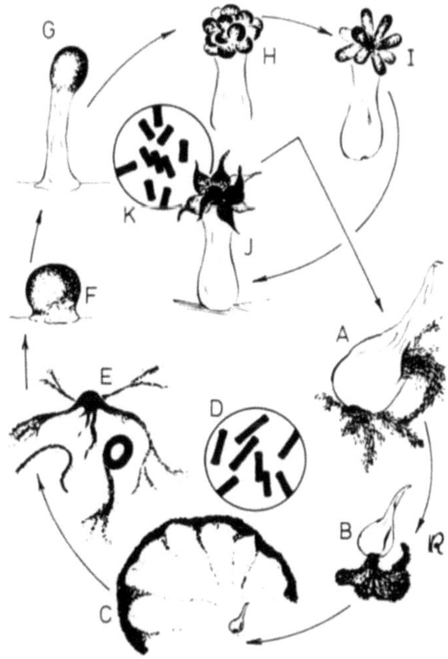


**Figure 1.4.** Cooperative morphogenesis in myxobacteria. Schematic presentation of fruiting body formation in *Chondromyces apiculatus.* D, Vegetative cells; K, myxospores from a sporangiole.

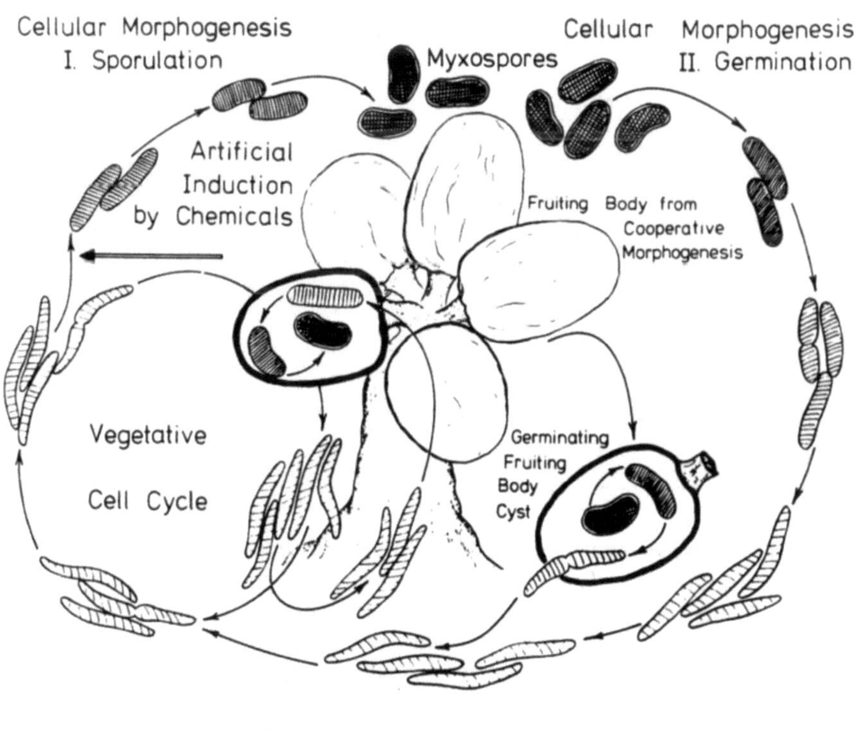


**Figure 1.5.** Cellular morphogenesis in myxobacteria. Schematic presentation of natural and experimentally induced myxospore formation in *Stigmatella aurantiaca.* Cells and sporangioles are not drawn to scale.

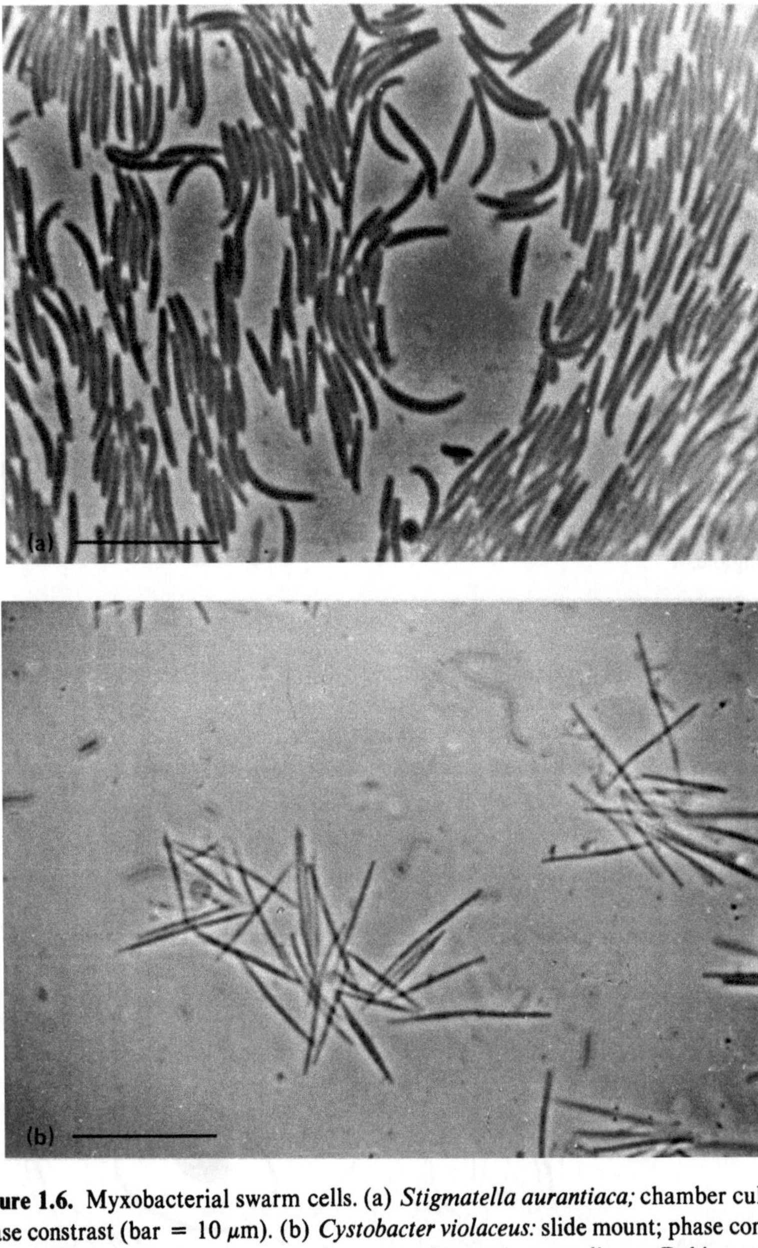


**Figure 1.6.** Myxobacterial swarm cells. (a) *Stigmatella aurantiaca;* chamber culture; phase constrast (bar = 10 μm). (b) *Cystobacter violaceus:* slide mount; phase contrast (bar = 12 μm). (c) *Cystobacter violaceus:* nuclear stain according to Robinow; most rods contain two nucleoids (bar = 21 μm). (d) *Myxococcus xanthus:* germinated myxospores; an occasional branched cell and empty spore capsules are recognizable. Stained with I-KI (bar = 12 μm). (e) *Chondromyces apiculatus:* chamber culture; phase contrast (bar = 11 μm). (f) *Nannocystis exedens:* chamber culture; phase contrast (bar = 11 μm).

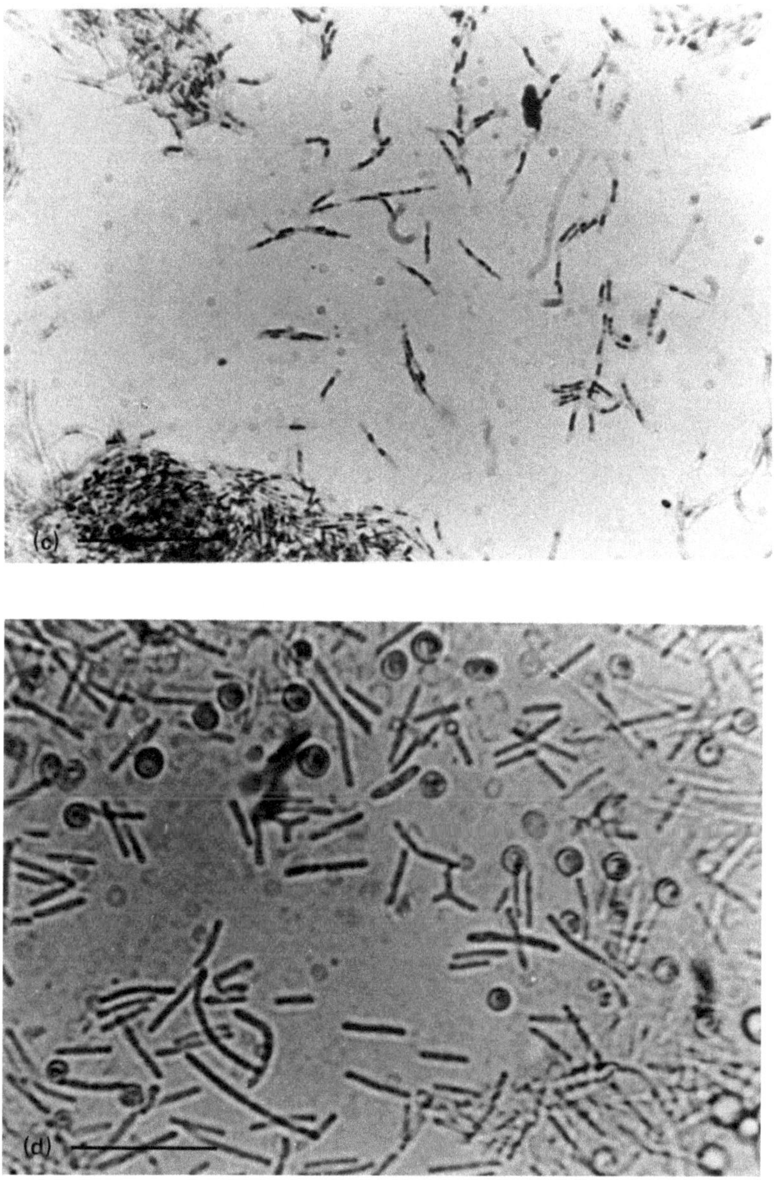


**Figure 1.6.** (Continued.)

ably by lyophilization or any other desiccation procedure, good results are easily obtained with myxospores. Myxospores within fruiting bodies that have been dried on filter paper in the laboratory may survive 6–16 yr, and this survival time seems to be surpassed by myxospores dried in their natural soil habitat. At least the more compact myxospores of the Cystobacterineae show also

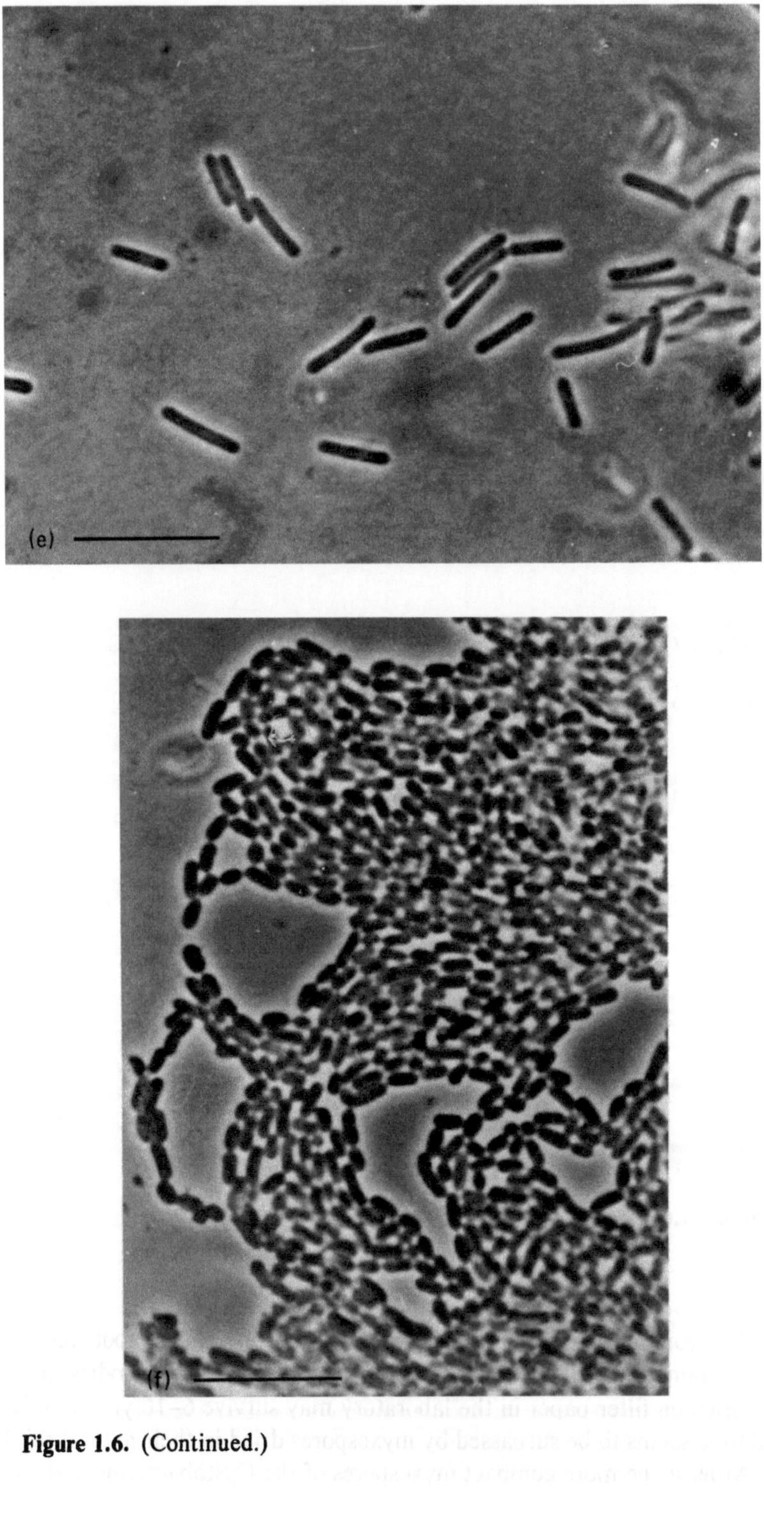


**Figure 1.6.** (Continued.)

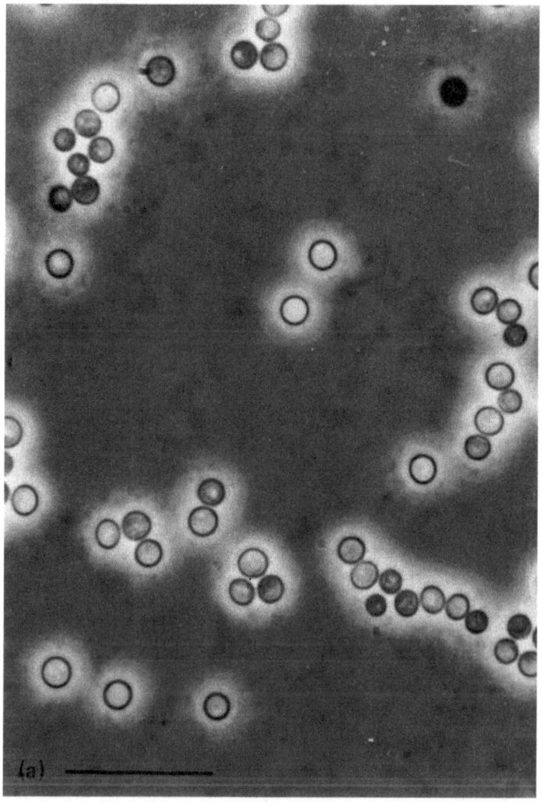

**Figure 1.7.** Myxospores. (a) *Myxococcus xanthus* from fruiting body (bar = 10 μm). (b) *Angiococcus disciformis:* crushed sporangiole (bar = 22 μm). (c) *Cystobacter violaceus* in chamber culture; vegetative cells, mature myxospores and intermediary stages (bar = 18 μm). (d) *Cystobacter ferrugineus* from degenerate *Archangium*-like fruiting body (bar = 9 μm). (e) *Stigmatella aurantiaca* in chamber culture on a thin layer of agar (bar = 16 μm). (f) *Cystobacter velatus:* mature myxospores from sporangiole (bar = 10 μm). All pictures in phase contrast. Although optically refractile, myxospores often appear dark in the microscope when observed under oil immersion.

considerable mechanical strength, as demonstrated, for example, by resistance to ultrasonic treatment. Further, they show increased resistance to ultraviolet (UV) light and to heat (Baur 1905; Reichenbach and Dworkin 1969; Sudo and Dworkin 1969). The heat resistance is moderate, but it is clearly above that of vegetative cells. The latter have their growth maximum at about 36 °C and are quickly killed above 45 °C, whereas mature myxospores heated in aqueous suspension usually survive at 58–60 °C for 10–30 min and longer. When heated in the dry state, myxospores resist even higher temperatures (70–90 °C for at least 15 min), so that they could probably withstand high temperature peaks

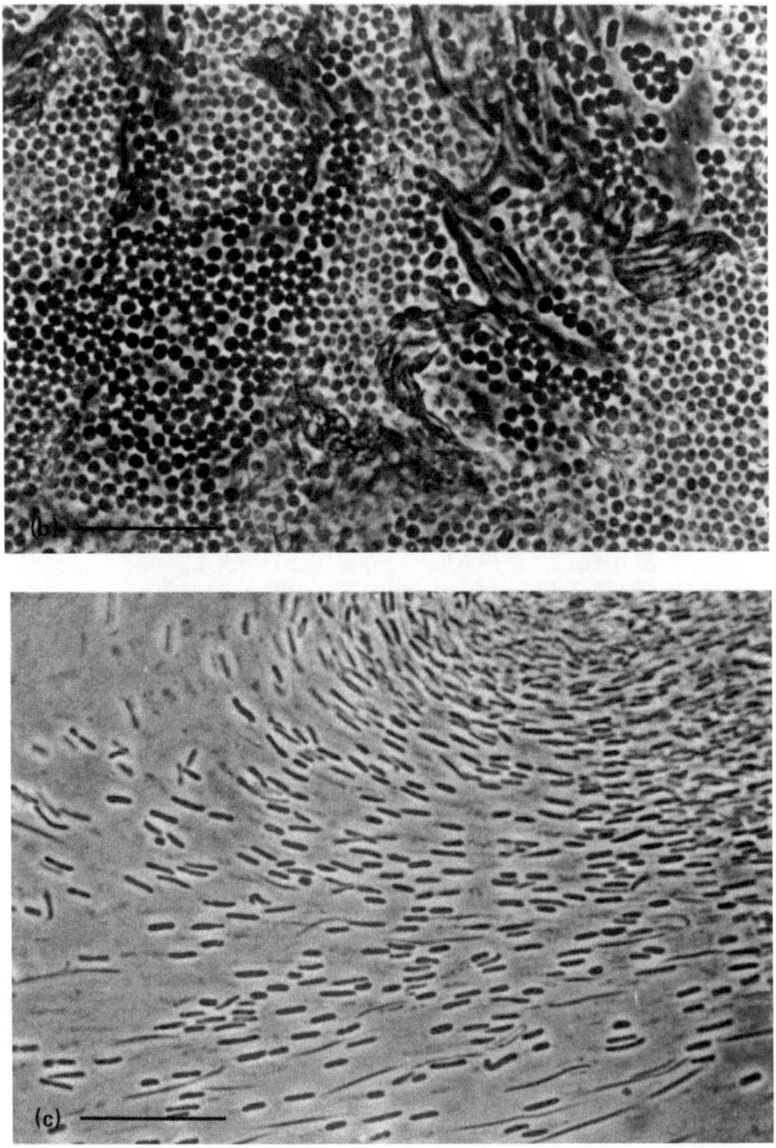


**Figure 1.7.** (Continued.)

occurring occasionally in the top layers of soil. Heat resistance of myxospores can be used to purify newly isolated strains from contaminating bacteria by suspending fruiting bodies in sterile water and heating them (Reichenbach and Dworkin 1981).

Physiologically, the myxospore is a dormant cell. This is obvious from the long storage periods that they survive and is experimentally supported by the

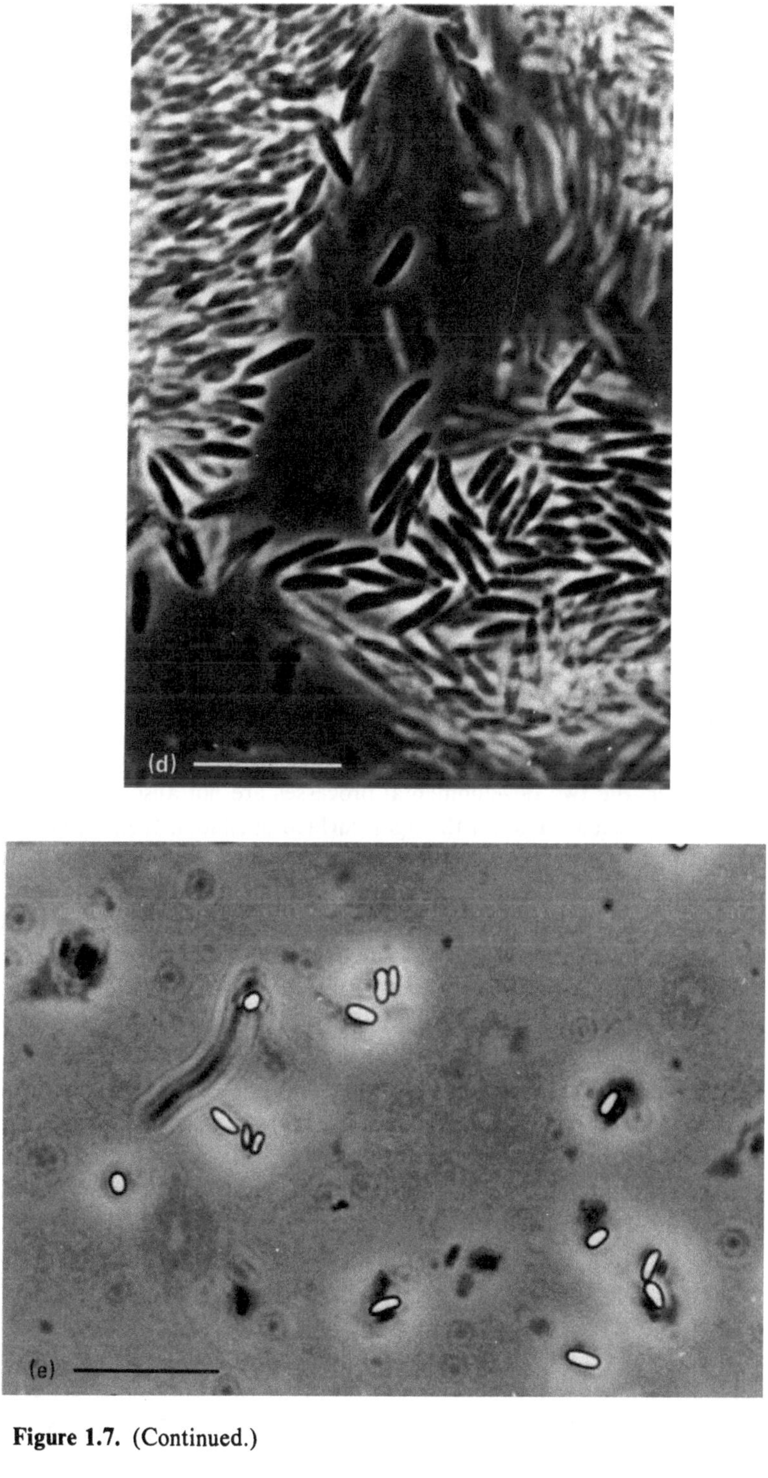


**Figure 1.7.** (Continued.)

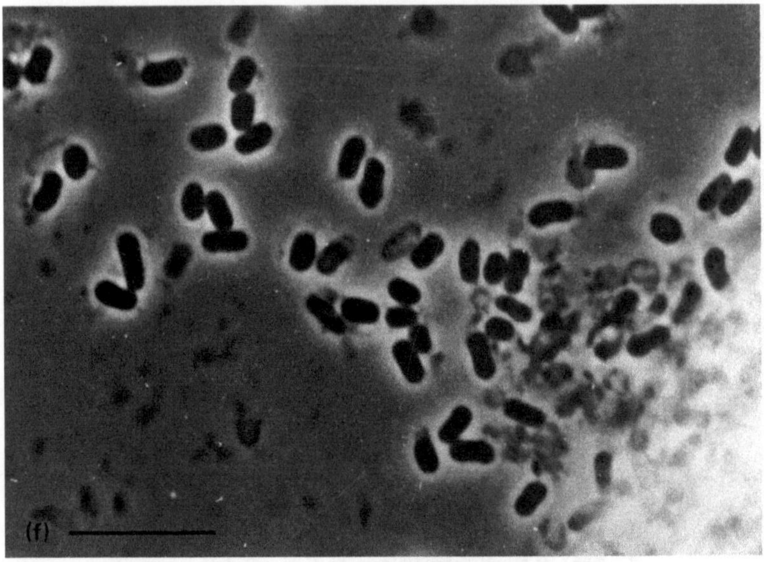


**Figure 1.7.** (Continued.)

extremely low respiratory activity in myxospores (Dworkin and Niederpruem 1964).

Ordinarily, myxospore formation takes place within the maturing fruiting body. However, the two developmental processes are not absolutely coupled. Myxospores are often found on the agar surface outside of fruiting bodies or even in complete absence of fruiting bodies. This may also happen in nature and emphasizes the fact that in essence the myxospore, and not the fruiting body, is the resting stage of myxobacteria. Furthermore, myxospore formation can be induced in suspension culture with glycerol and many other chemical compounds under controlled conditions (Dworkin and Gibson 1964). This discovery helped to draw the myxobacteria from near oblivion and make them attractive objects of research in developmental biology. Two model systems have been developed: *M. xanthus* (Dworkin and Gibson 1964; Dworkin and Sadler 1966; Sadler and Dworkin 1966; Rosenberg *et al.*, 1967; White *et al.*, 1968; Orlowski *et al.*, 1972; Foster and Parish 1973; Inouye *et al.* 1979a,b; and many other articles) and *Stigmatella aurantiaca* (Reichenbach and Dworkin 1970; Gerth 1975; Gerth and Reichenbach 1978), but it must be admitted that we are still far from a real understanding of the biochemistry of myxospore formation (see Chapter 2).

It should be stressed that when working on such a complex phenomenon as myxospore formation it may be wise not to concentrate entirely on one experimental system, important as it is not to split activities too much. Every organism has certain unique properties and its experimental limitations, and not every question may be answered with every system. Thus *Stigmatella auran-*

*tiaca* Sg a1 proved in some respects more versatile than *Myxococcus xanthus* FB, and many experiments could be done with the former that do not seem feasible with the latter.

Relatively little is known about germination of myxospores. The morphologic events are as follows: First, the myxospore loses optical refractility; then, the outer layer is penetrated from within by a stout rod cell. An empty shell is left behind (Voelz and Dworkin 1962). Activation by physical factors such as high or low temperatures or desiccation can be ruled out. Myxospores of *M. xanthus* obtained by glycerol induction will germinate in distilled water but only if present in high cell density ($5 \times 10^9$/ml). In this case, germination seems to be triggered by inorganic phosphate excreted by the cells (Dworkin 1973). These myxospores can germinate at low cell densities if suspended in 1–10 m*M* phosphate buffer. It has been speculated that this mechanism is designed to guarantee that germination takes place only in community. Unfortunately, myxospores from fruiting bodies do not seem to respond to the phosphate signal (White 1975).

## Fruiting Body Formation

The fruiting bodies of myxobacteria are quite peculiar natural objects. Measuring between 10 and 700 μm in diameter, they are just visible to the naked eye, particularly as they tend to appear in large numbers. In Minneapolis, I once found a rotting willow trunk that was covered with large orange patches consisting of thousands and thousands of fruiting bodies of *Stigmatella aurantiaca*. The structure of myxobacterial fruiting bodies may be quite complex, and it is thus understandable that for a long time they have been mistaken for fungi. Some representative fruiting bodies are shown in Figs. 1.8–1.14.

The most simple myxobacterial fruiting bodies are mounds of soft or hardened slime in which the myxospores are embedded (genera *Myxococcus, Corallococcus, Archangium*). In more advanced cases, a tough and very resistant wall of unknown chemical composition develops around the slime and myxospores. This walled structure is referred to as a sporangiole. The wall probably gives additional protection to the myxospores and prevents them from being easily washed away and the population becoming dispersed. Only few species form single, isolated sporangioles *(Haploangium, Nannocystis)*; usually a number of sporangioles are more or less tightly packed together and piled up, often surrounded by an additional, morphologically differentiated communal slime layer or envelope *(Angiococcus, Cystobacter, Polyangium, Sorangium)*. Finally stalks may be produced that lift the sporangioles into the air. The stalk may be topped by a single sporangiole *(Stigmatella erecta, Melittangium)* or by a cluster *(Stigmatella aurantiaca, Chondromyces, Polyangium thaxteri)*. In the most sophisticated case, *Chondromyces crocatus*, the stalk is branched, and each branch bears a bunch of sporangioles (Fig. 1.14). The stalk may exist entirely of excreted slime (*Chondromyces:* Fig. 1.12b,d; Reichenbach *et al.*

**Figure 1.8.** Myxobacterial fruiting bodies. (a) A small fruiting body of *Myxococcus fulvus* on a soil crumb in crude culture (bar = 35 μm). (b) *Myxococcus fulvus* on a piece of rotting wood. The soft slimy heads fuse when they touch each other (bar = 125 μm). (c) *Myxococcus stipitatus* on a soil crumb (bar = 125 μm). (d) *Angiococcus disciformis:* fruiting body, slightly squeezed under cover glass, showing chains of sporangioles (bar = 90 μm). (e) *Corallococcus coralloides* on agar surface (bar = 215 μm). (f) *Corallococcus coralloides* on agar surface. The shapes of the fruiting bodies of this organism are very variable (bar = 215 μm). (g) *Archangium serpens* on agar surface in enrichment culture; bright dots are soil amoebae (bar = 215 μm).

1965c, McCurdy, 1969b), or it may first be built out of cells and slime and the cells later leave or degenerate (*Stigmatella:* Voelz and Reichenbach 1969; Stephens and White 1980). Curiously, sporangioles and stalks seem to have been "invented" more than once; both subgroups of myxobacteria, which are not closely related to one another, produce these morphologic elements.

Often myxobacterial fruiting bodies are brightly colored, and at least in the more complex fruiting structures, this color is not simply due to the pigmentation of the encased myxospores. Rather, during the fruiting process, specific, very intensively absorbing pigments of as yet unknown chemical composition

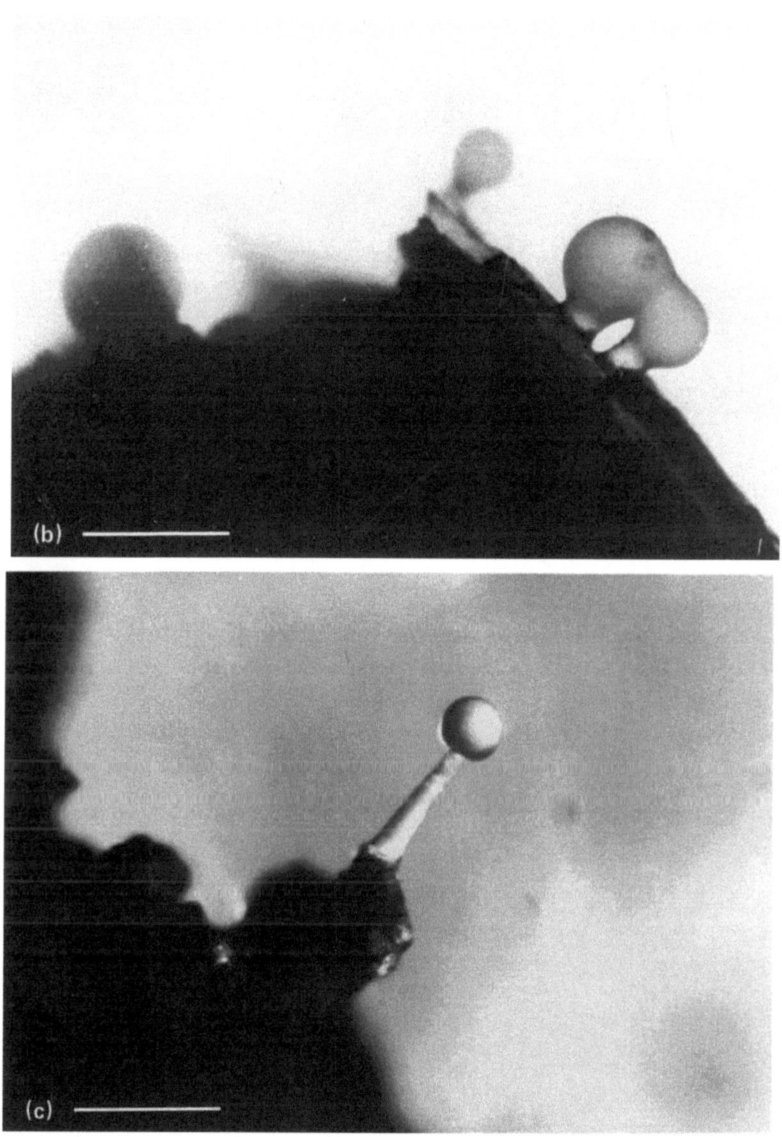


**Figure 1.8.** (Continued.)

are produced and deposited within the walls of the sporangioles and in the surrounding slime envelopes.

Fruiting body induction and formation are obviously under the control of environmental factors. Strains may be cultivated virtually indefinitely in the vegetative state without ever producing fruiting bodies. In fact, the ability to

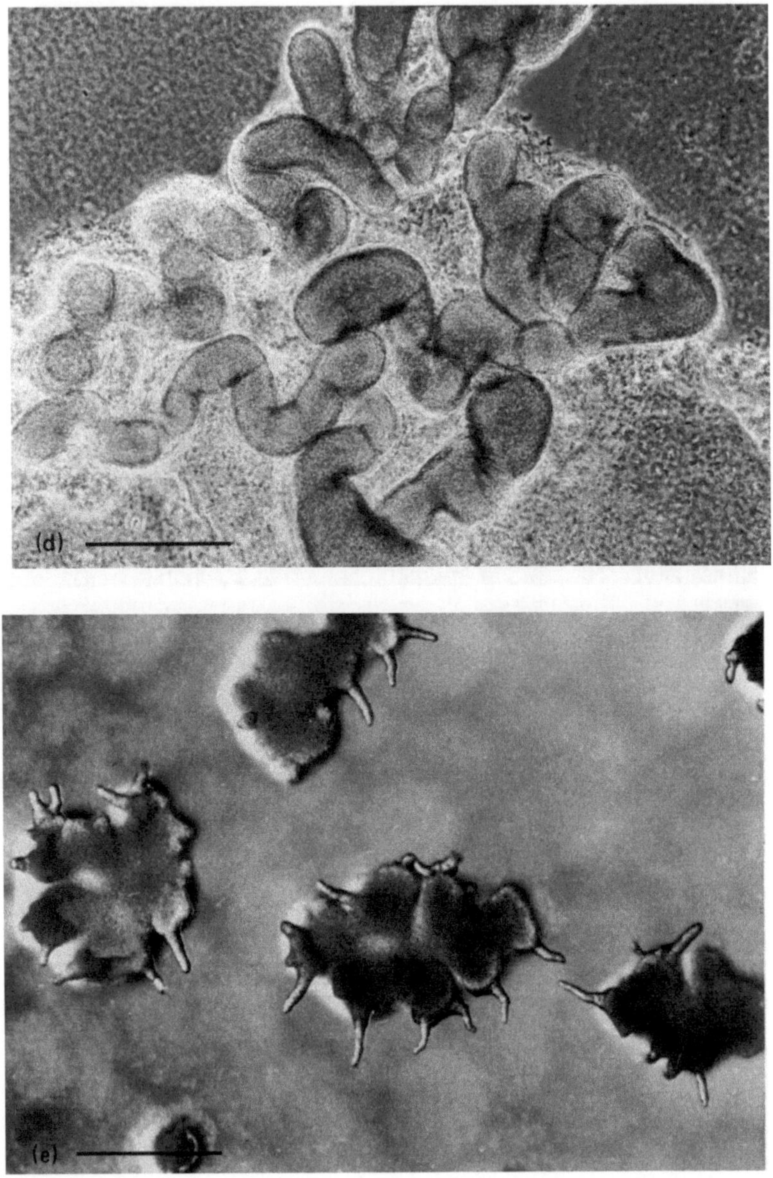


**Figure 1.8.** (Continued.)

form fruiting bodies is often lost when a newly isolated strain is transferred several times on a rich growth medium. In most cases we do not know whether and how the ability can be reestablished. Cultures started from fruiting bodies usually can easily be made to fruit again. Perhaps the degeneration of the fruiting capability is due to a selection of well-growing but nonfruiting mutants.

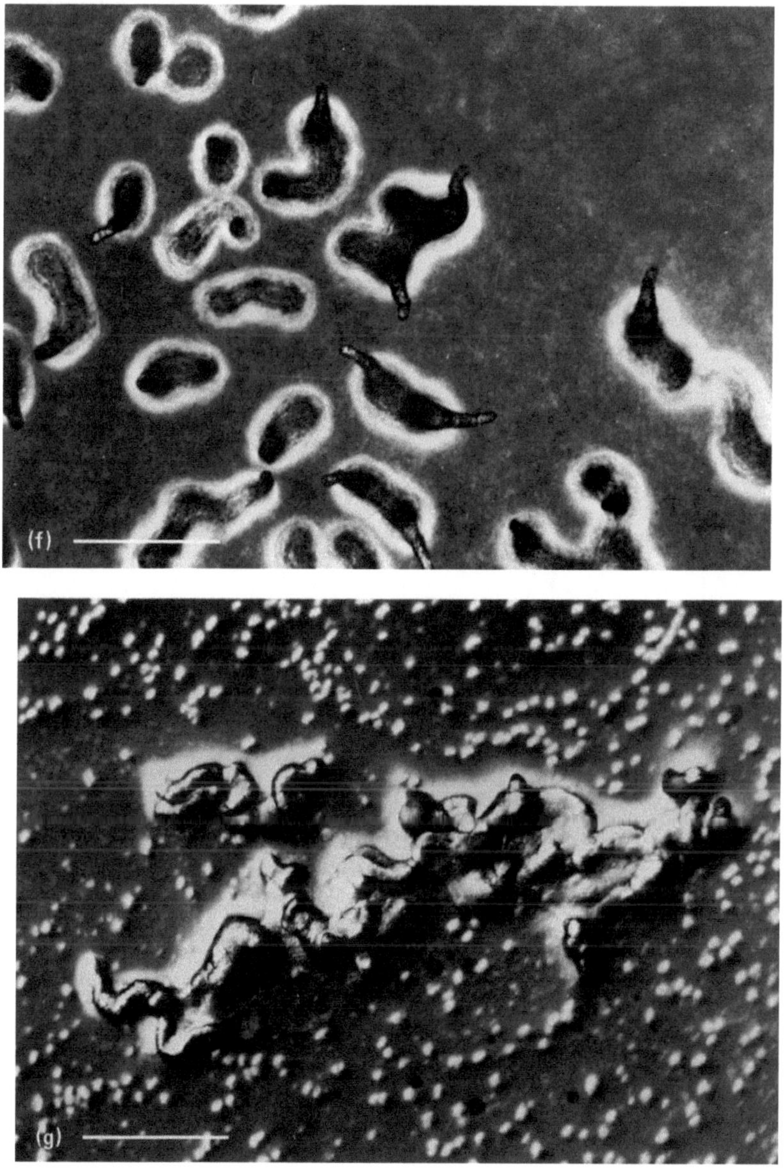


**Figure 1.8.** (Continued.)

The structure and physicochemical properties of the slime may play some role in this connection, for nonfruiting strains often have different looking swarms that show reduced spreading, a rudimentary surface texture, and a very soft slime. There are, however, also examples of strains that have been growing for hundreds of generations without ever producing fruiting bodies and that still

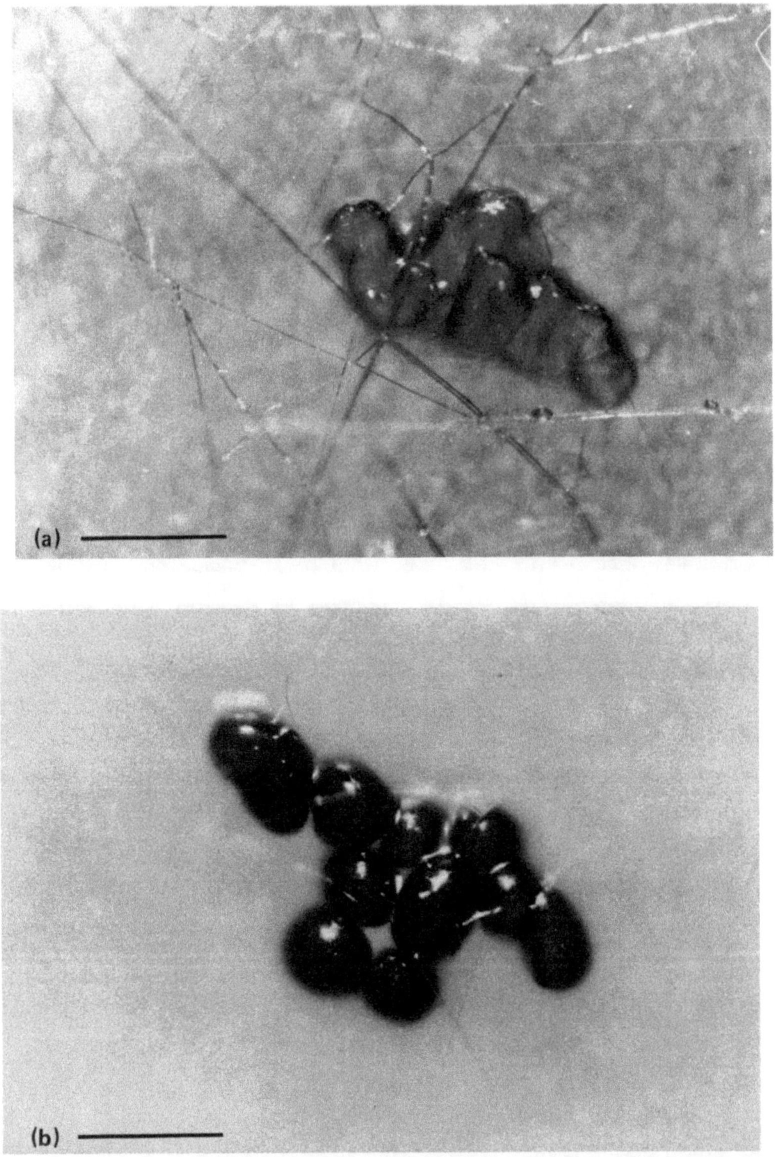


**Figure 1.9.** Myxobacterial fruiting bodies. (a) *Cystobacter fuscus:* early stage of fruiting body differentiation. On agar surface, crude culture (bar = 135 μm). (b) The same organism, differentiation into sporangioles completed (bar = 135 μm). (c) *Cystobacter velatus* on agar surface. The sporangioles are covered by a thin, finely plicated slime sheet (bar = 175 μm). (d) *Cystobacter ferrugineus:* sporangioles of very different size. Slide mount (bar = 60 μm). (e) *Cystobacter ferrugineus:* fruiting body on agar surface (bar = 140 μm). (f) *Polyangium* sp: fruiting body within the agar. Slide mount (bar = 45 μm).

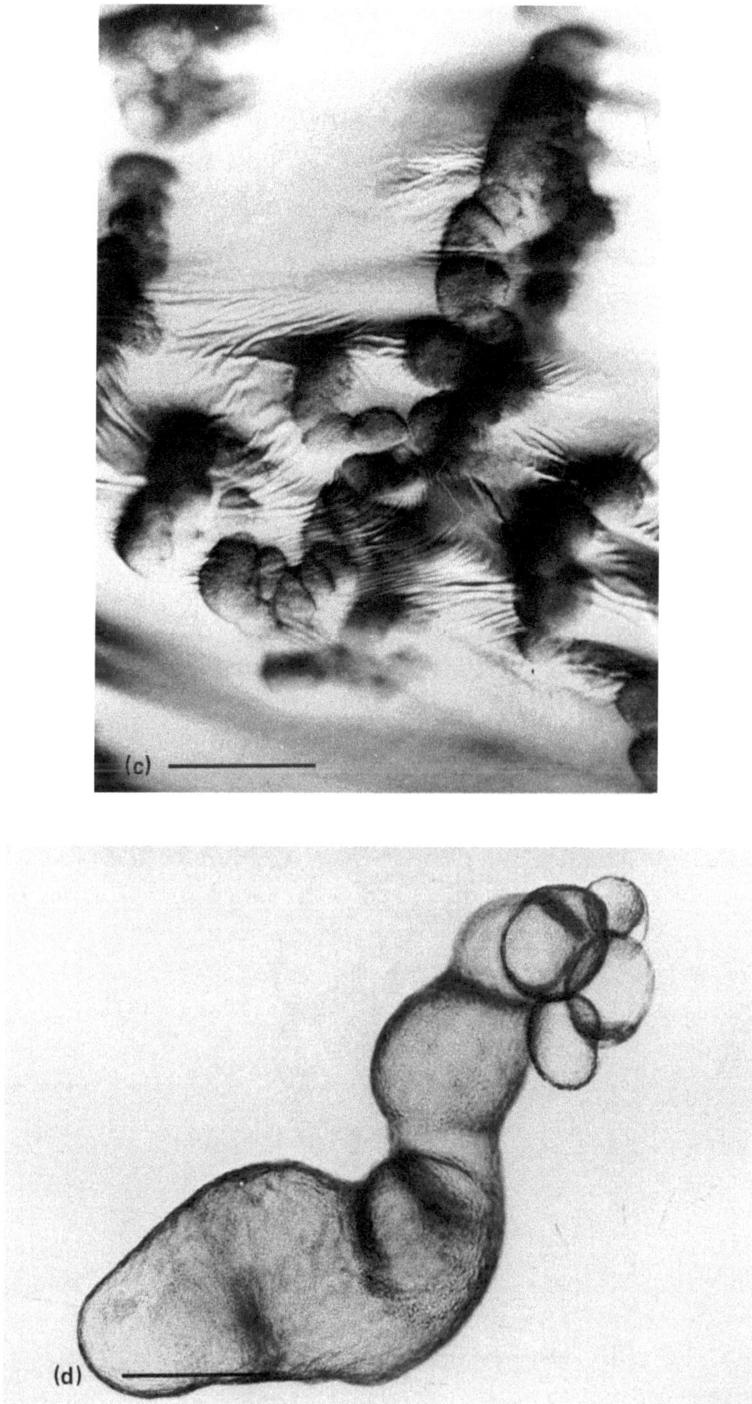


**Figure 1.9.** (Continued.)

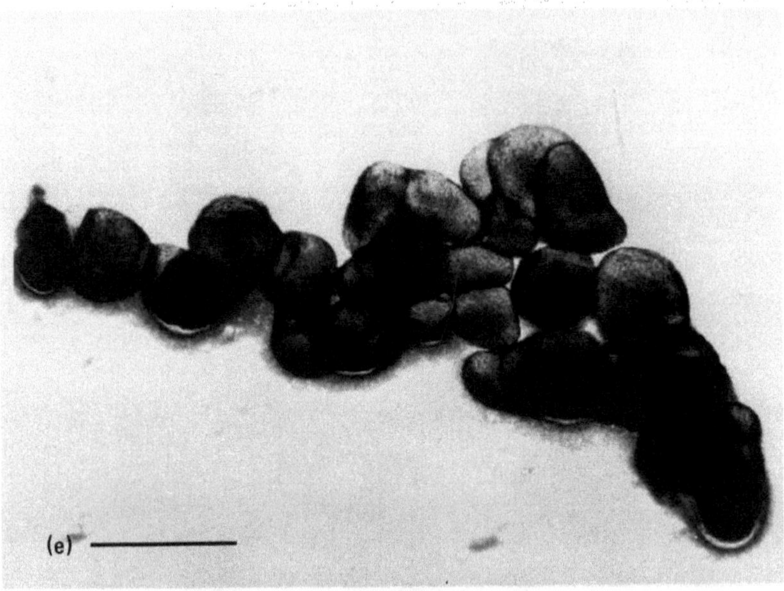


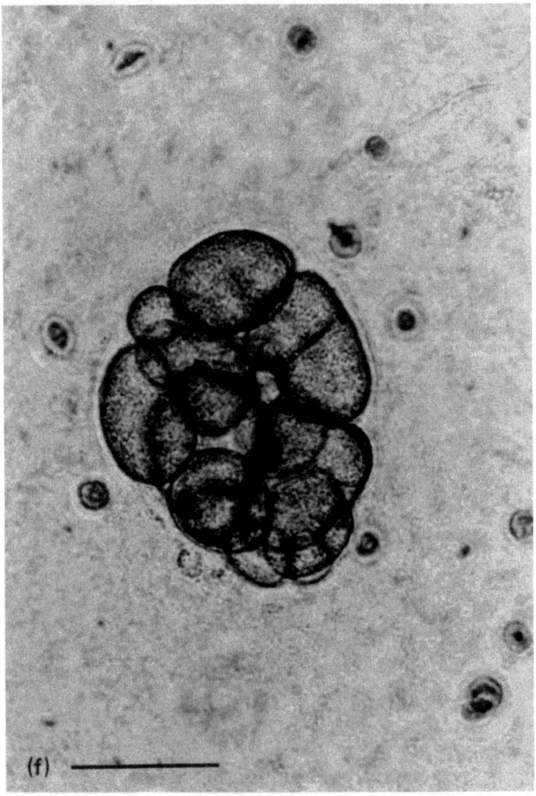


**Figure 1.9.** (Continued.)

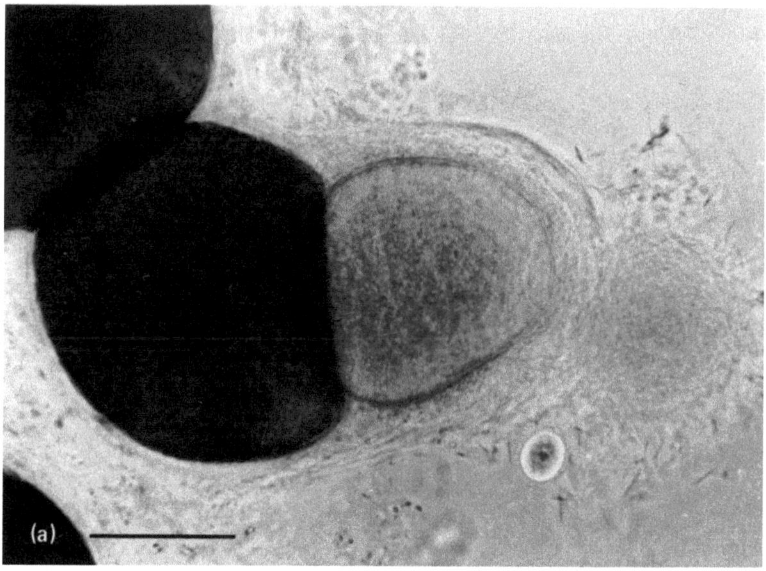


**Figure 1.10.** Myxobacterial fruiting bodies. (a) *Cystobacter fuscus:* fruiting body in the process of differentiation. The sporangiole at the right is at an early stage and consists of a tightly interwoven mass of vegetative cells; the one in the middle already shows a distinct wall and a layered slime envelope; the sporangiole at the left is fully differentiated with its wall impregnated with a dark brown pigment; slide mount (bar = 45 μm). (b) *Stigmatella aurantiaca:* fruiting body with its mirror image on an agar surface (bar = 120 μm). (c) *Stigmatella aurantiaca,* sporangioles: surrounding wall and the insertion point of the stalk are clearly recognizable; slide mount (bar = 25 μm). (d) *Stigmatella erecta* on agar surface (bar = 125 μm).

can be induced to form typical fruiting bodies under certain conditions (e.g., *Chondromyces apiculatus* strain Cm a2; Reichenbach 1974a,b).

In general, fruiting body formation is induced as a result of a depletion of nutrients. Transfer of cells from suitable growth media to plain water agar or to strips of filter paper on water agar is usually sufficient to start the fruiting process. If the filter paper is put into a vial and dried in a desiccator, the fruiting bodies retain the ability to germinate over a period of 6–16 yr; in this way the loss of the fruiting capacity of a strain can be avoided (Reichenbach and Dworkin 1981). Quite unexpectedly it was found that with two strains, *Chondromyces apiculatus* Cm a2 and *Stigmatella aurantiaca* DW 4, light plays an essential role in the aggregation of cells and in the differentiation of the primordial cell mass (Reichenbach 1974a,b; Qualls *et al.* 1978a,b). The biochemical details of the fruiting process will be treated in later chapters.

Fruiting bodies usually arise within the swarm in large numbers and often are distributed over the culture plate in striking patterns, such as parallel or

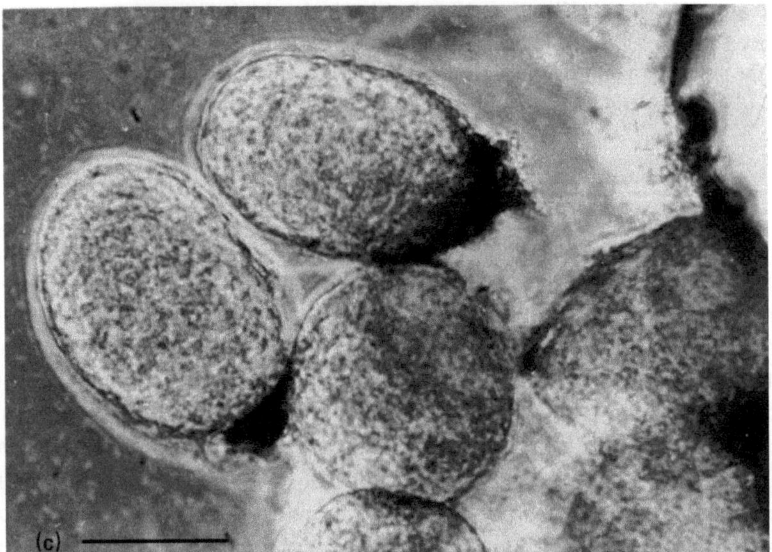


**Figure 1.10.** (Continued.)

**Figure 1.10.** (Continued.)

radial rows and (witches') rings (Finck 1950; Reichenbach 1962). Obviously, different mechanisms are responsible for the different patterns. The radial and parallel rows seem simply to reflect an uneven distribution of the vegetative cells within the swarm. Thus, the cells often tend to concentrate within the swarm along heavy radial veins, or migration roads, and in consequence also form a radial row of fruiting bodies. Parallel rows of fruiting bodies are often due to elasticotaxis (Fig. 1.15; Stainier 1942). The rings may possibly be due to diffusion gradients of factors that control the fruiting process.

There has been some speculation about the biologic meaning of fruiting bodies (Reichenbach, 1965b, 1974a; Dworkin 1973; Reichenbach and Dworkin 1981). Although the myxospore may indeed be somewhat better protected within a sporangiole, the isolated myxospore is still essentially resistant to harmful factors that play a role in the environment of myxobacteria: lack of water and nutrients, elevated or low temperatures, and intense UV irradiation. Rather, the function of the fruiting body seems to be to ensure that a new life cycle is started by a cell community and not by a single cell: The fruiting body

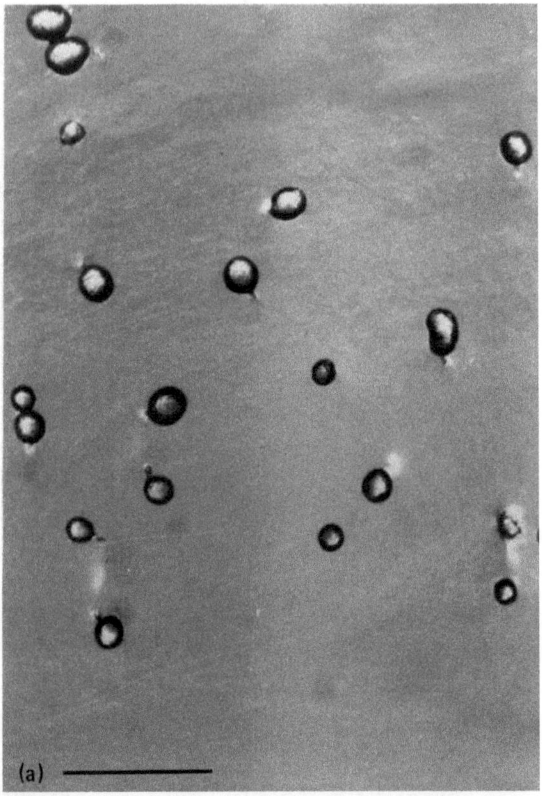


**Figure 1.11.** Myxobacterial fruiting bodies. (a)–(b) *Melittangium lichenicola.* (a) On agar surface (bar = 140 μm). (b) On a natural substrate, rotting wood. The tiny sporangioles are borne on minute stalks (bar = 75 μm). (c) *Melittangium boletus:* The characteristic caplike sporangiole is sitting on a short stalk; slide mount under oblique illumination (bar = 30 μm). (d) *Sorangium cellulosum* on agar surface (bar = 115 μm). (e) *Nannocystis exedens:* tiny sporangioles of widely varying size are embedded within the substrate; slide mount, phase contrast (bar = 70 μm).

is a resting colony. The explanation for this, as for probably all social phenomena in myxobacteria, may be found in the manner in which these organisms grow in nature. Typically, myxobacteria live from degradation products of biopolymers such as protein, cellulose, and all kinds of cell constituents. Many myxobacteria efficiently lyse other microorganisms, particularly bacteria and yeasts. In order to get access to such nutrients, the myxobacteria have first to invest into the production of exoenzymes. These enzymes diffuse away from the producer cell and may reach the substrate only at some distance from the cell. The small molecules produced by the enzymatic reaction diffuse into the direction of the enzyme-excreting cell, as well as into other directions, so that

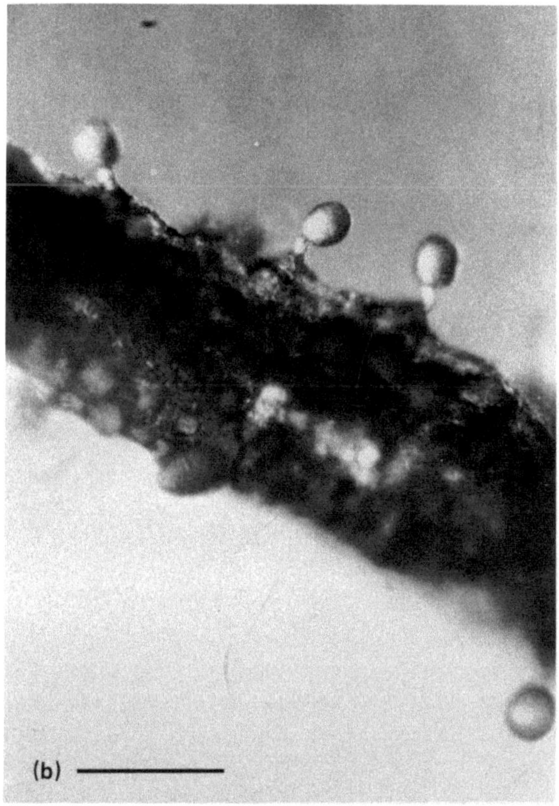


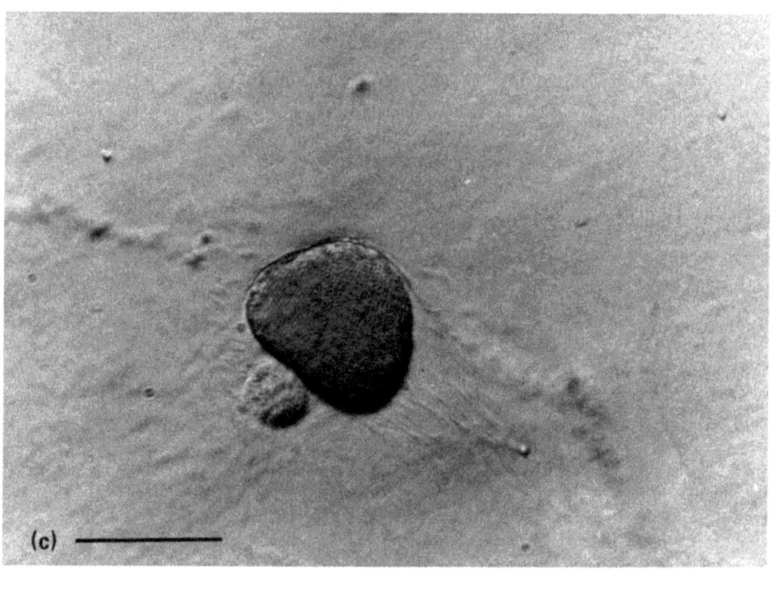


**Figure 1.11.** (Continued.)

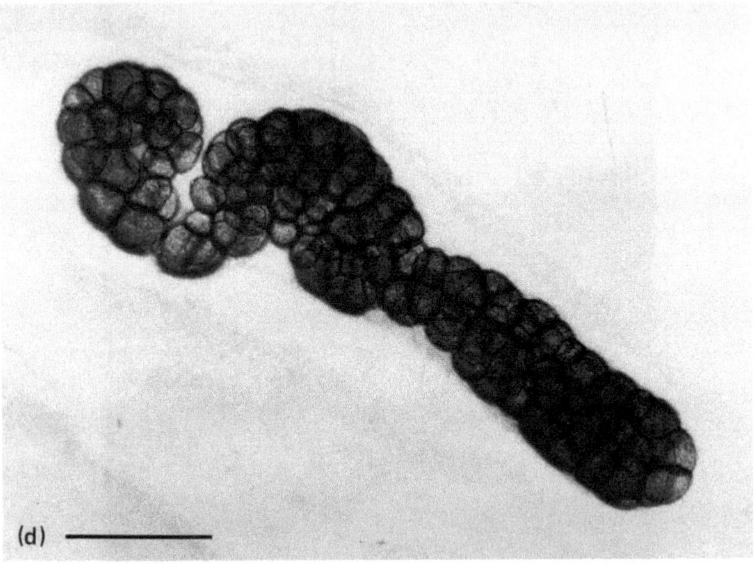


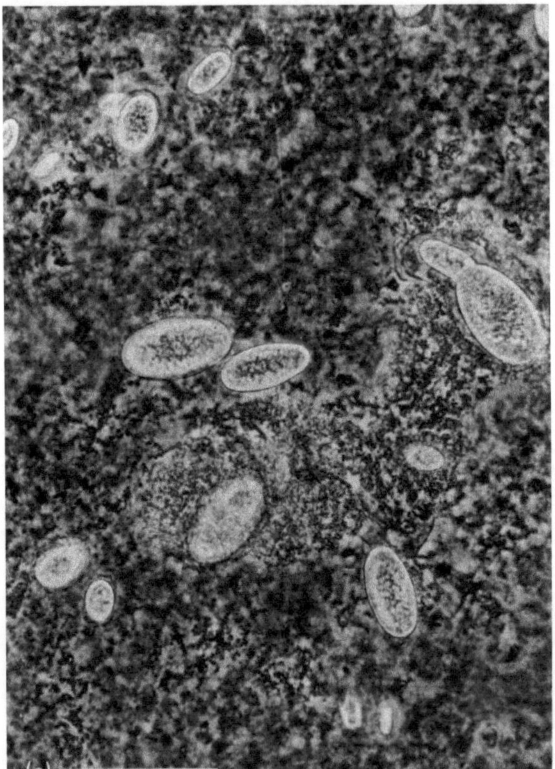


**Figure 1.11.** (Continued.)

**Figure 1.12.** Myxobacterial fruiting bodies. (a)–(e) *Chondromyces apiculatus.* (a) In situ (bar = 105 μm). (b) Piece of stalk: The stalk is acellular, and only few cells are embedded in its matrix. Slide mount, interference contrast (bar = 55 μm). (c) Late stage of differentiation. The developing sporangioles do not yet show the typical turnip shape. In situ (bar = 375 μm). (d) Earlier stage with the bacterial mass on top of the stalk still undivided. In situ (bar = 170 μm). (e) Mature sporangiole surrounded by a delicate but tough wall; slide mount, interference contrast (bar = 40 μm). (f)–(g) *Chrondromyces pediculatus.* (f) Late developmental stage: cell mass at the end of the stalk divided into several portions, each of which is going to become a sporangiole; slide mount, phase contrast (bar = 65 μm). (g) Mature fruiting body on a soil crumb with the characteristic bell-shaped sporangioles; slide mount under oblique illumination (bar = 120 μm).

**Figure 1.12.** (Continued.)

the return in energy is rather low. The situation is therefore more favorable when a cell community is at work. Not only can the cell community create higher enzyme levels and thus speed up the degradation, but it also minimizes loss by diffusion, because one cell may catch molecules set free by enzymes excreted by another cell and vice versa. In addition, a myxobacterial community may keep off competitors more effectively. In support of the ideas just discussed, it was shown that growth of *M. xanthus* on macromolecular casein takes place only if a relatively high inoculum is used (Rosenberg *et al.* 1977).

The shape and particular structure of the fruiting bodies have probably been selected to help distribution. Typically, fruiting bodies sit on top of the substrate, probably because light and slow desiccation of the substrate stimulate fruiting, but the biological meaning is to improve distribution. This is clearly suggested by the repeated evolution of stalks by several myxobacteria: The stalks further lift the sporangioles above the surface so that they can be carried away more readily by air currents or small animals. We can sometimes observe mites crawling over myxobacterial treelets and carrying sporangioles sticking to their backs.

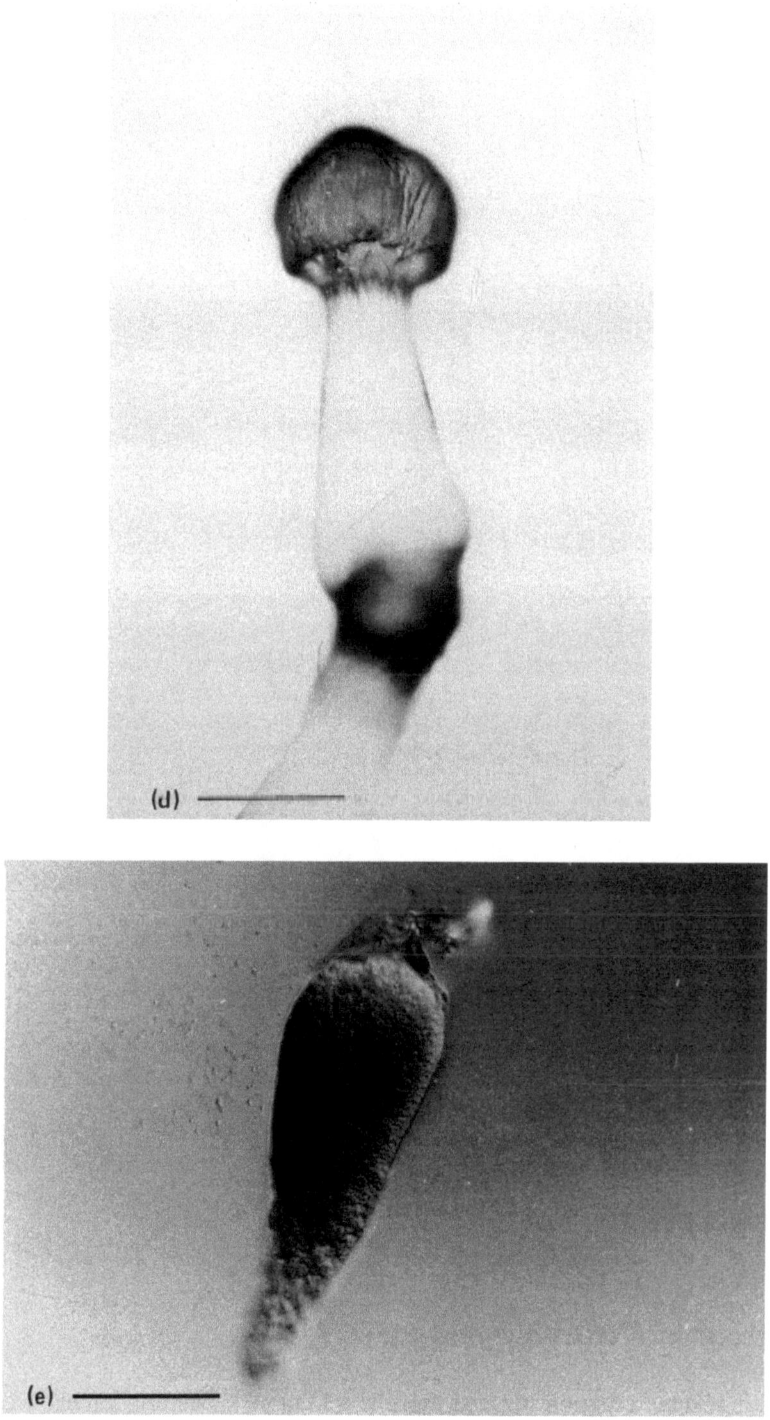


**Figure 1.12.** (Continued.)

**Figure 1.12.** (Continued.)

## Intercellular Communication

Fruiting body formation suggests that the cells of myxobacteria somehow communicate. Indeed, they must do that in a very precise manner because fruiting bodies have a distinct and species-specific shape (Figs. 1.8–1.14). Intercellular

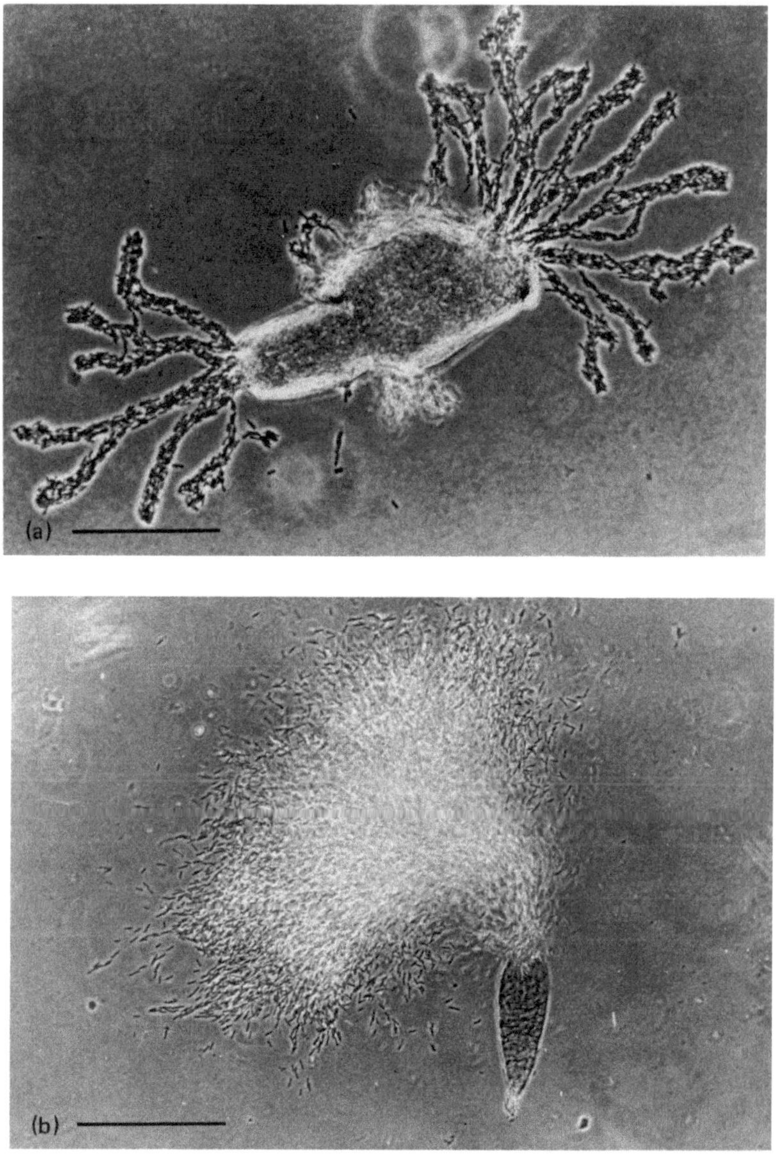


**Figure 1.13.** *Chondromyces apiculatus:* germination of fruiting body sporangioles. (a) Early stage with husk broken in several places and rod cells leaving in streams (bar = 40 μm). (b) Vegetative cells stay close together and form a little germination swarm beside the empty husk (bar = 60 μm). (c) Empty husk (bar = 16 μm). (d) End of a husk showing the hole through which the cells escaped (bar = 11 μm).

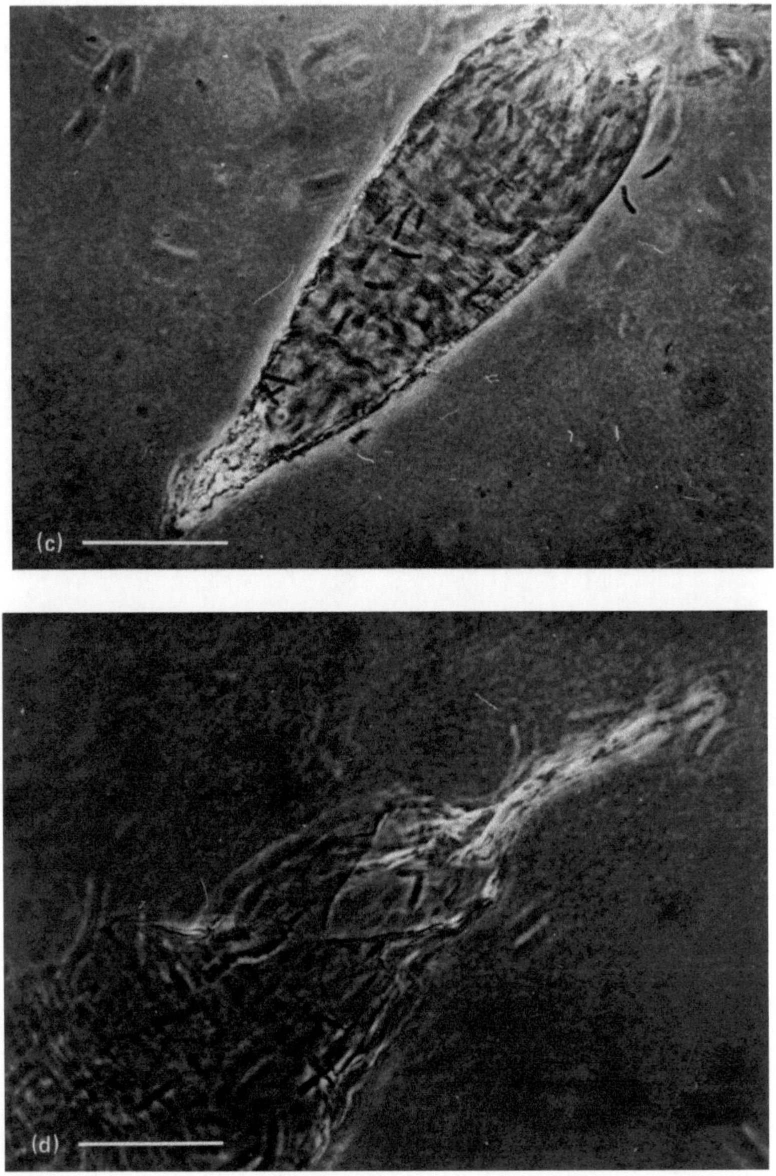


**Figure 1.13.** (Continued.)

communication is, however, by no means restricted to cooperative morphogenesis, so that myxobacteria must be regarded as truly social prokaryotes (Dworkin 1973). The social behavior of myxobacteria became apparent for the first time in a series of movies on several myxobacterial species (Reichenbach *et al.* 1965a–d, 1975–1976; Grimm *et al.* 1971; Kühlwein *et al.* 1971). It is striking to see that no cell will ever escape and move away from the expanding swarm,

**Figure 1.14.** Myxobacterial fruiting bodies: *Chrondromyces crocatus.* (a) In situ on agar surface (bar = 240 μm). (b) Slide mount, dark field (bar = 130 μm). (c) Subterminal developmental stage: stalk branched repeatedly and bears spherical cell masses at tips of the branches. The final step, the differentiation of the sporangioles, has not yet begun. Slide mount under oblique illumination (bar = 90 μm). (d) Tip of a branch with sporangioles and the insertion area of the latter on the stalk. The stalk shows a longitudinal striation and contains a few cells that have been lost from the terminal cell mass during stalk excretion. Slide mount, phase contrast (bar = 30 μm).

although individual cells can glide much faster than the rate at which the swarm edge advances. Whenever a cell moves beyond the swarm edge, which happens rarely, the cell either actively returns after it has covered a distance of a few cell lengths or it stops and waits until the community of the swarm has caught up. Very small swarms, for example, those that emerge from a single germinating sporangiole of a *Chondromyces* fruiting body, may even move as a whole (Fig. 1.13; Reichenbach *et al.* 1965b). Typically, in this case the cells at the leading edge move slowly and stick closely together, whereas those at the trailing edge move much more freely and faster. If such a migrating swarm approaches another swarm of the same strain, the cells at the respective front edges seem to sense one another as soon as they are 10–20 cell lengths apart, start to break up the formerly closed swarm edge, and move toward one another with increasing speed. These observations can be explained by assuming that a chemotactic substance is more or less uniformly distributed at the rear end of the migrating swarm, allowing the cells to move there freely in any direction, whereas at the front end there is a steep gradient that soon

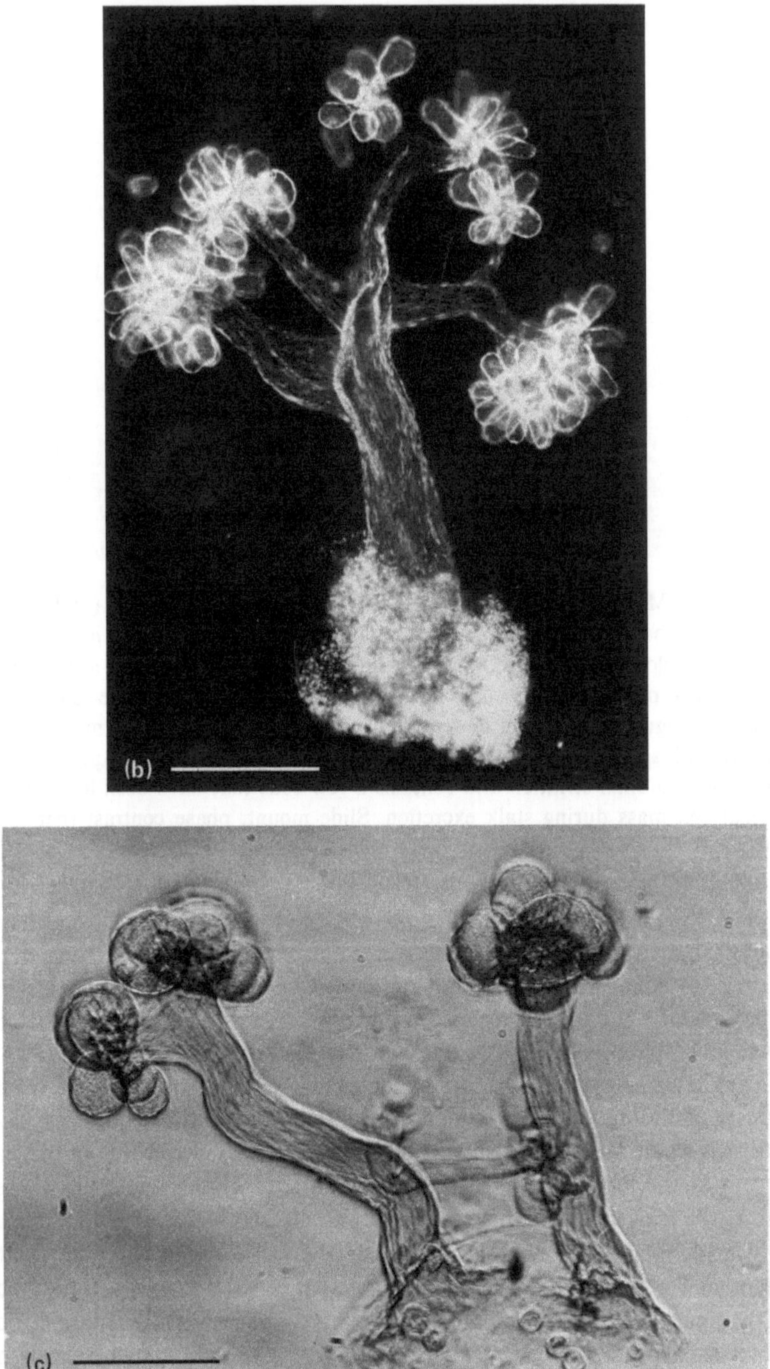


**Figure 1.14.** (Continued.)

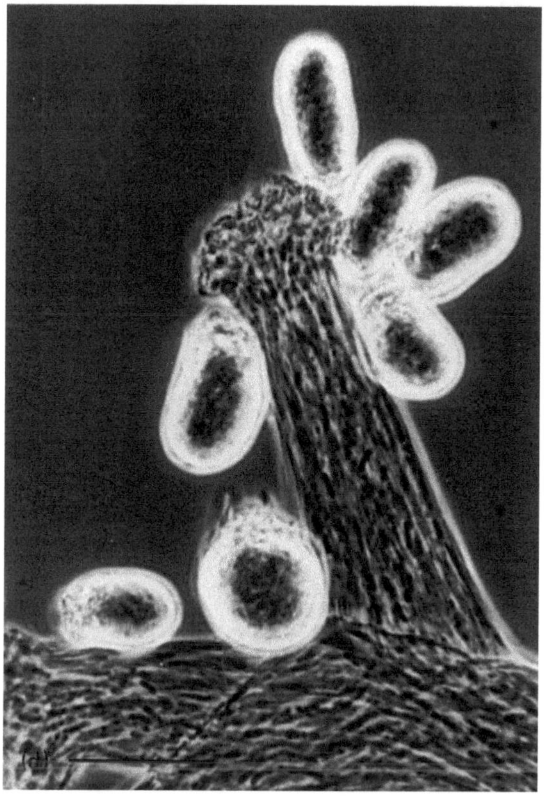

**Figure 1.14.** (Continued.)

stops any escaping cells. Recently a pheromone has been isolated that seems to play a role in fruiting body development of *Stigmatella aurantiaca* (Stephens *et al.* 1982) and gives experimental support to the hypothesis of chemical communication over some distance.

Cell clusters that migrate as packs are found regularly with members of the genera *Sorangium, Polyangium,* and *Nannocystis.* On poor substrates the swarms of these organisms tend to break up into a multitude of tiny cell masses that then migrate by themselves away from the swarm, usually leaving long radial ditches or tunnels within the agar behind them and often forming fruiting bodies at the ends of their paths (Fig. 1.1f, 1.2f). Presumably in all of these cases the cells are held together by some type of communication system.

The movies show another fascinating social behavior of the swarm cells: the phenomenon of the oscillating waves (Reichenbach 1965a). Within the swarm there often appear more or less extended fields of tiny ridges or waves. The waves are arranged one behind the other in long radial rays and also move along these lines. The whole field strikingly resembles the ripple marks at the

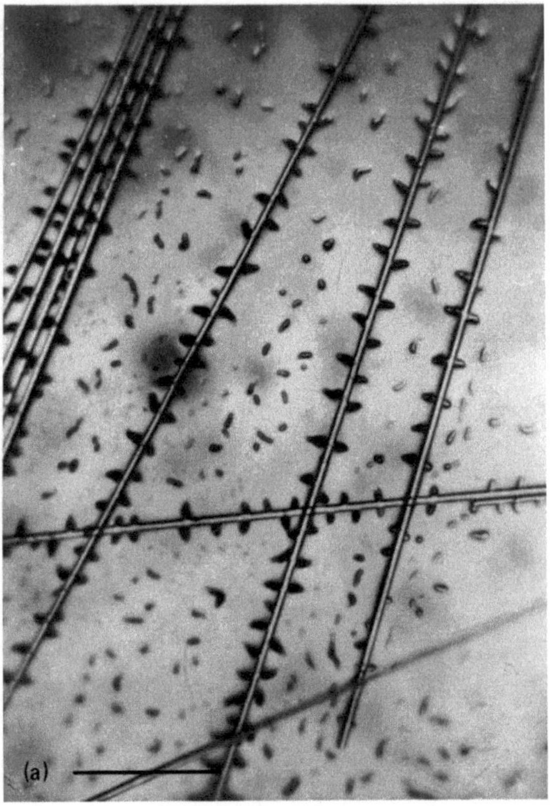


**Figure 1.15.** Guiding effect by structures on the substrate surface ("elasticotaxis"). (a) *Corallococcus coralloides:* along glass fibers placed on the surface of the culture plate, fruiting bodies are all oriented perpendicular to the fiber (bar = 650 μm). (b) Caused by a parallel alignment of the migrating cells close to the glass fiber (bar = 45 μm). (c) *Myxococcus xanthus:* cells move strictly parallel to (invisible) scratches in a cellophane surface (bar = 45 μm).

seashore. It is not yet clear what is really going on. Each wave consists of a large number of cells piled on top of one another and aligned more or less perpendicular to the axis of the wave or in the direction of the wave movement. In the areas in between the waves relatively few cells are found. The movements of the waves are fast enough to be directly observed under the microscope, but an exact analysis of the movements requires time-lapse photography. It can then be seen that all waves move with uniform speed, but the speed of the movement may become rhythmically faster and slower (Reichenbach 1965a). The movements of the cells thus seem not only coordinated within each wave ray, but pulses spread through the swarm, which activate or inactivate cell movement. The exact pattern of cell movement within the wave ray is not yet understood. Either the cells piled up in one wave remain always together

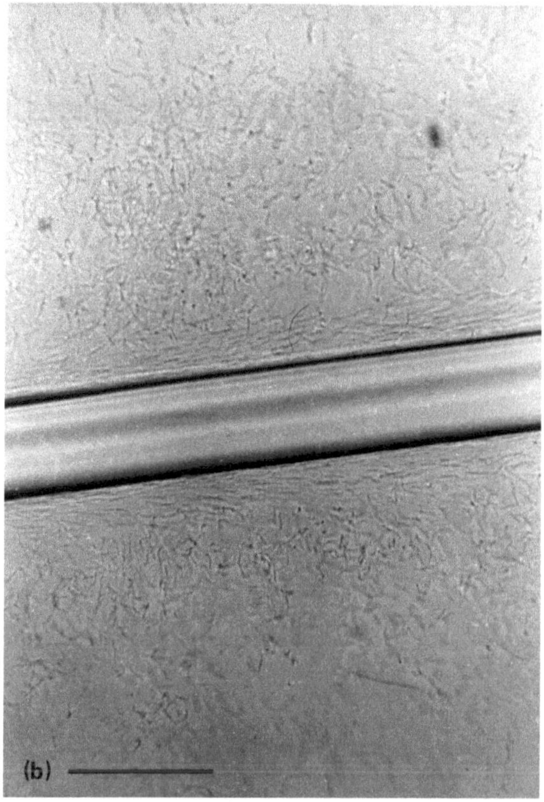


**Figure 1.15.** (Continued.)

and move as a whole—always in one direction or rhythmically back and forth—or the population of each wave is constantly rearranging itself—part of it moving forward, part of it backward, each part forming displaced new waves with cells coming from neighboring waves. In some cases two wave systems move on the same track, or ray, in opposite directions, so that the waves penetrate through each other (Reichenbach *et al.* 1975–1976). Are there two independent cell populations, each forming its own wave system? Unfortunately, it is very difficult optically to resolve at the same time the wave and the individual cell within it. For an analysis of movements, it would be essential that the phenomenon could be studied in chamber cultures with very thin cell layers. Perhaps the recent discovery that in *M. xanthus* wave formation can be induced by components of the peptidoglycan (Shimkets and Kaiser 1982b) will allow such experiments. This observation may also give some clues as to the origin of the wave phenomenon. Most myxobacteria are "micropredators" (Singh 1947) (i.e., they live from degradation products of other microorganisms, particularly bacteria). The peptidoglycan components that are set free during

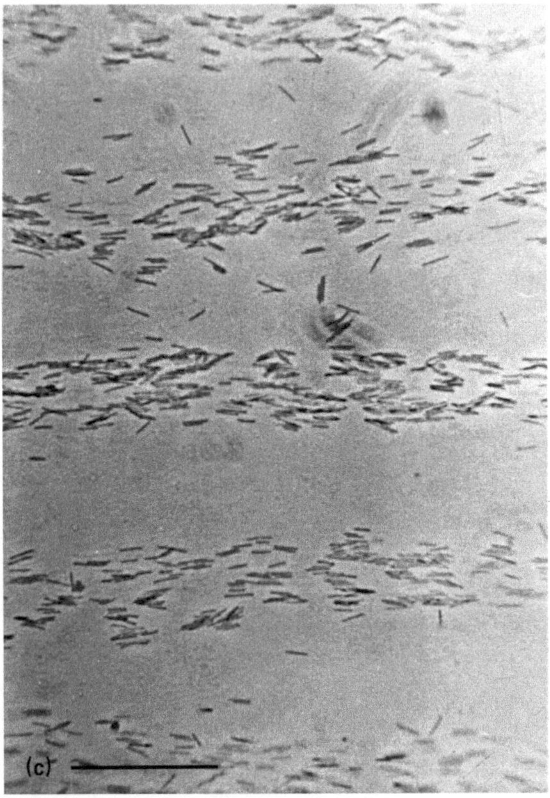

**Figure 1.15.** (Continued.)

breakdown of bacteria may serve as signals attracting the myxobacteria to sites where food is available. The wave phenomenon as we observe it in cultures is perhaps artificial and a degeneration of a behavior that is meaningful under natural conditions. The waves seem to have nothing to do with fruiting body formation; at least they can be observed in swarms that do not fruit, and fruiting bodies may arise in absence of waves.

An obvious prerequisite for cell–cell interactions in myxobacteria is that the individual cell is physiologically prepared to respond to the signals of its companions. The cell population in the swarm seems, however, to be heterogeneous in this respect. Thus, under ordinary culture conditions not all cells participate in wave formation. Immediately beside the wave rays are many cells that move randomly: The wave signal must reach them, but they do not respond. Similarly, in a fruiting swarm not all cells participate in fruiting body formation: In the direct neighborhood of developing fruiting bodies, there are many cells that move freely and are seemingly uninfluenced by the fruiting centers (Reichenbach *et al.* 1965a–d). Therefore, doubts arise as to whether the cells really pile up under chemotactical control (i.e., whether there is active *aggregation*) or whether the cells simply become trapped whenever they happen to chance into a fruiting center (i.e., whether there is a more passive *accumulation*).

A prerequisite for the study of chemotaxis and fruiting body formation in myxobacteria is to produce biochemically defined and uniform cell populations. This has now been achieved for *M. xanthus* and *Stigmatella aurantiaca,* but it is still an unsolved problem for the fascinating *C. apiculatus* system (Reichenbach 1974a,b).

## Ecology

In spite of their exotic appearance and complex developmental cycle, myxobacteria are very common organisms. In the neutral pH range between 5 and 8 and in aerated surface layers, virtually every pinch of soil throughout the world, regardless of climate, contains at least some myxobacteria. The occurrence and distribution of myoxbacteria has been studied repeatedly, and there are several surveys available (Krzemieniewska and Krzemieniewski 1927a,b; Norén 1952; Sabados-Saríc 1957; Jeffers 1964; Rückert 1979; Dawid 1979).

Basically, myxobacteria are soil organisms; however, they are able to leave the soil and colonize other habitats, particularly those that harbor rich microbial communities, for example, the rhizosphere of plants (Agnihothrudu *et al.* 1959), the dung of herbivorous animals, rotting wood and decaying organic material in general, the bark of living and dead trees (Peterson 1959), and even the phyllosphere (Rückert 1981). A few species appear to have become specialized for particular niches (e.g., *Haploangium,* which has only been observed on bark and rotting wood). From the soil myxobacteria are easily washed into water bodies in which they can survive and multiply (Jeffers 1964; Gräf 1975; Hook 1977). One species, *Polyangium parasiticum,* has even been described as an (obligate?) parasite of the green alga, *Cladophora* (Geitler 1925). Myxobacteria have apparently not adapted to marine environments.

Because myxobacteria tend to stick together, it is very difficult to determine their actual numbers in the environment. Sabados-Sarić (1957) estimated that rich soil contains up to $8 \times 10^4$ cells per gram; Singh (1947) reported about the same number for soil and more than $5 \times 10^5$ for decomposing compost. In most studies only the number of different species occurring in a certain soil was recorded. Since almost always several species are isolated from one small sample, the cell numbers given above probably still underestimate the real numbers. One single fruiting body contains between $10^3$ and $10^6$ tightly packed myxospores.

The studies on the occurrence of species also give some idea about distribution in different climate zones. Particularly rich myxobacterial populations are observed in warm semiarid and arid climates; in these climates the desiccation resistance of the myxospores obviously gives them a strong selection advantage (Singh and Singh 1971; Brockman 1976; Rückert 1978a). Rich floras are also found in warm temperate and in tropical climates (Krzemieniewska and Krzemieniewski 1930; Kühlwein 1960; McCurdy 1969a; Dawid 1978). In tropical areas, however, poor quality of the soil (e.g., acid pH, depletion of organic material) may become limiting. In cold temperate, subarctic, and arctic environments, including high mountain ranges, the spectrum of species becomes more uniform, consisting mainly of *Myxococcus* and *Corallococcus,* although abundance may still be high (Kühlwein 1960; Brockman and Boyd 1963; Rückert 1973). There are other curious differences in the distribution of species. Although *Stigmatella aurantiaca* is a very common organism in the temperate climate zone in North America, where it is found regularly on bark and rotting wood (McCurdy 1969a; Reichenbach and Dworkin 1969), it has been reported in Europe only once, and in that case from soil (Dawid 1979). *Chondromyces crocatus* seems to be an omnipresent rhizosphere organism in India (Agnihothrudu *et al.* 1959); however, in 20 years of isolation from samples coming from all over the world, we obtained this species only once from a soil sample collected in Madeira. Thus, it seems that some species are adapted to very special niches. In not even one single case do we understand the reasons for the localization of species. Why, for example, does *Stigmatella aurantiaca* prefer bark and rotting wood? It is not a cellulose decomposer nor does it have special growth requirements compared with other myxobacteria. Here is a ripe field for the physiologically oriented ecologist.

Some myxobacteria, such as the *Myxococcus* and *Corallococcus* species and *Nannocystis exedens,* are apparently very flexible with respect to the habitats they can populate. Perhaps with the exception of cold climates, *N. exedens* is the most common of all myxobacteria, although it is never mentioned in studies on the ecology of myxobacteria.

Although most myxobacteria are rather sensitive to sodium chloride, *Myxococcus virescens* (Rückert 1978b) and some *N. exedens* strains are still able to grow reasonably well in presence of 0.8% salt. Many myxobacteria were originally isolated from dung, mainly of herbivorous animals; the classical source of myxobacteria is rabbit dung. While it was believed for a long time that these organisms would depend on certain factors contained in this complex substrate, all myxobacteria obtained from dung are regularly found in soil as well and can easily be grown on the standard media for myxobacteria based on peptones or yeast cells. Although myxobacterial sporangioles and perhaps myxospores apparently pass undamaged through the intestinal tract of animals (Kühlwein 1950; Reichenbach and Dworkin 1981), it is not likely that they live permanently within animals, being strictly aerobic organisms. Rather, myxobacteria colonize the dung later from the surrounding soil, being attracted by the rich supply of organic material with its microbial flora. This hypothesis is corroborated by one of the classic isolation methods for myxobacteria: baiting soil samples with sterilized rabbit dung (Krzemieniewska and Krzemieniewski 1926).

## Physiology, Isolation, and Cultivation

Our knowledge of the physiology of myxobacteria is rather patchy. It has been assumed that myxobacterial metabolism is more or less identical with that of other aerobic Gram-negative bacteria. In general, this may be true, but myxobacteria still seem to have some pecularities that are important to know when working with the organisms. Since later chapters will go into the details of myxobacterial physiology, in this section I shall only discuss some general aspects.

Myxobacteria are strictly aerobic mesophilic organisms. They grow in the range between 6 and 38 °C, with the optimum at about 34 °C. Generation times are generally about 4–12 hr.

There are two distinct physiologic types of myxobacteria. By far the majority are "micropredators" (Singh 1947) with an essentially bacteriolytic–proteolytic metabolism. *Myxococcus xanthus* falls into this class. A few species, which apparently all belong to the genus *Sorangium,* are cellulose decomposers. The cellulose degraders may be cultivated either on filter paper or glucose and use inorganic nitrogen as the only nitrogen source (Krzemieniewska and Krzemieniewski 1937a,b; Imschenezki 1959; Pronina 1962; Couke and Voets 1967). They are not really dependent on cellulose. In fact, they grow very slowly on filter paper, usually producing swarm colonies of only 1–2 cm in diameter in 6–8 weeks. They grow much faster on media containing peptone plus starch (Ringel *et al.* 1977) or with autoclaved yeast cells. Many strains also grow on streaks of autoclaved *Escherichia coli* on water agar, and some even grow on pure peptone media. Thus, it is not clear whether cellulose is really their natural substrate.

The organisms of the bacteriolytic–proteolytic type excrete a host of exoenzymes degrading many kinds of biopolymers (see Chapter 5). Many strains do not use mono- and disaccharides (Watson and Dworkin 1968), nevertheless oligo- and polysaccharides may have a growth-stimulating effect (McCurdy 1963; Norén 1955). There are, however, also bacteriolytic myxobacteria that metabolize glucose quite efficiently, either with (*Stigmatella aurantiaca:* Reichenbach and Dworkin 1969; Gerth and Reichenbach 1978) or without (*Cystobacter fuscus:* unpublished) production of acid. The bacteriolytic organisms can be cultivated on peptone-containing media; apparently they rely heavily on the peptides within the peptone. Usually they do not grow on amino acid mixtures such as acid hydrolysates of casein. It has been possible to develop defined media for a few strains, consisting mainly of mixtures of amino acids (Dworkin 1962; Mayer 1967; Witkin and Rosenberg 1970; Bretscher and Kaiser 1978). Each medium can only be used with that one strain for which it has been developed. The problem is obviously that there are essential amino acids that differ from strain to strain, that the quantitative relationship between the amino acids may become very critical, and that alternative energy and carbon sources are often not available or, to be more exact, not known. Surprisingly, there are also bacteriolytic myxobacteria that grow on relatively simple media. One strain of

*Cystobacter fuscus* grows on a liquid medium consisting of 0.5% glucose, 0.1% glutamine, 0.004% phenylalanine 0.004% isoleucine, 0.1% $MgSO_4 \cdot 7H_2O$, 0.5 m*M* phosphate buffer pH 6.5, and standard trace elements (Reichenbach unpublished data). Another strain of *Cystobacter fuscus* grows on the same simple medium but requires thiamine in addition. Vitamin requirements have only rarely been demonstrated in myxobacteria. Another example is a strain of *Corallococcus macrosporus* that requires vitamin $B_{12}$ (Reichenbach and Dworkin 1981).

The isolation, cultivation, and conservation of myxobacteria have been summarized elsewhere (Kühlwein and Reichenbach 1965; Peterson 1969; Reichenbach and Dworkin 1981). I should, however, mention some recently introduced improvements in the purification of myxobacteria (Reichenbach 1983). Purification can become a considerable problem because myxobacteria do not grow too vigorously, spread into contaminated areas, and deposit a slime sheet that gives shelter to contaminants and prevents easy suspension for plating. When isolating cellulose degraders on pieces of filter paper, nematodes may become a considerable nuisance. By storing the contaminated plate for 1–3 days at −80°C and then, after thawing, making a transfer from the lysis zone to a fresh plate, we can usually get rid of the nematodes. According to Dr. Martin Aschner, soil amoeba, another serious probelm in enrichment cultures, can selectively be killed by exposing the culture plate for 30–120 sec to the vapors of a 5% ammonia solution; after another 60–120 sec, the myxobacteria are transferred to a fresh plate. Bacterial contamination can be eliminated by shaking fruiting bodies in a rich medium, which prevents germination of the myxospores but allows good growth of the contaminants, and to which is added a mixture of antibiotics at high concentrations. After 15–20 hr at 30°C, the fruiting bodies are washed and transferred to a suitable agar medium that allows germination and growth of the myxobacteria. Swarms appear within 1–6 days. In this way we often get pure strains in one step.

## Phylogeny and Taxonomy

Until recently the phylogenetic origin of the myxobacteria was totally in the dark. It has been speculated that there might be connections to the cyanobacteria because of their gliding motility (Jahn 1924) or to the actinomycetes because of some correspondence in their chemical composition (Reichenbach 1981). Both hypotheses have been shown to be incorrect by 16 S ribonucleic acid (RNA) studies (Ludwig *et al.* 1983). The myxobacteria form a clearly interrelated group of organisms that can be connected to the large branch of the purple bacteria among the Gram-negative bacteria. Thus, the next relatives of the myxobacteria are the sulfur and nonsulfur purple bacteria, the enterobacteria, and the pseudomonads (Stackebrandt and Woese 1981).

Surprisingly, from the 16 S RNA studies there also emerged the fact that there is no relationship between the myxobacteria and *Cytophaga johnsonae.*

It had been proposed in several taxonomic schemes that the *Cytophaga*-like bacteria and the myxobacteria form a common phylogenetic branch, the class (or subclass, or order) *Flexibacteriae*. (Soriano and Lewin 1965; Soriano 1973; Reichenbach 1974a, 1981). There seemed to be good reasons for such an assumption. Not only are essentially all unicellular representatives of the gliding bacteria found within these two groups, but they also share two biochemical peculiarities: their fatty acid patterns are dominated by isobranched and 2- and 3-hydroxy fatty acids (Walker 1969; Schröder and Reichenbach 1970; Ware and Dworkin 1973; Fautz *et al.* 1979, 1981; Oyaizu and Komagata 1981) and their respiratory chains contain exclusively menaquinones (Kleinig *et al.* 1974; Callies and Mannheim 1978; Oyaizu and Komagata 1981). Both properties are unusual for aerobic Gram-negative bacteria.

The 16 S RNA data clearly support the idea that all fruiting body forming gliding bacteria belong phylogenetically together and can thus taxonomically be united, perhaps as an order, *Myxobacterales*. In accordance with this is the narrow GC range of the deoxyribonucleic acid (DNA)-base composition of myxobacteria, which is between 66 and 72% (Mandel and Leadbetter 1965; McCurdy and Wolf 1967; Behrens *et al.* 1976). Within the order *Myxobacterales* there are two distinct subgroups, which differ in their cell shape (Krzemieniewska and Krzemieniewski 1928), in their colony form (Reichenbach and Dworkin 1981), in the staining behavior of their slime with Congo red (McCurdy 1969a), in their fatty acid patterns (Fautz *et al.* 1979, 1981), and in their pigment patterns (see Chapter 6). Therefore, a proposal has been made to subdivide the *Myxobacterales* into two suborders, the *Cystobacterineae* and *Sorangineae* (Reichenbach 1974a). This proposal is supported by the 16S RNA studies that show a deep division line between *Myxococcus, Cystobacter* and *Stigmatella,* on the one side, and *Sorangium* and *Nannocystis,* on the other (Ludwig *et al.* 1983). The similarity between *Sorangium* and *Nannocystis* is only slightly higher than that between the two suborders, so that two families, *Polyangiaceae* and *Nannocystaceae,* could be assumed in this branch, whereas there is apparently only one family, *Cystobacteraceae,* in the other. The difference between *Sorangium* and *Nannocystis* is also indicated by the fact that *Nannocystis* contains very unusual pigments (see Chapter 6) and appreciable amounts of steroids (Kohl *et al.* 1983).

There are several treatises on the taxonomy of myxobacteria on the genus and species level (Jahn 1924; Krassilnikov 1959; McCurdy 1974; Reichenbach and Dworkin 1981). This taxonomy rests almost exclusively on morphologic characteristics: the shape, structure, and size of fruiting bodies, swarm colonies, myxospores and vegetative cells. In some cases (e.g., *Myxococcus xanthus* and *M. virescens*), pigmentation has been used as the distinguishing character. *Myxococcus stipitatus* seems to be characterized by a compound with strong yellow fluorescence at 366 nm (Lampky and Brockman 1977), which is very useful in its otherwise difficult differentiation from *Myxococcus fulvus.* In general, pigmentation is not a reliable character because it strongly depends on growth conditions (e.g., illumination) and there may be a (genetically determined?) variability within the strains. The only physiologic characteristic of taxonomic relevance is the ability to decompose cellulose (genus *Sorangium*). Efforts to find diagnostically useful physiologic characteristics have failed so far (McCurdy 1969a). With better knowledge of the metabolism of myxobacteria perhaps new possibilities will emerge (e.g., a differentiation of *Archangium* and *Cystobacter* on the basis of glucose utilization). As already mentioned, there are several biochemical properties that help to classify larger groups within the myxobacteria; they are Congo red reaction, fatty acid, and pigment patterns.

There are many unresolved problems in myxobacterial taxonomy and diagnosis, and activity in this field would be welcome. These problems have three sources: (1) The first source is an overemphasis in the past on fruiting body morphology. Thus, *Stigmatella aurantiaca* and *Stigmatella erecta* were wrongly classified until recently, because the treelet-shaped fruiting bodies of the former suggested a relationship to *Chondromyces,* which is in fact only remotely related, whereas the appearance of a stalk with one single sporangiole in the latter was regarded as so essential that *Stigmatella erecta* was united with the *Melittangium* species (McCurdy 1969a; Reichenbach and Dworkin 1969; McCurdy 1971). (2) The second source is the tendency of strains under cultivation to form fruiting bodies with an atypical degenerate morphology. So far we do not know how to restore the typical morphology of fresh isolates. Thus, *Cystobacter violaceus* (Kühlwein and Gallwitz 1958) changed in culture to such a degree that it was renamed *Archangium violaceum* (Kühlwein and Reichenbach 1964). (3) The third source is the lack of specific and reliable biochemical and physiologic characters that would also allow for classification of "atypical" strains. Presently, major problems with taxonomy and diagnosis are seen in the *Cystobacter–Archangium–Melittangium* complex, with *M. xanthus* and *M. virescens,* the *Corallococcus* group, the *Cystobacter* and *Polyangium* species, and the *Sorangium cellulosum* complex. Further, we can be certain that there are still undescribed organisms related to *N. exedens.* These problems could efficiently be attacked today with the methods of DNA–DNA and DNA–RNA hybridization. A first attempt in this direction has been made (Johnson and Ordal 1968). However, careful comparative studies on the morphology, biochemistry, and physiology of representative strains are also badly lacking and would indeed be an essential complementation to the molecular taxonomic investigations.

# 2

# Structure and Function of Myxobacteria Cells and Fruiting Bodies

David White*

The structure of vegetative myxobacterial cells is generally similar to that of other Gram-negative bacteria. The cells are surrounded by a polysaccharide glycocalyx (slime) through which fimbriae sometimes protrude at the cell poles. The slime forms a common matrix for the cell population when growth occurs in swarms on solid surfaces. Underneath the slime layer is a cell envelope consisting of an outer membrane and an inner membrane (plasma or cell membrane). Sandwiched between the outer and inner membranes is a third layer, presumably the peptidoglycan, responsible for much of the strength of bacterial cell walls. The cytoplasm is typically prokaryotic and contains no obvious membrane-bound organelles. The deoxyribonucleic acid (DNA) appears to be similar to other bacterial DNA (see Chapter 3). The cells move by gliding motility, but it has not been possible to find any intracellular organelles or filaments that are specifically involved in motility (Burchard 1981). As discussed in this chapter, polar fimbriae are involved in the gliding of groups of cells. Fimbriae, however, are not required for the gliding of single cells.

*Myxobacteria are not slime molds.* The myxobacteria are alone among the known prokaryotes in that the individual cells aggregate to form large multicellular fruiting bodies. When one discusses the phenomenon of single-cell microorganisms aggregating to form a multicellular structure, the comparison between the cellular slime molds and the myxobacteria is inevitable. However, such comparisons can be misleading. The myxobacteria are distinct from the cellular slime molds not only in being typical prokaryotes in cell structure and chemistry, as described in this chapter, but also in the dynamic interactions within the cell community. Whereas the cellular slime mold in the feeding

*Department of Biology, Program in Microbiology, Indiana University, Bloomington, Indiana 47405 U.S.A.

stage is a solitary organism, myxobacteria grow as a community of interdependent cells in a common slime matrix.

When the cellular slime mold aggregates, individual cells emit positive chemotactic signals (e.g., adenosine 3′:5′-cyclic phosphate—cAMP) that attract other amoebae over large distances (in terms of cell length) to the aggregation centers (Rossier *et al.* 1980). There is no necessity for the myxobacteria to begin aggregation in this way since the cells are already relatively close together and contact between cells is frequent. A change either in motility pattern or in cell adhesiveness alone could possibly initiate aggregate formation. Other aspects of fruiting body development as the construction of stalks and sporangia (e.g., in *Stigmatella*) may bear very little resemblance to events that take place during slime mold fruit development.

It would be more profitable to compare events that take place during myxobacterial development with the behavior of other prokaryotes. Bacteria often live in communities attached to plant, animal, or inanimate surfaces, as well as to each other (Corpe 1980). They also attach to each other transiently, for example, during mating (Achtman and Skurray 1977). In at least one case (*Streptococcus* plasmid transfer) it is known that chemical signal compounds are secreted by the cells before cell–cell attachment presumably to promote that attachment (Dunny *et al.* 1979). These signals promote changes in the surface properties of the cells. Myxobacteria attach to each other during fruiting body formation and almost certainly exchange signals that influence gene expression and cell differentiation. A great deal can be learned from myxobacteria about how bacteria recognize other cell surfaces, how they attach, the nature of intercellular signals, and how they influence cellular behavior and differentiation. As the attention of microbiologists begins to shift from the physiology and genetics of the single cell to the biology of cell communities, the myxobacteria will be an invaluable source of information. An approach to investigating myxobacteria development and cell interactions is to focus on programmed changes that occur in the structure of the developing cells and fruiting bodies and how these changes are induced or maintained by exogeneous stimuli.

## Structure of Vegetative Cells and Myxospores

### Two Cell Types

As described in Chapter 1, myxobacteria are capable of cellular differentiation and exist as two distinct cell types—vegetative cells and myxospores (Figs. 2.1 and 2.2). The vegetative cell is the growing cell. Vegetative cells are either long thin rods with tapered ends or short fat rods with blunt ends. Myxospores can be spheric, kidney-bean shaped, or short fat rods. Often a myxospore coat (capsule) surrounds the myxospore, but this is not universal. It is presumed that myxospores are physiologically different from vegetative cells, but this has been

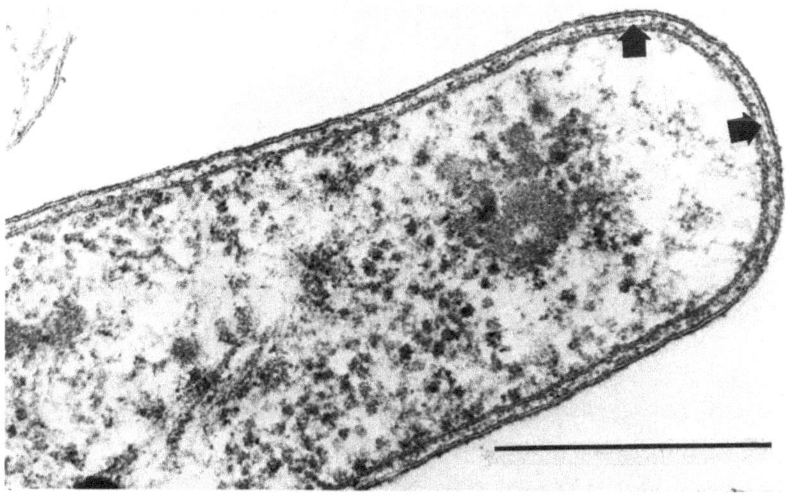

**Figure 2.1.** Cell pole of a vegetative cell, *Stigmatella aurantiaca*. The arrow points to the putative peptidoglycan layer between the inner and outer membranes. The bar is 0.5 μm. Photograph courtesy of Dr. Herbert Voelz. Reichenbach *et al.* (1969). J. Bacteriol. 97:905–911.

investigated in detail only with *Myxococcus xanthus* (Dworkin and Niederpruem 1964; Sudo and Dworkin 1969). *Myxococcus* myxospores have a very low metabolic rate and are much more resistant to desiccation than are the vegetative cells. The process of converting from a vegetative cell to a myxospore is an encystment and grossly resembles the formation of several other prokaryotic resting cells such as *Sporocytophaga* microcysts, *Azotobacter* cysts, and cyanobacterial akinetes (Sudo and Dworkin, 1973).

## Fimbriae and Their Possible Roles in Cell–Cell Adhesion and Gliding Motility

Vegetative cells of myxobacteria have slender threadlike appendages extending from the cell poles (Fig. 2.3). These were first reported by MacRae and McCurdy in the myxobacterium *Chondromyces crocatus* and subsequently found by McCurdy and colleagues in other myxobacteria, including *Myxococcus* and *Stigmatella* (MacRae and McCurdy 1975; MacRae, Dobson and McCurdy 1977). Dobson and McCurdy purified the appendages from *M. xanthus* and concluded that they were glycoprotein (Dobson and McCurdy 1979). As pointed out by these authors, the appendages resemble fimbriae or pili that are widespread in bacteria, especially Gram-negative bacteria.

It is likely that the fimbriae in myxobacteria are involved in cell–cell adhesion within the myxobacterial community. Dobson *et al.* (1979) pointed out

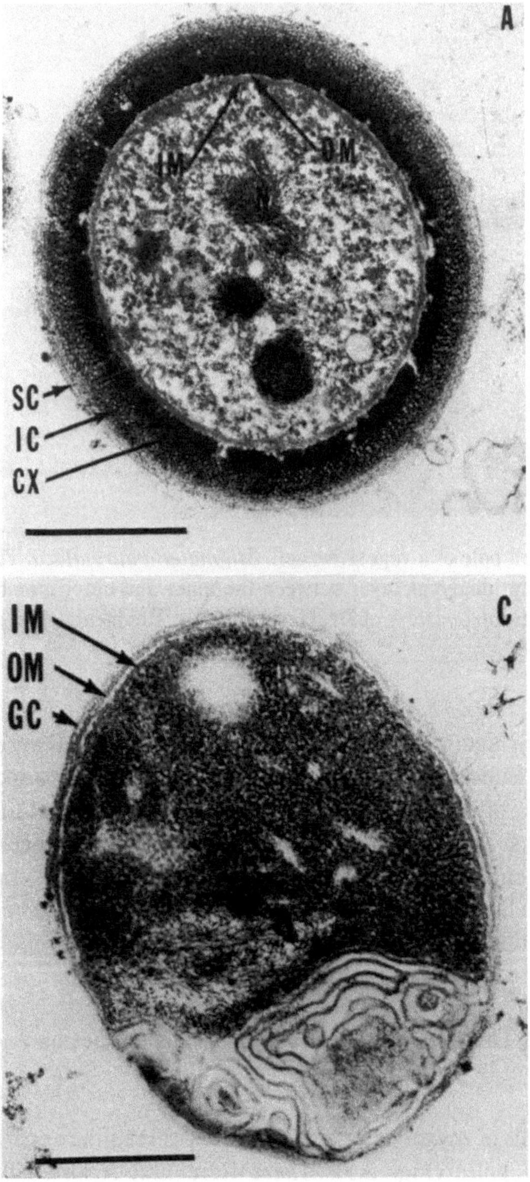


**Figure 2.2.** Myxospores of *Myxococcus xanthus.* (A) Fruiting body myxospore. (C) Glycerol-induced myxospore. I, Inner membrane; OM, outer membrane; CX, cortex; IC, intermediate coat; GC, glycerol spore coat. The bar represents 0.5 $\mu$m. From Inouye *et al.* (1979). Proc. Natl. Acad. Sci. 76:209–213.

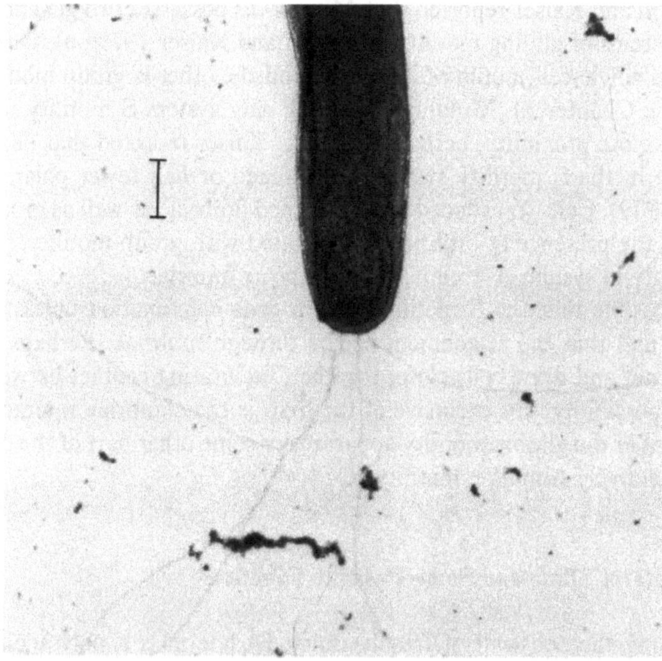

**Figure 2.3.** *Myxococcus xanthus* with polar pili. The cells were negatively stained with uranyl acetate. The bar is 0.5 μm. Photograph courtesy of Dr. Dale Kaiser. From Kaiser (1979). Proc. Natl. Acad. Sci. 76:5952.

that the tendency to form cell clumps in liquid cultures correlated with the presence of fimbriae (Dobson, McCurdy and MacRae 1979). Kaiser also made the observation that fimbriated strains of *M. xanthus* grow as clumps in liquid and attach to the glass, whereas nonfimbriated strains grow dispersed (Kaiser 1979). The bacterial receptors for the myxobacterial fimbriae are not known. The fimbriae may attach to a polysaccharide or oligosaccharide. This conclusion follows from the report by Dobson *et al.* (1979) that isolated fimbriae from *M. xanthus* agglutinated red blood cells and the agglutination was prevented by mannose or *N*-acetylgalactosamine. Furthermore, these authors reported that the addition of mannose or *N*-acetylgalactosamine to liquid meida caused fimbriated cells to grow dispersed. The fimbriae may attach to the polysaccharide in the slime since this material consists of mannose and sometimes hexosamine (Sutherland and Thomson 1975). Alternatively, the fimbriae may protrude through the slime and attach to the oligosaccharide portion of the lipopolysaccharide. The latter possibility is attractive because it provides an opportunity for direct contact of the fimbriae with the cell envelope of adjacent cells. This conceivably could influence cooperative gliding motility as discussed below.

Hodgkin and Kaiser reported that *M. xanthus* possesses two genetically separable systems of gliding motility (Hodgkin and Kaiser 1979a,b). One type of motility is single-cell motility (system A) and the other is group motility (system S; see Chapter 8). Mutants possessing only system S motility will move only if in close proximity, perhaps touching. Kaiser reported that all mutants examined in the S motility system ($S^-$) lacked or had fewer polar fimbriae (Kaiser 1979). Cells transduced to $S^+$ regained fimbriae as well as group motility. Thus the presence of fimbriae is correlated with group motility. Cells can move singly in system A even if they lack polar fimbriae.

It is possible that the S motility system does not function unless cells are attached and that the attachment occurs through fimbriae. Perhaps the fimbriae retract and draw cells closer together, facilitating contact between cells. Another possibility, not exclusive of the first, is that fimbriae interact with a component of the gliding motility apparatus or some other part of the cell envelope and actively stimulate motility.

## Myxobacteria Slime and Some Possible Functions

Surrounding the cell wall of freshly isolated bacteria is a polysaccharide or glycoprotein layer sometimes called the glycocalyx (Costerton and Irvin 1981). In Gram-negative bacteria the glycocalyx is external to the outer membrane. The form of the glycocalyx can vary from a regular array of glycoprotein subunits or a well-defined thick capsule to a loose "slime" that is shed into the medium. Myxobacteria are surrounded by a loose glycocalyx commonly referred to as a slime layer.

The most thorough chemical analysis of myxobacterial slime was published by Sutherland and Thomson (1975). They analyzed acetone-precipitable material found in culture filtrates of cells grown in suspension or acetone-precipitable material that could be dislodged by vortex mixing from vegetative cells grown on a agar surface. Their methods would not have detected material tightly bound to the cell surface or soluble in acetone. The material analyzed was neutral polysaccharide. The yields were very high (i.e., 5–10% of the dry weight of vegetative cells grown on solid media). The most abundant sugars were mannose, D-glucose, and D-galactose. One strain also had substantial amounts of hexosamine. Sutherland and Thomson (1975) pointed out that the absence of uronic acids in the exopolysaccharide is uncommon for bacterial exopolysaccharides.

There are several possible functions for the myxobacteria slime. One is cell–cell adhesion, or adhesion to substrata on which the bacteria grow. Other possible roles for the slime include the prevention of dehydration and protection against phagocytosis and toxins. Burchard (1981) discussed the possibility that the slime may assist in gliding motility.

Does the myxobacterial slime perform any or all of these functions? There are no definitive answers. A more thorough analysis of the chemical structure of the slime in both the wild type and mutants that are unable to adhere or to glide would be helpful. Since at least a portion of the slime can be removed from the cells by vortex agitation, it might be possible to raise specific antisera to it and use an immunologic approach to investigate the role that the slime might play in adhesion and motility.

## Outer Membrane

Gram-negative bacteria have a discrete membrane external to the peptidoglycan surrounding the cell membrane. It is called the outer membrane. The cell membrane (inner membrane) and the outer membrane are referred collectively to as the cell envelope. The outer membrane consists of lipopolysaccharide, protein, and phospholipid. The properites of the outer membrane were reviewed by Nikaido (1973, 1979). The proteins of the outer membrane were recently reviewed by Osborn and Wu (1980). Both protein and the oligosaccharide portion of the lipopolysaccharide are exposed to the outside of the cell and are thus potentially available for interaction with the environment and other cells. Several functions have been attributed to the outer membrane. These include a role as a permeability barrier to toxic compounds including antibiotics, hydrolytic enzymes, and surfactants. Several of the proteins in the outer membrane in *Escherichia coli* are involved in the specific transport of nutrients and metals into the cell (Nikaido 1979). At least one of the outer membrane proteins, *pOmp A,* may be involved in the formation of stable mating aggregates in cultures of *E. coli* (Achtman and Skurray 1977). The lipopolysaccharide in the outer membrane of *Agrobacterium* has been postulated to bind to a polysaccharide in wounded portions of plant cell walls. This event possibly precedes the establishment of the crown gall tumor (Whatley *et al.* 1976; Lippincott and Lippincott, 1978; Banerjee *et al.* 1981). There is also a report that the lipopolysaccharide in *Rhizobium* attaches to the root hair lectin (Wolpert and Albersheim 1976). It should be pointed out that the phenol extraction procedures for lipopolysaccharides will also extract nonlipopolysaccharide polysaccharides. Such polysaccharides can contaminate the lipopolysaccharide preparations as reported for certain *Rhizobium* preparations (Planque and Kijne, 1977). Also, certain strains of *E. coli* have a nonlipopolysaccharide glycan apparently covering the lipopolysaccharide and resembling it chemically and immunochemically (Goldman *et al.* 1982).

The first report of the isolation of lipopolysaccharide from myxobacteria was by Weckesser *et al.* (1971). The lipopolysaccharide was extracted frym *Myxococcus fulvus* using aqueous phenol and purified by centrifugation. The sugars identified were mannose, galactose, rhamnose, glucose, ribose, xylose, arabinose, and 3-*O*-methyl-D-xylose. A short time afterward, Smith and Sutherland (1973) analyzed the sugars in the lipopolysaccharide preparations from several myxobacteria including *M. xanthus* and *Stigmatella aurantiaca.* The monosaccharides varied from strain to strain and included ribose, mannose, galactose, rhamnose, glucose, glucosamine, and galactosamine. Notable was the absence of heptose, a sugar common to many bacterial lipopolysaccharides. A more thorough examination of the lipopolysaccharide from *M. fulvus* by Rosenfelder *et al.* (1974) showed that the lipopolysaccharide was essentially the same as the lipopolysaccharide from other Gram-negative bacteria, except for the absence of heptose.

Orndoff and Dworkin (1980) succeeded in separating the outer and inner membranes from *Myxococcus* by isopycnic density centrifugation. The outer membrane was less dense than the inner membrane, a situation that is the reverse of what is found in *E. coli*. This may be due to the very high phospholipid content in the *Myxococcus* outer membrane. The polypeptide content of the outer membrane was complex. Approximately 26 polypeptides were revealed by one-dimensional gel electrophoresis. At least 10 of these were present in major amounts.

There have been no reports of the involvement of the lipopolysaccharide or outer membrane proteins in cell–cell adhesion, signaling, aggregation, or fruiting body development. The investigations of these matters must await a more thorough chemical analysis of the outer membrane and the isolation of mutants defective in outer membrane components. Such mutants would be expected to include those resistant to certain phages and to antibiotics and may include mutants defective in gliding motility.

## Peptidoglycan Layer

Most bacteria contain within their cell walls a heteropolymer called peptidoglycan (murein, mucopeptide, or glycopeptide), which is in large part responsible for the mechanical strength of the cell wall. Peptidoglycan consists of glycan chains of alternating residues of *N*-acetylmuramic acid and *N*-acetyglucosamine. The glycan chains are cross-linked to each other through short peptide chains. The structure of peptidoglycan was reviewed by Ghuysen and Shockman (1973). In thin sections of Gram-negative bacteria, viewed with a transmission electron microscope, the peptidoglycan can sometimes be observed as a darkly stained line 30–80 Å thick situated between the outer and inner membranes (Glauert and Thornley 1969). The identity of this layer as peptidoglycan can be confirmed by demonstrating its destruction by lysozyme (Murray *et al.* 1965). In *E. coli* and certain other Gram-negative bacteria, the peptidoglycan is difficult to visualize in thin sections apparently because it sometimes adheres to the outer or inner membrane (Glauert and Thornley 1969). Most published electron micrographs of myxobacteria do not show a peptidoglycan layer. A putative peptidoglycan layer was observed by Reichenbach *et al.* (1969) in *Stigmatella* and by MacRae and McCurdy (1975) in *Chondromyces* (Fig. 2.1). In Weidel's pioneering publications, he suggested that the peptidoglycan was solely responsible for the strength and rigidity of bacterial cell walls (Weidel *et al.* 1960; Weidel and Pelzer 1964). In fact, the peptidoglycan from *E. coli* and some other Gram-negative bacteria can be isolated as a huge bag-shaped macromolecule of approximately the size and shape of the cell and held together entirely by glycosidic and peptide bonds (Weidel *et al.* 1960; Heilmann 1972). It is called a sacculus. The sacculus from *E. coli* is resistant to trypsin and detergent but is sensitive to lysosyme, an enzyme that catalyzes the hydrolysis of the glycan portion. It has been suggested that the covalently bonded structure is necessary to provide the tensile strength required to withstand the large hydrostatic pressures within the bacterial cell (Koch 1983).

Myxobacteria often appear to be flexous when gliding. At one time it was not clear whether they possessed a peptidoglycan-containing cell wall as did the more rigid bacteria. However, Mason and Powelson (1958) published a report in which they described for the first time the presence of the peptidoglycan component, diaminopimelic acid, in the cell wall of vegetative cells of *M. xanthus*. This was followed by a report from Adye and Powelson (1961) that a substance tentatively identified as the peptidoglycan component, muramic acid, was present in myxospores of *M. xanthus*. White *et al.* (1968) reported the isolation and chemical composition of the peptidoglycan from vegetative cells and myxospores of *M. xanthus*. *Myxococcus* vegetative cells readily autolyze, and to minimize autolysis sodium lauryl sulfate (SLS) was added to a final concentration of 10% before the cells were harvested by centrifugation. Acid hydrolysis of the cell walls revealed the presence of the peptiodoglycan components. These were glutamic acid, diaminopimelic acid (DAP), alanine, glucosamine, and muramic acid in molar ratios of 1.0, 1.0, 1.7, 0.75, and 0.75. In addition, there were substantial amounts of glycine and some serine. All other amino acids were present in insignificant amounts.

Trypsin treatment of the cell walls had a surprising result. In contrast to *E. coli* sacculi, trypsin treatment of *Myxococcus* walls fragmented them into smaller pieces that could be sedimented by centrifugation. This suggested that the *Myxococcus* cell walls may consist of peptidoglycan patches held together by trypsin-sensitive material. The trypsin solubilized most of the glycine, implicating glycine in the putative interpatch material. The serine present in the cell wall preparation was not solubilized by trypsin. Lysozyme treatment of the trypsin fragments solubilized the peptidoglycan but not the serine, indicating that it was not covalently bound to the peptidoglycan. It should be emphasized that the evidence for a patchlike arrangement of peptidoglycan is indirect. It has not been ruled out that limited autolysis was not prevented by the detergent treatment. However, it was later reported that the addition of detergent to a culture of a marine pseudomonad did prevent autolysis and allowed the isolation of a complete peptidoglycan sacculus when previous attempts had failed (Forsbert *et al.* 1972).

Apparently there is considerable glucose covalently bound to the lysozyme-solubilized fragments of peptidoglycan from *M. xanthus*. This was discovered by Johnson and White (1972) who separated the fragments by electrophoresis and determined that in one of the fragments the glucose was bound to a moiety containing muramic acid, alanine, trace amounts of lysine, and possibly glutamic acid. Glucosamine and DAP were absent from the fragment (Johnson and White 1972). The molar ratios of glucose to DAP in the entire peptidoglycan was 2.6.

## Myxospore Coat

Myxospores are resting cells usually formed within fruiting bodies. The physiologic properties of *Myxococcus* myxospores have been reviewed (Sudo and Dworkin 1973; White 1975). Dworkin and Gibson (1964) described a method to induce myxospore formation rapidly and synchronously in cell suspension by adding large amounts of glycerol (Dworkin and Gibson 1964). This method, called glycerol induction, became the basis for a number of investigations of the physiology and chemistry of myxospore formation. The properties of glycerol-induced myxospores are not exactly the same as those of fruiting body myxospores (White 1975).

When myxospores are induced with glycerol, a myxospore coat is synthesized. This coat was isolated and its chemical composition determined (Kottel *et al.* 1975). The coat preparations were treated with detergent and lysozyme before analysis. Thus, if any peptidoglycan or lipid were present in the coat it probably would not have been detected. One preparation of coats isolated after detergent and lysozyme treatment contained on a dry weight basis 55.5% galactosamine, 22.6% glucose, 13.5% protein, 6.8% glycine, and 0.34% nonlipid phosphorus. Most of the glycine was probably not in the protein as discussed below. Coats remained morphologically intact despite boiling in 10 *M* urea, sodium lauryl sulfate, plus $\beta$-mercaptoethanol or extraction with warm phenol. The deposition of the galactosamine, protein, glycine, and phosphorus was complete by 6 hrs, but a glucose polymer was added to the coat over a 24-hr period.

Coats from fruiting body myxospores have not been adequately analyzed. However, the available information suggests some differences from glycerol-induced myxospore coats. Fruiting body myxospore coats are thicker and appear to consist of multiple layers in thin section (Fig. 2.2; Voelz and Dworkin 1962). Furthermore, fruiting body myxospore coats have an external protein layer, protein S, not found in glycerol-induced myxospores (Inouye *et al.* 1979b; Inouye *et al.* 1981). The protein can be removed from the myxospores with a high salt wash. Protein S will reassemble on the surface of salt-washed fruiting body myxospores if calcium ion is present. Protein S does not attach to glycerol-induced myxospores. It is possible that the myxospore coats from glycerol-induced cells reflects an incomplete myxospore coat, perhaps an internal layer. Kottel *et al.* (1975) isolated a coat from fruiting body myxospores whose polysaccharide composition resembled that of the coat from glycerol-induced myxospores. Protein and glycine were not analyzed.

### Physiologic Changes Accompanying Production of the Myxospore Coat

Most physiologic studies on myxospore formation have been performed with glycerol-induced myxospores. Since the glycerol-induced myxospore coat is 75% polysaccharide by weight, we might expect considerable gluconeogenesis to occur from the amino acids that are the carbon and energy sources for *Myxococcus*. In fact, during glycerol-induced myxospore formation, there occurs a dramatic shift of carbon flow in the direction of polysaccharide synthesis. Over a 24-hr period the amount of neutral polysaccharide increased approximately 200% and accounted for 20% of the dry weight of the cells (Bacon *et al.* 1975). However, 50% of the polysaccharide was recovered from the soluble portion of the cell and probably represented reserve material.

Gluconeogenesis by whole cells during glycerol-incuced myxospore formation was investigated by Filer *et al.* (1977) using radioactive acetate as a precursor to polysaccharide. The acetate was incorporated into *N*-acetyl-glucosamine-6-phosphate and uridine-5′-diphosphate-*N*-acetygalactosamine *(UDPGalNac)* plus uridine-5′-diphosphate-*N*-acetylglucosamine *(UDPGlcNac)* at an increased rate during myxospore formation. These intermediates would be expected to be coat polysaccharide precursors. Acetate was also incorporated into trichloroacetic acid insoluble material, which was probably coat. In vitro assays of enzyme activities revealed significant increases in specific activities of the enzymes required to convert fructose diphosphate to UDP-galactosamine (Filer *et al.* 1977). The activities of the enzymes began to increase just before the beginning of coat polymer deposition. It is interesting that the activities of the enzymes remained high for only a few hours and then decreased. Orlowski *et al.* (1972) also observed transient increases in enzyme activities during glycerol-induced myxospore formation. The mechanism of enzyme inactivation was not determined.

Net synthesis of carbohydrate from acetate is often associated with the enzymes of the glyoxylate bypass—isocitrate lyase and malate synthase. In vitro assays showed that both of these enzymes increased in activity during the early stages of myxospore formation (Orlowski *et al.* 1972). The acetate that feeds into the glyoxylate bypass probably originates from the catabolism of amino acids.

The glycine in the myxospore coat deserves special comment. Filer *et al.* (1977) showed that glycine incorporation into the coat fraction was inhibited 75% by bacitracin. Valine incorporation into was inhibited only 15%. This suggests that much of the glycine may not be part of the coat protein and that it is incorporated via a precursor containing bactoprenol, the $C_{55}$ isoprenoid phospholipid. The glycine itself could be synthesized from glyoxylate generated in the glyoxylate bypass using transaminases reported to be present in *M. xanthus* (Kottel *et al.* 1974). The biochemical and topographic details of coat synthesis have not been elucidated and remain subjects for further study.

## Problem of Shape Change during Myxospore Formation in *Myxococcus*

One of the most intriguing aspects of myxospore formation in *Myxococcus* is the fact that the cell changes from a cylinder to a sphere. During glycerol induction the change in shape is rapid. After 30 or 40 min the cells begin to shorten, and by 50 min they are ovoids. Within 70–80 min they are spheres (Fig. 2.4; Dworkin and Gibson 1964). The surface area of the myxospore is no more than one-third that of the vegetative cell (unpublished observations). The amount of peptidoglycan per cell in *Myxococcus* remains the same during glycerol-induced myxospore formation despite the smaller surface area (White *et al.* 1968). However, glucose decreases by two-thirds and the cross linking between diaminopimelic acid and alanine increases 11% (Johnson and White 1972).

During myxospore formation there is increased turnover of the peptidoglycan as measured by the loss of radioactivity from peptidoglycan labeled with [$^{14}$C] mesodiaminopimelic acid (Dawson and Jones 1979). Bonds must be broken with the peptidoglycan and new bonds formed as the restructuring from a cylinder into a sphere takes place. In *E. coli,* at least, it is known that the peptidoglycan sacculus from rod-shaped cells is rod shaped and the sacculus from mutants that grow as spheres is round (Henning *et al.* 1972). (This does not necessarily mean that murein *determines* cell shape, merely that it reflects the shape of the cell.) How, then, does *Myxococcus* restructure its peptidoglycan into a sphere without severely weakening the wall?

One possible mechanism lies in the putative patchlike composition of the peptidoglycan suggested by White *et al.* (1968). Perhaps the interpatch material evaginates bringing the peptidoglycan patches closer together, allowing the patches to be connected by cross-linking via the diaminopimelic acid. This would account for the increased cross-linking in myxospore peptidoglycan. After the cross-links are formed, the interpatch material could be severed and allowed to enter the newly synthesized coat. The coat contains substantial amounts of glucose and glycine, and these might originate in part from the interpatch material between the peptidoglycan. This might explain the decrease in glucose in the peptidoglycan. Parts of the model are testable. For example, it may be possible to label the glucose and glycine in the vegetative wall with radioisotopes and then determine whether these same molecules are incorporated into the coat. The model has the virtue of allowing the cell to change the shape of the peptidoglycan layer and even remove parts of it without necessarily weakening the structure. This follows because the nonpeptidoglycan portions would not be removed before the peptidoglycan patches are cross-linked to each other. Similar considerations of the need to cross-link

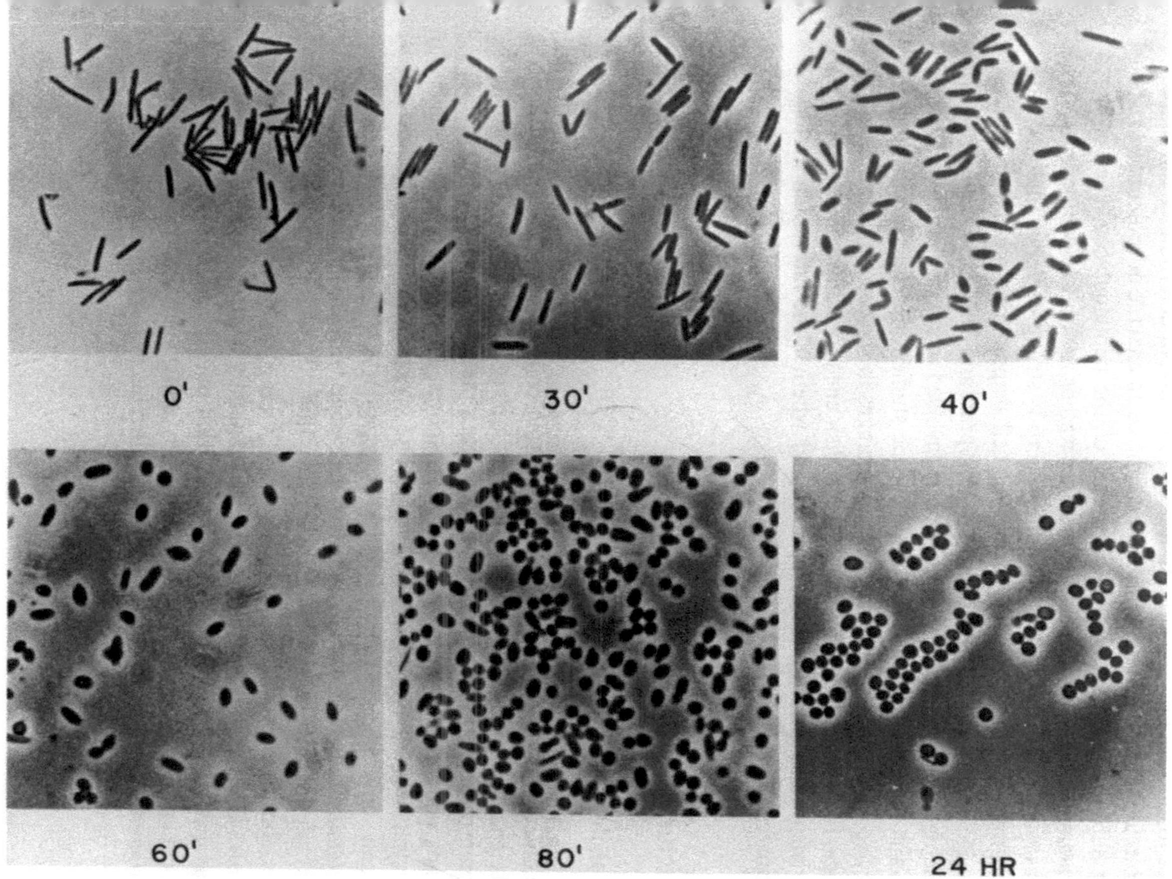


**Figure 2.4.** Morphologic changes during glycerol induction of *M. xanthus*. Glycerol was added to a growing culture of *M. xanthus* at a final concentration of 0.5 *M*. At the times indicated in the figure, samples of cells were fixed in glutaraldehyde and photographed with a phase microscope. The average length of the vegetative cells (0′) was 5 μm. From White (1975). Spores VI, American Society for Microbiology, pp. 44–51.

newly added peptidoglycan before cleaving existing bonds were discussed by Koch (1983) in a theoretical analysis of growth of the *E. coli* sacculus.

The discussion of cell shape change in *Myxococcus* has thus far omitted reference to any specific molecules that might function in determining cell shape. A molecule that *determines* cell shape might be an enzyme or protein (or some other substance) that alters the structure of the peptidoglycan layer or the interaction of the peptidoglycan with either wall or membrane components. As a consequence the cell becomes round. There is strong evidence that at least three proteins are involved in determining cell shape during normal growth of *E. coli* (Sonntag *et al.* 1978; Spratt 1975; Spratt and Pardee 1975).

The nature of the proteins that may be involved in shape change in *Myxococcus* is not known. Before the issue can be studied in detail, it will be necessary to learn more about the chemical changes in the cell wall during shape change. It is particularly important to establish unequivocally whether the peptidoglycan does indeed exist as patches and whether the interpatch material evaginates to become part of the coat.

## Structure of Fruiting Bodies

### Morphology

A detailed description of myxobacteria fruiting bodies was recently published by Reichenbach and Dworkin (1981). The simplest fruiting bodies are formed by members of the family *Myxococcaceae*. The fruits are either spheric mounds, columns, or ridges filled with myxospores. There is usually no distinct outer wall to the fruiting body or a stalk. The *Myxococcus* fruiting bodies are soft and slimy mounds that frequently become flat. Other myxobacteria such as *Stigmatella* form more complex fruiting bodies with distinct outer walls (or coats) of unknown chemical compositions (Fig. 2.5). Depending upon the genus, the fruits are either sessile or consist of sporangia (sporangioles) born on stalks as in *Stigmatella*. The stalks are frequently referred to as slime stalks, but as discussed below there is some indication that the stalks of *Stigmatella* are cellular. Myxobacterial fruiting bodies are quite small. For example, fruits of *Stigmatella* are approximately 100 $\mu$m tall. This can vary, however, and the source of variability is not known.

### Cellular Stalks

Fruits of *Stigmatella* and *Chondromyces* were examined in detail by Grilione and Pangborn (1975) using the scanning electron microscope. They concluded that the stalks of both myxobacteria were cellular in 2-week-old fruiting bodies. Eight-week-old fruits of *Condromyces* had stalks composed of empty tubules that may have been lysed cells.

**Figure 2.5.** Fruiting body of *Stigmatella aurantiaca*. Magnification is 1600×. Courtesy of Dr. Karen Stevens.

Stephens and White (1980) found that under certain conditions *Stigmatella* formed fruits without an outer coat. Under these circumstances it was possible to see unequivocally, using the scanning electron microscope, that the stalks were cellular. Voelz and Reichenbach (1969) examined thin sections of fruiting bodies of *Stigmatella* with the transmission electron microscope. They concluded that the stalks of mature fruits consisted of empty tubules, approximately the dimensions of cells, and that perhaps these tubules arose from lysed cells. They also concluded that there was little justification in referring to the stalks of *Stigmatella* as slime stalks. A cellular stalk suggests that the fruiting body may form by morphogenetic movements of the cells rather than by the secretion of slime at the base of a rising mass of cells as proposed earlier (Dworkin 1973; Parish 1979). In older fruits the cells in the stalk may lyse.

### Influence of Light and Chemicals

In 1975 Reichenbach reported at the Second International Symposium on the Biology of Myxobacteria that light was required for fruiting body formation in *Chondromyces apiculatus*. Later, Qualls *et al.* (1978b) reported that light strongly stimulated the development of fruiting bodies in *Stigmatella* and that the cells tended to aggregate into ridges in the dark. When cell densities are low, light causes the cells to aggregate faster and to produce more aggregates. The most effective wavelengths of light are in the blue range, but far red is also active (White *et al.* 1980b).

The shapes of *Stigmatella* fruiting bodies are also influenced by the cation composition of the medium and the inclusion of guanine compounds (White *et al.* 1980a; Stephens and White 1980). In addition, *Stigmatella* secretes a pheromone into the medium. When present at sufficient concentrations, the pheromone can substitute for light (Stephens *et al.* 1982). The data suggest that light stimulates fruiting by increasing the sensitivity of the cells to the pheromone (Stephens *et al.* 1982). Limited fruiting sometimes takes place in the dark at high-cell densities, and this may be due to the occasional production of sufficient pheromone. An interesting parallel occurs in the cellular slime mold, *Polysphondylium violaceum* (Teta and Hanna 1981). *Polysphondylium* aggregates faster and produces more aggregates in the light than in the dark. As with *Stigmatella, Polysphondelium* produces an aggregation factor (D factor) that stimulates aggregation of cells incubated in the dark in a similar manner to the stimulation of aggregation by light. A further parallel is that the *Polysophondelium* aggregation factor is more effective in the light than in the dark. It is interesting that despite the unrelatedness of the myxobacteria to the cellular slime molds, there may be similarities in the signaling systems in *Stigmatella* and *Polysphondelium.* Since *Myxococcus* aggregation does not appear to be stimualted by light and since the pheromone produced by *Stigmatella* does not enhance aggregation in *Myxococcus,* it is probable that the signaling systems differ in these two genera of myxobacteria. This is not surprising and is paralleled in the cellular slime molds that also have genus-specific signaling systems (Bonner 1983).

## Conclusions

There are several reports of cell–cell interactions in prokaryotes other than the myxobacteria; these interactions result in specific changes in cell structure and chemical composition. For example, *Streptococcus* cells signal each other before transfer of conjugative plasmids (Dunny *et al.* 1979). This is accompanied by changes in the chemistry of the cell surface. The cyanobacterium *Cylindrospermum* apparently produces a diffusible intercellular sporulation signal during sporulation in certain media (Hirosawa and Wolk 1971a,b). It has been reported that *Streptomyces griseus* produces a substance capable of promoting conidium formation in certain strains of *streptomyces* (Biro *et al.* 1980). Cell–cell signaling in prokaryotes is probably more widespread than now appreciated. The fact that myxobacteria development is strongly influenced by chemical and light signals (in certain genera) introduces the possibility of investigating the effect of light and chemicals, including cations, on chemical composition and organization of the cell surface and its appendages.

# 3

# Genome of *Myxococcus xanthus*

Thomas Yee and Masayori Inouye*

Is the relatively complex life cycle of myxobacteria reflected in a corresponding complexity in the physical organization of their DNAs? Do distinct differences exist between chromosomal replication in myxobacteria and nondevelopmental bacteria? Can these differences be correlated with the characteristic life cycle of myxobacteria? Such questions have been the focus of ongoing investigations in our laboratory and those of others. Most of the studies have concentrated on *Myxococcus xanthus*. The results of these investigations indicate that the genome of *M. xanthus* exhibits several unusual and characteristic features. This chapter is not a comprehensive review of all aspects of the myxobacterial genome; instead, it concentrates on those aspects of myxobacterial chromosome structure and dynamics that appear uniquely related to the myxobacterial life cycle.

## Coordination of DNA Replication with Cell Division

### Cell Cycle of *Escherichia coli*

First, we will briefly discuss general features of the cell cycle in *Escherichia coli;* then, we will compare this information with what is known about the cell cycle of *M. xanthus*. The cell cycle of *E. coli* is markedly different from that of eukaryotes. In all eukaryotes, it appears that the growing cell *must* proceed through an orderly sequence of clearly separated stages, G1, S, G2, and mitosis (Lewin 1980). In contrast, it has been noted that *E. coli* can engage in chro-

*Department of Biochemistry, State University of New York at Stony Brook, Stony Brook, New York, 11794 U.S.A.

mosomal replication throughout 100% of the cell cycle (Pachler *et al.* 1965). Under various conditions, DNA replication can be initiated before the previous round of synthesis has been completed, resulting in chromosomes with multiple replication forks. Only in slowly growing cultures can gaps in DNA synthesis be detected (Helmstetter 1967). A second point of difference between *E. coli* and eukaryotes is that whereas in eukaryotes, nuclear division generally precedes cell division by a few minutes at the most, in *E. coli* a round of chromosomal replication may be completed well before cell division. The data of Cooper and Helmstetter (1968) suggested a constant average delay of about 20 minutes between the completion of chromosome replication and the onset of cell division (the D period), regardless of the growth rate, although it must be noted that individual cells exhibit wide variations around this average period (Chai and Lark 1970). Furthermore, the time required for one round of DNA replication (the C period) is nearly constant at 40 min over a wide range of growth rates (Chandler *et al.* 1975), decreasing to a limit of 37 minutes only at high growth rates (Churchward and Bremer 1977). Hence, the major control of the repplication rate in *E. coli* must be the rate of chromosomal initiation. A variety of experimental evidence has been shown to be consistent with the Cooper–Helmstetter model (Helmstetter *et al.* 1979; Bremer and Chuchward 1977a; Mendelson 1982). Measurements at different growth rates of the DNA per cell (Cooper and Helmstetter 1968), the DNA per cell mass ratio (Pritchard and Zaritsky 1971), and the C/(C + D) ratio (Buckley and Anagnostopoulos 1976) are in agreement with this model. Bird *et al.* (1972) and Chandler and Pritchard (1975) found that this model correctly predicted the relative gene frequencies of various markers situated at different positions in the *E. coli* chromosome. This model has also proven useful in the analysis of transient accumulations of DNA during various perturbation experiments including the following: (1) step-up experiments in which additional thymine is added to cultures of thymine-requiring bacteria growing in low thymine, resulting in decrease of the C period (Bremer *et al.* 1977); (2) experiments in which protein synthesis is inhibited, which prevents new initiations while not affecting the elongation rate of already initiated molecules (Bremer and Churchward 1977b); (3) experiments in which DNA elongation but not initiation is temporarily halted by nalidixic acid or thymine starvation (Zaritsky 1975; Ward and Glaser 1970).

Although the Cooper–Helmstetter model has been markedly successful in many applications, Helmstetter *et al.* (1979) cautioned against possible misapplication of the model to inappropriate systems. The model was developed using a particular strain (B/r) of *E. coli,* but the conclusions reached are not necessarily applicable even to different strains. For example, in cultures of *E. coli* growing at very slow growth rates, whether the C period is invariant or not is a strain-dependent phenomenon. Furthermore, although the Cooper–Helmstetter model has frequently been interpreted as a deterministic model, they did not believe the cell cycle to be necessarily entirely deterministic in the sense that all cells proceed in a fixed step-by-step sequence toward division; instead, they felt that several steps in the cycle could involve probablistic components (Smith and Martin 1973).

Since the major point of control of the replication rate in *E. coli* must be in the initiation event, it is of great interest to study the relationship of chromosome initiations to the rest of the cell cycle. Several experiments have yielded data that support the idea that initiation occurs, on the average, at a remarkably constant ratio of cell mass to the number of chromosome origins and that cell mass and chromosome content are continually adjusted to a characteristic ratio (Donachie *et al.* 1976; Cooper and Helmstetter 1968; Pritchard 1974). For example, Donachie (1969) and Donachie *et al.* (1968) reported that cells that were starved for thymine showed, when the thymine was restored, an increased rate of DNA synthesis due to new initiations, and this increased rate of synthesis persisted until the DNA content reached normal levels. Cell division was delayed until the mass/DNA ratio returned to a normal value. Their data suggested that when cell mass exceeds a certain critical amount, DNA synthesis initiates at all available origins.

This precise correlation between cell mass and chromosome initiations, which is apparent in studies dealing with populations of cells, becomes somewhat less clear when techniques are used that allow monitoring of individual cells. Chai and Lark (1970) pulse-labeled slowly growing *E. coli* with a doubling time of 120 min and examined by electron microscopy autoradiograms of individual cells. With this slow growth rate, gaps between rounds of chromosome duplication were discerned. Although initiation of DNA synthesis was strongly correlated with certain periods in the cell cycle, it is interesting to note that no cell size class was completely free of replicating cells, nor was any cell size class comprised totally of replicating cells. Indeed, the only consistent way of interpreting the data is to assume that, even in slowly growing cells, initiation of chromosome synthesis may take place at a significant frequency at virtually *any* cell mass (Koch 1977). Hence, it is not possible to claim that accurate phasing exists between cell mass and chromosome initiation.

Recently, experimental conditions have been found whereby the initiation of DNA synthesis can be uncoupled from cell mass (Grossman and Ron 1980). A synchronous culture of newly divided cells of *E. coli* B, obtained by the membrane elution technique of Helmstetter (1967), was starved for an amino acid. Upon restoration of the amino acid, the cells not only divided at a size that was smaller than normal, but also initiated DNA replication long before they could increase their masses to reach the expected ratio of mass/DNA presumably required for initiation.

Theories on the regulation of chromosome initiation must therefore explain not only the observed constancy of the cell-mass chromosome-origin ratio at different growth rates, but also the cell-to-cell variability in the timing of initiation and the perturbability of the mass/DNA ratio after amino acid starvation. Both positive and negative regulatory mechanisms have been proposed to explain the observed coupling between chromosome initiation and growth rate. Positive regulatory models postulate that initiator molecules are synthesized during the cell cycle and that accumulation to a critical level triggers the initiation (Donachie 1968; Helmstetter *et al.* 1968; Jacob *et al.* 1963). Negative regulatory models, however, propose that the synthesis during the cell cycle of an inhibitor molecule supresses initiation, until cell growth dilutes the cellular concentration of inhibitor to less than a critical value (Pritchard *et al.* 1969; Sompayrac and Maaløe 1973). Alternatively, negative regulation may involve the continuous synthesis of an antiinhibitor (Rosenberg *et al.* 1969). Still another class of models propose associations between the chromosome and the cell surface as possibly being involved in controlling the timing of initiation (Helmstetter 1974; Marvin 1968; Nicoladis and Holland 1978). However, Craine and Rupert (1979) have argued against any role for membrane–DNA interactions in initiation of synthesis or elongation. At present there is insufficient evidence to select positively between these general classes of models.

On a molecular level, little is known about the factors regulating the initiation of replication in *E. coli.* It is known that blockage of protein synthesis inhibits the initiation of new cycles of replication but allows initiated chromosomes to continue replication until completion (Ward and Glaser 1969). RNA transcription appears to be necessary in the initiation process (Lark 1972; Zyskind *et al.* 1977). Some role for polyamines in initiation seems to be suggested by the fact that the addition of spermidine to starved *E. coli* cells will stimulate initiation (Geiger and Morris 1980).

Normally, cell division is closely coordinated with chromosome replication. What factors regulate the timing of cell division? During balanced growth, cell division in *E. coli* takes place over a narrow size range (Donachie 1969). It was initially believed that the time between the completion of DNA replication and cell division (the D period) is a constant (Cooper and Helmstetter 1968). Subsequent work revealed, however, that the D period is neither constant at different growth rates (Helmstetter *et al.* 1979) nor is even necessarily fixed among individual cells in a population (Koch 1977; Chai and Lark 1970). Hence, chromosomal termination by itself does not serve as the timer for subsequent cell division. However, there is considerable evidence that chromosome termination is a necessary condition for cell division in *E . coli* (Dix and Helmstetter 1973). Jones and Donachie (1973) reported a requirement for synthesis of proteins specifically associated with the termination of chromosome replication. Tormo *et al.* (1980) proposed that the *ftsA* gene product had the properties of such a termination protein.

Studies of specific replication and cell division mutants in *E. coli* have yielded considerable information concerning the role and expression of genes involved in these processes. Discussion along these lines, however, is necessarily outside the scope of this chapter, since no comparable genetic studies are available for *M. xanthus* against which we may make comparisons. Comprehensive reviews of DNA replication and the cell cycle are available which cover these topics, including reviews by Kornberg (1980, 1982), Lloyd *et al.* (1982), Mendelson (1982), and Helmstetter *et al.* (1979).

## Cell Cycle of *Myxococcus xanthus*

Several distinctive differences are seen between the pattern of chromosome replication and cell division in *M. xanthus* and *E. coli.* As with *E. coli,* Zusman and Rosenberg (1970) found that in exponential cultures, cell division in *M. xanthus* took place over a narrow range of cell sizes. In contrast to *E. coli,* however, when they pulse-labeled cells with thymidine and counted the number of silver grains associated with cells of various sizes, they found the period of DNA synthesis to be positioned within a relatively constant interval within the cell division cycle. It should be especially noted that they found DNA synthesis in *M. xanthus* to be preceded and followed by well-defined gaps. Unlike what Chai and Lark (1970) found in *E. coli,* they observed size classes almost completely free of replicating cells, as well as size classes comprised almost totally of replicating cells. The only comparable situation described in *E. coli* has been certain B/r strains growing in poor growth media (Helmstetter and Peirucci 1976). Despite this relatively precise phasing, Zusman and Rosenberg (1970) did not consider that their data supported any model predicting DNA synthesis to be initiated at a particular cell mass or mass to DNA ratio. Those cells that were synthesizing DNA did so at a constant rate independent of cell size or cell age, that is, the authors found no evidence for multiple rounds of initiation such as would be associated with a dichotomous replication pattern.

In subsequent work, Zusman *et al.* (1978) studied DNA synthesis in *M. xanthus* in three different media permitting a twofold variation in doubling time. They confirmed the existence of a simple DNA cycle within the cell division cycle, observing clearly defined gap, synthesis, and second gap periods. An interesting finding was that the rate of chromosome replication was variable: Synthesis of DNA occurred during 85% of the cell division cycle, even though generation times were varied from 5 to 11 hr. The first gap period before DNA synthesis was also variable in length as a function of the generation time, but the second gap period was virtually constant.

Rosenberg *et al.* (1967) studied DNA synthesis during glycerol-induced spore formation. Glycerol apparently halted new initiations but allowed DNA synthesis to continue until chromosome completion. After glycerol induction, DNA continued to be synthesized, but at a decreasing rate until the final increase was only 19%. This was a particularly difficult result to explain: Since under their growth conditions DNA synthesis occupied 80% of the cell division cycle, their observed increase was only about half of the value we might predict on an a priori basis. To distinguish among several possible explanations for this result, Rosenberg *et al.* (1967) performed various autoradiographic studies; in particular, they studied the total uptake of labeled thymidine into glycerol spores and pulse-labeled uptake into exponential cells. They found that the DNA made after glycerol induction was not evenly distributed among the myxospores. Their observed distribution of grains supported the following hypothetical model: In exponential phase, each daughter cell receives two chromosomes that replicate during 80% of the cell division cycle. However, these two chromosomes do not replicate simultaneously, as might be expected; instead, they replicate in a sequential fashion, one after the other. They found no evidence of any period in the cell cycle when both chromosomes are engaged in simultaneous replication. Such a replication scheme was not unprecedented; a similar sscheme was proposed by Lark and Lark (1965) for DNA synthesis in *E. coli* 15T$^{-}$ growing in a succinate medium after amino acid starvation. Hence, after glycerol induction, individual myxospores may contain from two to four completed chromosomes depending on the replicative state of their individual precursor cells at the time of induction. This heterogeneity among individual myxospores accounts for the observed uneven grain distribution.

Additional indication that glycerol-induced myxospores contain completed chromosomes comes from Komano, Inouye, and Inouye (unpublished data), who germinated myxospores in the presence of labeled thymidine and established that virtually every restriction fragment that was labeled during the initial stages of germination mapped within a single 250-kb region. Hence, elongation apparently proceeds to the very end of the chromosome during myxospore induction. Furthermore, the general pattern of labeling supported a bidirectional mode of replication.

The hypothesis that new chromosomal initiations are blocked during glycerol-induced sporulation received support from the work of Kimchi and Rosenberg (1976). In this work, a different strain of cells and different growth conditions were used than those in the Rosenberg *et al.* (1967) study; possibly because of these differences, Kimchi and Rosenberg (1976) found the increase of DNA during glycerol induction to be 39%, close to the predicted value for chromosome completion by an exponentially growing population. Hence, sequential replication of multiple chromosomes is not a universal pattern among different *M. xanthus* strains. Chloramphenicol treatment also blocked chromosomal initiations while allowing chromosome completion. Termination of chromosome replication was found to be a necessary condition for cell division in *M. xanthus,* since exponentially growing cells failed to divide in the presence of nalidixic acid, whereas cells pretreated with chloramphenicol for a time sufficient for chromosomes to complete replication divided even in the presence of nalidixic acid. Germinating myxospores, however, did not divide in the presence of nalidixic acid in spite of the fact that they contained completely replicated chromosomes. It appears that myxospores containing completed chromosomes require additional DNA synthesis during germination for subsequent cell division. The critical period of DNA synthesis required during germination for subsequent cell division involved 5% of the *M. xanthus* chromosome and involved RNA and protein synthesis. The data suggest that replication of the origin region triggers formation of a protein product necessary but not sufficient for subsequent cell division, this postulated protein being destroyed during myxospore induction. *Myxococcus xanthus* differs from *E. coli* with regards to the effect of amino acid starvation on DNA synthesis: In *E. coli,* protein synthesis inhibitors and amino acid starvation both appear to allow chromosome completion while blocking initiation. In *M. xanthus,* however, chloramphenicol treatment allows for chromosome completion, but amino acid starvation does not allow sufficient DNA synthesis for chromosome completion.

Zusman and Rosenberg (1968) studied DNA synthesis during germination of glycerol-induced spores by labeled thymidine incorporation, autoradiography, and chemical analysis. Glycerol spores contained an average of 6.6 conserved units of DNA corresponding to three to four chromosomes per cell. This figure is consistent with the hypothesis of Rosenberg *et al.* (1967) that exponential cells contain from two to four chromosomes in various stages of sequential replication. At a gross level, DNA synthesis is seen to begin 3.5–4 hr after induction of germination and is synchronous through subsequent replication cycles. Cell division begins at 4.5 hr and continues exponentially until 9 hr. The cell number quadruples during this 4.5-hr period. This is surprising since under normal conditions the particular strain of *M. xanthus* that Rosenberg *et al.* (1967) used has an optimal doubling time of 4.5 hr.

Hence, their results indicate an unusual synchrony of cell division. Individual spores containing only three completed chromosomes appear to have a different pattern of DNA synthesis during germination than do spores containing four completed chromosomes. It appears that spores containing three completed chromosomes begin DNA synthesis 3.5–4 hr after induction. All three chromosomes are replicated simultaneously. On the other hand, spores containing four completed chromosomes do not begin any DNA synthesis until 6 hr after induction.

## Base Composition

The DNA base composition of bacterial species, expressed as the percentage of guanine plus cytosine (G + C), ranges widely from about 24 to 74% (Normore 1976). The few determinations lying outside this range are generally not confirmed when alternate methods of measurement ($T_{m}$, chemical analysis, bouyant density) are used. For the greater part of this range of G + C contents, roughly 34–64%, it can be argued with considerable degree of plausibility that selection for any particular level of G + C is largely a neutral process (Bak *et al.* 1972; Lesk 1973).

Outside of this broad range of G + C ratios, it becomes difficult to argue in favor of neutral selection. On theoretical grounds, G + C contents of 27 and 71% have been identified as genetic code limits (Woese and Bleyman 1972; Elton 1973). In species with such extreme G + C contents, the redundant positions are almost entirely filled with one type of base pair (G − C in the case of species with high G + C ratios; A − T in the case of low G + C content), and the remaining content in the DNA of the other type of base pair stems almost entirely from the need in the coded proteins for minimum percentages of amino acids with this base pair in their nonredundant codon positions. As G + C contents approach within 10% of either end of the code limit range, underlying pressures toward extreme base compositions reflects itself in grossly altered amino acid compositions (Elton 1973). Organisms with G + C contents outside the code limit are presumably forced to have a large fraction of noncoding DNA.

Myxobacteria are genetic code limit organisms, with measured G + C percentages of 67–71% (Johnson and Ordal 1968; McCurdy and Wolf 1967; Mandel and Leadbetter 1965). In this connection, it should be noted that in the past various gliding organisms not forming fruiting bodies have been classified among the myxobacteria. For example, the seventh edition of *Bergey's Manual* (Breed *et al.* 1957) lists as myxobacteria *Cytophaga* and *Sporocytophaga.* However, members of these two genera have measured G + C contents in the range of 35–41% (Normore 1976). It is now generally agreed that inclusion of these two genera among the myxobacteria is a misclassification (Reichenbach and Dworkin 1981).

What sorts of selective pressures would favor the adoption of extremes of G + C content? The code limits are represented at the low G + C extreme by, for example, *Mycoplasma, Clostridium,* and in the eukaryotes, *Tetrahymena.* At the high G + C extreme, in addition to myxobacteria, there are various genera of the actinomycetes and actinoplanetes, the streptomycetes, *Micrococcus, Cellulomonas,* and certain *Corynebacterium* species.

Woese and Bleyman (1972) suggested nutritional factors as possibly being responsible for the extremely low G + C content of *Tetrahymena;* these organisms require guanine for growth, and hence during evolution of this organism selective advantage could have favored low G + C individuals. Ashmarin *et al.* (1969) noted a correlation between decreasing content of GC pairs and increasing nutrient dependence, whereas Kharchenko (1975) suggested an evolutionary trend toward increasing contents of AT pairs. Extreme thermophiles usually possess DNA of very high G + C percentages (Oshima and Imahori 1974), this high G + C content presumably being required for greater DNA stability at high temperatures.

The general nutritional and evolutionary arguments of Woese and Bleyman (1972), Ashmarin *et al.* (1969) and Karchenko (1975) are lacking in predictive power; the relationship between thermophiles and high G + C content seems obvious and strong, but it is limited in applicability. The most ambitious and specifically predictive theory attempting to explain variations in G + C content is the theory of Singer and Ames (1970). A strong correlation was noted between the amount of sunlight to which a bacterium is exposed and its G + C content. Bacteria presumed to be exposed to high levels of sunlight (bacteria with aerial reproduction, aquatic bacteria, and carotenoid-containing bacterial) were noted as almost universally having a high G + C content. Conversely, bacteria presumed to be exposed minimally to sunlight in their life cycles (nonphotosynthetic obligate anerobes and obligate parasites) were noted as almost all having low G + C contents. High G + C contents, they argued, have a photoprotective function.

Precisely because the theory of Singer and Ames (1970) was so specifically predictive, it was possible for Bak *et al.* (1972) to challenge it by presenting a number of exceptions to their generalization. In particular, Bak *et al.* (1972) pointed out a number of parasitic, nonpigmented, nonsporulating bacteria that are presumably exposed minimally to sunlight but that nevertheless have high G + C contents and, conversely, a number of marine and fresh-water bacteria that are presumably exposed heavily to sunlight but that have G + C contents that are not exceptionally high.

The rather obvious point that must be made is that variability in G + C content is by no means the sole source of DNA photoprotection; effective alternative mechanisms of photoprotection via screening and repair mechanisms allows lower G + C contents in various instances, thus providing exceptions to Singer and Ames' generalization (Mennigmann 1979; Shakhova 1978). The correlation between high G + C content and a high sunlight exposure, despite puzzling exceptions, is a strong one (Singer and Ames 1972). The remainder of this section will seek to demonstrate that the known facts concerning the life cycle of myxobacteria are consistent with an exceptional requirement for photoprotection, although of course no claim can be made for any absolute relationship between this requirement and their high G + C contents.

In nature, myxobacteria are usually found in the surface layers of the soil. They are thus subjected to intermittent ultraviolet (UV) irradiation, elevated temperatures, and dessication, and many aspects of the myxobacterial life cycle can be interpreted as adaptations to a periodically harsh and variable environment.

Fruiting myxobacteria may be expected to experience high sunlight exposures because of the suspension of their fruiting bodies above the solid substrate. Furthermore, the presence of characteristic red, yellow, and orange carotenoid pigments in myxobacteria (Kleinig 1972), whose synthesis is photoinducible and that appear to function as photoprotectants (Burchard and Dworkin 1966a), also argues that myxobacteria experience, at least intermittently, high levels of UV exposure.

The UV radiation in sunlight is capable of causing considerable damage to DNA. The major product of UV damage is considered to be pyrimidine dimers. Singer and Ames (1972) calculated that in direct sunlight at sea level, a chromosome with the size and base composition of the *E. coli* chromosome would suffer 10 dimers per minute. Although bacteria in general appear to have evolved efficient means of repairing dimerization, it should be noted that certain other forms of photodamage, such as DNA interchain cross-linkage (Cole *et al.* 1976) and DNA–protein cross-linkage (Shetlar 1980), may not be repairable except by error-prone mechanisms (Hanawalt *et al.* 1979). If all other things were equal, myxobacteria with their larger chromosomes would be even more susceptible to UV damage.

The mechanisms that bacteria have evolved to protect their DNA against UV damage may be classified as (1) screening methods, (2) repair methods, and (3) the evolution of base ratios. Aside from carotenoid pigmentation, additional important sources of screening include the RNA and tryptophan contents of the cell. Various species of myxobacteria exhibit acriflavin-sensitive systems for the reapir of UV damage (Grimm 1978; Grimm and Herdrich 1980) and show UV-reactivation and host-cell-reactivation phenomena, which would indicate the presence of an error-prone repair mechanism. The presence of multiple chromosomes in myxobacteria (Rosenberg *et al.* 1967) may itself be an adaptation facilitating recombinational repair (Hanawalt *et al.* 1979). Additional resistance to UV exposure would be derived from having a lowered amount of thymine. In DNA with a 50% G + C content, it has been observed that thymine dimers and thymine–cytosine dimers together constitute 90% of the photoproducts, whereas cytosine dimers constitute only 10% (Swenson and Setlow 1966). Experimentally, it has been shown that UV sensitivity does indeed decrease with increasing G + C content in bacteria.

The hypothesis that myxobacteria have evolved their high G + C content in response to exposure to UV radiation, like most evolutionary arguments, suffers somewhat from an inability to be directly tested. Nevertheless it is a plausible hypothesis, especially since it is evident that myxobacteria are adapted to dealing with a periodically harsh environment.

## Genome Size

The genome size of *M. xanthus* was measured by Yee and Inouye (1982) to be approximately 5690 + 510 kb in confirmation of earlier results (Yee and Inouye 1981). The method in this work used a technique of two-dimensional DNA electrophoresis using in situ enzymatic redigestions, which enabled the nearly complete resolution of all DNA restriction fragments on two separate gels, as seen in Fig. 3.1. The distribution in sizes of all the resolved spots was mathematically analyzed to yield an unambiguous total genome size estimate that is almost certainly correct within its error limits. *Myxococcus xanthus* therefore has one of the larger measured genome sizes among bacteria (Wallace and Morowitz 1973), approximately 62% greater in size than *E. coli* K-12 (3520 ± 250 kb, measured using the same technique of two-dimensional DNA electrophoresis).

For several years previous to the work of Yee and Inouye (1982), however, it was believed that the genome size of *M. xanthus* was even greater than this,

→

**Figure 3.1.** Two-dimensional separations of *M. xanthus* DNA on agarose. The first dimension separations are from left to right; the second dimension separations are from top to bottom. The horizontal bands are molecular weight markers from a *Hin*dIII digest of phage lambda cI857 DNA. (a) First dimension, *Bst*EIII, second dimension after in situ digestion by *Sal*I. (b) The same restriction enzymes as in (a) but used in reverse order.

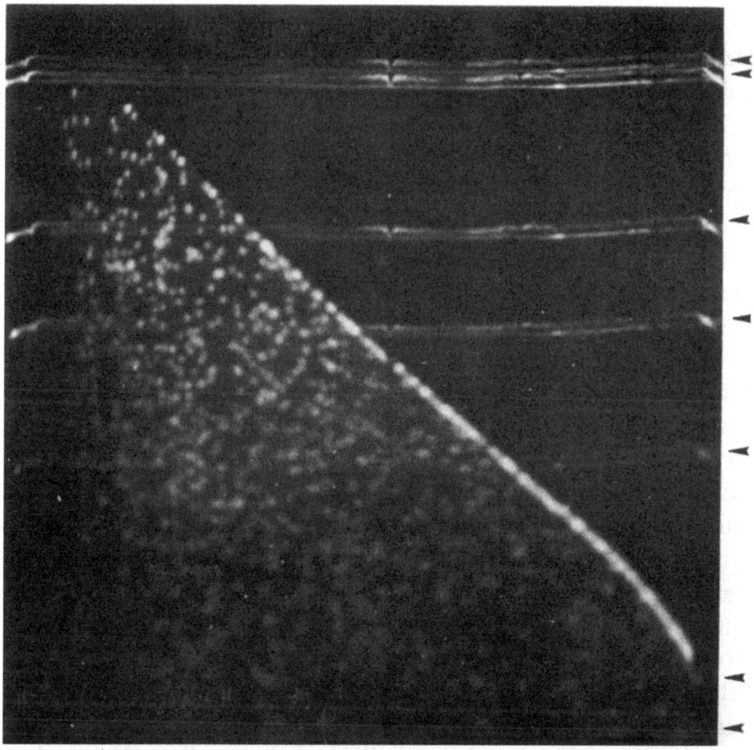

(a)

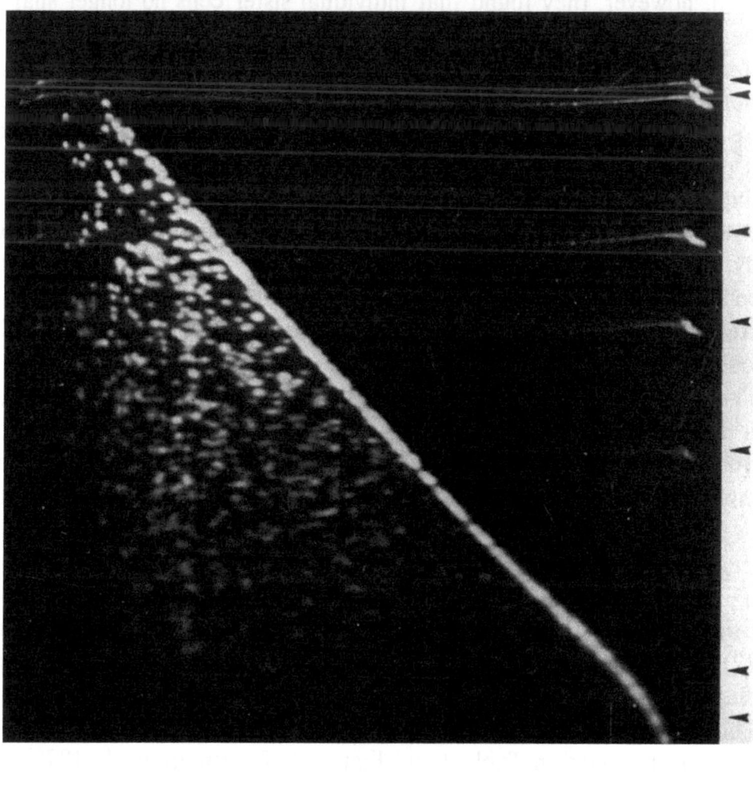

(b)

indeed, over twice as great (Kaiser *et al.* 1979; Zusman 1980). This belief was based on measurements by Zusman *et al.* (1978), who obtained a figure of 8.4 $\pm$ 1.2 $\times$ $10^9$ $M_r$ (approximately 12, 700 kb) as part of a study on regulation of DNA synthesis during the cell division cycle. We have examined carefully the published data of Zusman *et al.* (1978) and are most impressed with the care and accuracy with which these workers performed their experiments; their only fault was an easily understandable misinterpretation of their results. In reinterpreting their data, we find clear evidence for an unusual coordination of chromosome segregation in exponentially growing cells of *M. xanthus.*

To arrive at their genome size estimate, Zusman *et al.* (1978) continuously labeled an exponential phase culture for four generations with tritiated thymidine and then suspended the cells in a nonradioactive conditioned medium. After precisely zero, one, and two generations of growth in the nonradioactive medium, they removed samples and prepared them for autoradiography. Only the silver grains above dividing sister cells were scored in each sample. By scoring sister cells at precise integral generation periods after suspension into the nonradioactive medium, they avoided the complications that would arise from scoring partially labeled chromosomes, since *M. xanthus* clearly exhibits a gap in DNA synthesis around the period of cell division.

After zero or one generation of growth, they found that individual sister cells shared identical mean numbers of grains per cell. After two generations of growth, however, they found that individual sister cells no longer shared the same mean number of grains; instead, the radioactivity appeared to segregate preferentially into one sister, the other sister receiving, on the average, little radioactivity. Zusman *et al.* (1978) concluded that the radioactivity in the more heavily labeled sister cell represented one single conserved replication unit of DNA (i.e., one single strand) and that nascent cells therefore contain a single, completely replicated chromosome. Furthermore, their analysis of Giemsa-stained cells showed that nascent cells contained a single nucleoid body, whereas aged cells contained two nucleoids. Knowing that the average exponential cell has replicated 50% of its DNA and having measured the DNA content of exponential phase cells to be approximately 20 $\times$ $10^{-9}$ $\mu$m/cell, they calculated their molecular weight value.

Their analysis of their data was logical and straightforward but was subject to a deep and nonobvious flaw: They assumed that if nascent cells contained not one, but multiple, chromosomes, the chromosomes would segregate randomly. In other words, their radiolabeling experiment could not distinguish between one chromosome and multiple chromosomes with a linked pattern of segregation, such as has been described for F′ and the *E. coli* chromosome (Cuzin and Jacob 1967). They themselves acknowledged this as a possibility, but they erroneously dismissed it as unlikely, as we also did (Yee and Inouye 1982). Furthermore, it was discovered by Chai and Lark (1970) that nucleoid bodies cannot necessarily be equated with completed chromosomes.

The keys to understanding what goes on in exponentially growing *M. xanthus* are to be found in Table 3 and Figure 3 of Zusman *et al.* (1978). Three salient points must be noted: First, the mean grain count per sister cell after one generation is 2.47 grains per cell, but after the second generation, the mean grain count in the more heavily labeled member of a sister pair is slightly less, at only 2.22 grains per cell. Second, the mean grain count in the less heavily labeled member of a second generation is low, but well above background at 0.20 grains per cell. Third, the grain counts in the less heavily labeled sister cells deviate markedly from the expected Poisson distribution.

These facts can be explained if dividing sister cells contain two chromosomes segregating in neither a perfectly linked nor in a random fashion. We have performed a mathematical analysis of the data of Zusman *et al.* (1978) in which we assume that the single strands in two chromosomes segregate together 80% of the time and segregate apart 20% of the time. Also essential to our analysis was the realization that since grain counts follow a Poisson distribution, errors will occasionally be made in identifying the more heavily and less heavily labeled members of a sister pair (i.e., although one member of a sister pair may in fact have been labeled to a greater extent than or to equal extent as the other member of the pair, fluctuations in the grain counts will result in a finite number of cases in which the more heavily labeled member will be misidentified as a less heavily labeled member). Using these assumptions, we find we can model extremely well the non-Poisson distribution of grain counts seen in second-generation sister cells. In Fig. 3.2 we replotted the data of Zusman *et al.* (1978) from their Figure 3d, superimposing on their data the results of our theoretical calculations. The frequency function $P(n) \times n!$ is plotted on a logarithmic scale in the ordinate, and $n$, the number of grains per cell, is plotted on the abscissa; $P(n)$ is the fraction of cells containing $n$ number of grains. In this form of plot, a straight line indicates a Poisson distribution; furthermore, a given $y$ intercept implies a given slope, since both are functions of the mean number of grains per cell. Zusman *et al.* (1978) separately plotted grain counts from apparently more heavily labeled sister cells

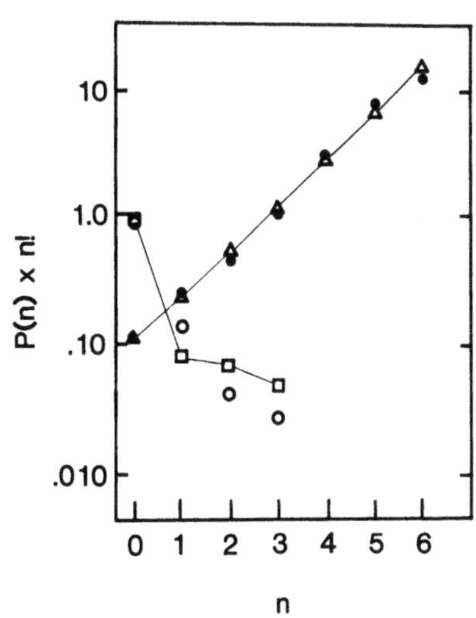


**Figure 3.2.** Theoretical and observed autoradiographic grain distributions of radioactive DNA in dividing cells after growth and division for two generations in nonradioactive media, with dividing cells considered as one heavily labeled sister and one lightly labeled sister. Symbols: ●, observed grain counts for the heavily labeled sisters; △ theoretical grain counts for the heavily labeled sisters; ○ observed grain counts for the lightly labeled sisters; □ theoretical grain counts for the lightly labeled sisters; Data for observed grain distributions was adapted from Zusman *et al.* (1978).

and apparently more lightly labeled sisters. The apparently more heavily labeled sisters seemed to fit a Poisson distribution reasonably well, except for a small unexplained concavity; however, the lightly labeled sisters deviated drastically from a Poisson distrbution. Our calculated points for the more heavily labeled sisters and lightly labeled sisters demonstrate that our model fits the data very well, explaining not only the drastic non-Poisson distribution of grains in the apparently lightly labeled members, but also the slight concavity of the plot of the apparently heavily labeled members. The small apparent discrepancy between our theoretical predictions and the observed grain distributions in the plot for the less heavily labeled members is insignificant compared with the large discrepancy between the observed counts and a Poisson distribution.

Our conclusion therefore is that the data of Zusman *et al.* (1978) imply a molecular weight of $4.2 \times 10^9$ $M_r$ for the *M. xanthus* genome, or approximately 6360 kb, which corroborates our own results to within 12%.

It is interesting to note that Zusman and Rosenberg (1968), 10 years previously in their analysis of germinating myxospores, obtained a value of $4.9 \times 10^9$ $M_r$ for the molecular weight of *M. xanthus,* which is only 30% above our best estimate. In this earlier work they uniformly labeled exponentially growing cells, induced with 0.5 *M* glycerol to obtain myxospores with completed chromosomes, then germinated the cells in fresh medium with a cold chase. After 25 hr they counted 4.9 grains per labeled cell, which, when compared to an initial level of 32.4 grains per myxospore, allowed them to calculate that the myxospores originally contained an average of 3.3 chromosomes each. The accuracy of this earlier result compared with the later result of Zusman *et al.* (1978) is probably due to their simply having grown the cells for a longer period of time, so that individual chromosomal strands had a chance to segregate apart. This pattern of oriented nonrandom strand segregation in *M. xanthus,* which is biased toward comigration of labeled DNA from the two chromosomes into daughter cells, may be compared with the pattern of strand segregation of *E. coli.* On the one hand, Chai and Lark (1970), working with rapidly growing cells, reported strand segregation in *E. coli* to be mostly random except for a bias *against* comigration of labeled DNA from dichotomous forkings of the single chromosome into daughter cells (Chai and Lark 1970). On the other hand, Cooper and Weinberger (1977), working with methylcellulose (Methocel)-immobilized cells, reported that segregating DNA strands tend to segregate toward the cell poles to which they have previously segregated. The degree of "strand inertia" that they observed depended on growth conditions: Segregation became increasingly oriented as the growth rate decreased. The pattern of strand segregation in *M. xanthus,* therefore, may be similar to that of slowly growing *E. coli* cells.

Multiple chromosomes are not unique to *M. xanthus. Bacillus subtilis* cells seem generally to contain two sets of genomes (Paulton 1971; Nanninga *et al.* 1979); this situation may be related to the sporulation properties of this organism. Various species of cyanobacteria contain multiple genome equivalents per cell even when growing very slowly, whereas rapidly growing cells may contain as many as 20 genome equivalents per cell (Doolittle 1979).

## DNA Methylation

In prokaryotes there has been considerable evidence suggesting the importance of DNA methylation in replication, recombination, and mutation (Razin and Riggs 1980). In the absence of methylation, Billen (1968) reported that replication of the *E. coli* chromosome becomes abnormal after a single round of synthesis. Increased rates of spontaneous mutation and heightened senstivity to mutagenic agents are associated with deficiencies in the *E. coli dam* gene, coding for the (G–A–T–C) methylase (Marinus and Morris 1975). The origin of replication contains a high frequency of the sequence G–A–T–C, methylated in most strains of *E. coli* (Sugimoto *et al.* 1979; Meijer *et al.* 1979).

In eukaryotes, methylation has been implicated in the regulation of gene activity and cellular differentiation. Specific undermethylation has been found to be associated with active transcription of such genes as the chicken betaglobin gene (McGhee and Ginder 1979), the chicken ovalbumin gene (Mandel and Chambon 1979), the rabbit betaglobin gene (Waalwijk and Flavell 1978), and the human gamma–delta–beta–globin genes (Van der Ploeg and Flavell 1980). Sites of cytosine methylation appear to be hot spots for mutation (Bird 1980). Specific methylation patterns were found to be clonally inherited in mouse cells and conserved over many generations, but only when located at C–G residues (Stein *et al.* 1982) indicating that the maintenance methylase (Riggs 1975; Holliday and Pugh 1975) is specific for these sites. A diverse range of chemical carcinogens were reported as capable of disrupting cytosine methylation (Wilson and Jones 1983) indicating that carcinogens may induce demethylation and activation of endogenous transforming genes (Cooper *et al.* 1980; Murray *et al.* 1981).

Are specific DNA methylation or demethylation events associated with cellular differentiation in *M. xanthus?* Using two-dimensional thin-layer chromatography, we previously detected the presence of 5-methylcytosine in the DNA of this organism, as well as variations in the overall level of methylation apparently associated with cell growth and development (Yee and Inouye, unpublished results). However, we felt that a more discriminatory method was needed to ascertain the possible function of these variations, and to this end we developed a two-dimensional DNA electrophoretic method based on use of the *MspI* and *Hpa*II isoschizomers (Yee and Inouye 1982). Both of these enzymes recognize the C–C–G–G sequence, but *Msp*I cuts regardless of the state of methylation of the internal cytosine (Waalwijk and Flavell 1978). We separated *Hpa*II digests of *M. xanthus* DNA isolated from various stages of growth on polyacrylamide gells, then redigested it in situ with *Msp*I before running in a second dimension. In this fashion we simultaneously probed the entire genome for specific methylation events. The results of this study were interesting. The amount of DNA used, 40 μg, was such that were any *Hpa*II sites completely methylated at the internal cytosine, recutting by *Msp*I should have resulted in bright spots below the diagonal. Instead, although we observed increasing spot intensities with increasing times of development as well as in aging cultures, the level of spot intensities never approached the levels that we expected for single-copy amounts. The conclusion we reached is that during development as well as during late stages of vegetative growth, *M. xanthus* DNA becomes methylated at highly specific *Hpa*II sites, but only a fraction of the population of DNA molecules becomes methylated at any particular site. Developmental cells of *M. xanthus* have varying fates: Some cells are destined to become spores, whereas by far the greater proportion remain vegetative and eventually lyse. The heterogeneity that we observed in DNA methylation may reflect this heterogeneity in the population of developmental cells.

## Phase Variation and DNA Rearrangement

Recombinational switching events play important roles in gene expression in a variety of prokaryotic and eukaryotic systems. For example, in the bacterium *Salmonella typhimurium,* the production of two distinct antigenic types of flagella is controlled by the orientation of a 995-bp segment carrying a promoter for transcription of the flagella gene (Simon *et al.* 1980). In bacteriophage Mu, inversion of the G loop segment within the structural region of the adaptor protein alters the host specificity of the phage particle (Van de Putte *et al.* 1980). The mating type of the yeast *Saccharomyces cerevisiae* is controlled by the copying of unexpressed alternative mating type genes into an expression locus (Nasmyth *et al.* 1981). In the immune system, linkage of constant-region sequences to variable-region sequences involves gene rearrangement with deletion of the DNA between the sequences chosen for expression (Tonegawa *et al.* 1980). African trypanosomes evade the immune response of their host by moving silent copies of the surface antigen gene to a locus, resulting in the expression of antigenically varying sets of their major surface glycoprotein (Bernards *et al.* 1981). These and other examples illustrate the importance of gene rearrangement as a mechanism for the generation of diversity.

*Myxococcus xanthus* exhibits phase variation phenomena that might be explained as a result of recombinational switching. Cultures of *M. xanthus* are heterogeneous. Burchard and Dworkin (1966a) first observed the existence of two colonial types in this organism, which they referred to as "yellow" and "tan," distinguished by the presence in yellow cells of a highly polar pigment (Burchard *et al.* 1977). Furthermore, heterogeneity with respect to motility has been described along with linkages between color and motility phenotypes (Wireman and Dworkin 1975). Interconversion rates between phenotypes can be very rapid, in some strains as high as $10^{-2}$–$10^{-3}$ per generation. In addition, pleiotropic changes in protein patterns between yellow swarmer and tan nonswarmer cultures have been observed (Zusman 1980).

The existence of these alternate states in *Myxococcus* and their high frequency of conversion suggested that the process may be due to recombinational switching (Kaiser *et al.* 1979; Zusman 1980). The isolation and cloning of the switching genes in most of the well-characterized examples was facilitated by the existence of abundant messenger RNA (mRNA) or structural protein gene products or, in the case of yeast, a relatively well-characterized genetic system including the availability of shuttle vectors replicating in both yeast and *E. coli*. In *M. xanthus*, however, the yellow pigment would be expected to be the end product of a chain of enzyme synthesis, unassociated with abundant mRNA or structural protein gene products. Furthermore, transformation protocols do not yet exist for this organism. Plasmid tranfers from *E. coli* and *Pseudomonas fluorescens* to *M. xanthus* have been reported but at very low efficiencies (Parish 1975; McCann and Clarke 1981). Efficient transduction methods are available (Kuner and Kaiser 1981). It was not clear, however, how a selective transductional scheme could be devised to isolate a presumed recombinational control element.

In view of these difficulties in attempting to design genetic methods for selecting switch element clones, we decided to join our technique of two-dimensional DNA electrophoresis (Yee and Inouye 1982) with S1 nuclease heteroduplex mapping (Shenk *et al.* 1975) to yield a technique for physically detecting mismatches between total bacterial genomes as well as detecting repetitive elements within a bacterial genome (Yee and Inouye 1984). To detect DNA rearrangement between two variant bacterial strains, total chromosomal DNA preparations from the two strains are digested with four-base restriction enzymes, mixed together, denatured, renatured, and separated on first-dimension polyacrylamide slab gels. Gel strips are cut out and soaked in a buffer containing S1 nuclease, which diffuses into the strips and digests the DNA fragments at sites of heteroduplex mismatch. The digested DNA is then electrophoresed in a second dimension. This method was developed and tested on a wide variety of model systems before use on *M. xanthus* and is still under development. Workers desiring to use this technique should contact the authors of this chapter for the latest improvements in protocol.

Total chromosomal DNA from yellow and tan strains of *M. xanthus* strain DZ-2 was digested with *Hin*fI or *Rsa*I and subjected to two-dimensional S1 analysis. Using the *Hin*fI digests, no significant differences were seen between the yellow–tan heteroduplex mapping pattern and the yellow–yellow and tan–tan homoduplex patterns. However, when *Rsa*I digests were subjected to S1 mapping, a prominent cluster of high molecular weight spots was easily discerned in the yellow–tan heteroduplexing, as seen in Fig. 3.3. These spots were cut out and eluted, and we used the material, nick-translated, to probe total chromosomal digests of yellow and tan strains of DZ-2. Figure 3.4 illustrates the result of hybridizing this material to *Rsa*I and *Sal*I digests of chromosomal DNA from yellow and tan cultures. With either digest, the majority of hybridizing bands are shared by both the yellow and tan lanes. Whether this represents nonspecific binding, contamination, or the presence of homologous

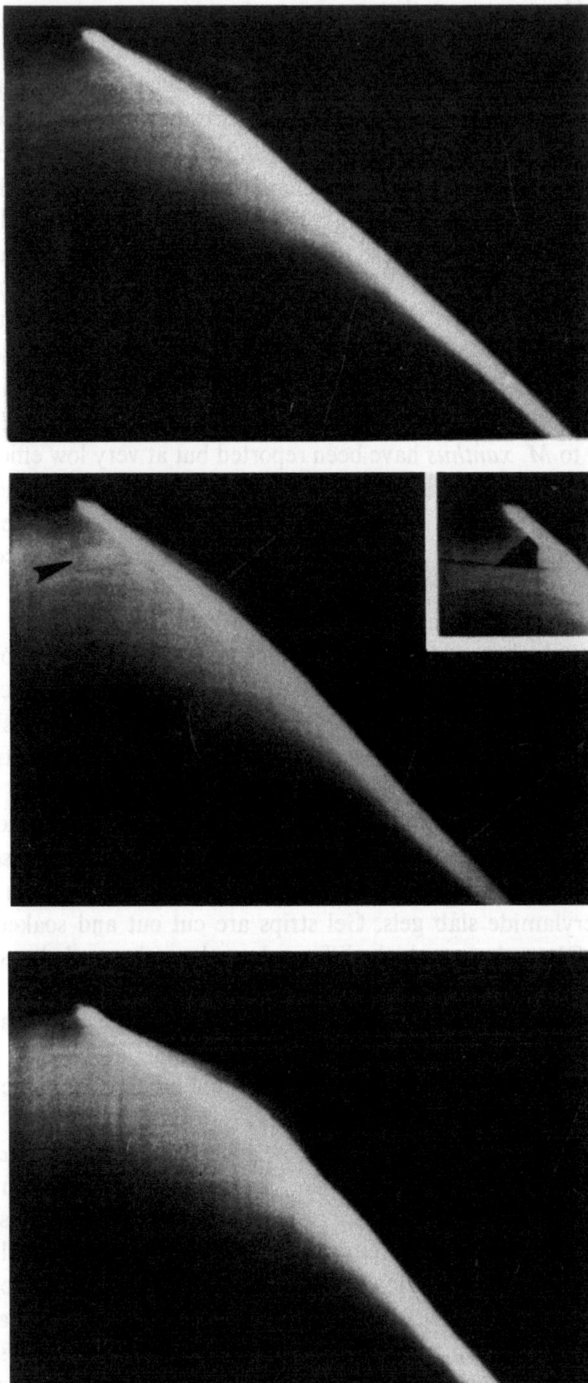

**Figure 3.3** The S1 mapping of *Rsa*I digests of mixed and unmixed DNA from yellow and tan variants of *M. xanthus* strain DZ2. (a) Analysis of homorenatured DNA from yellow cells only. (b) Analysis of mixed, heterorenatured DNA from yellow and tan cells. The inset shows the region of gel that was cut out and electroeluted. (c) Analysis of homorenatured DNA from tan cells only.

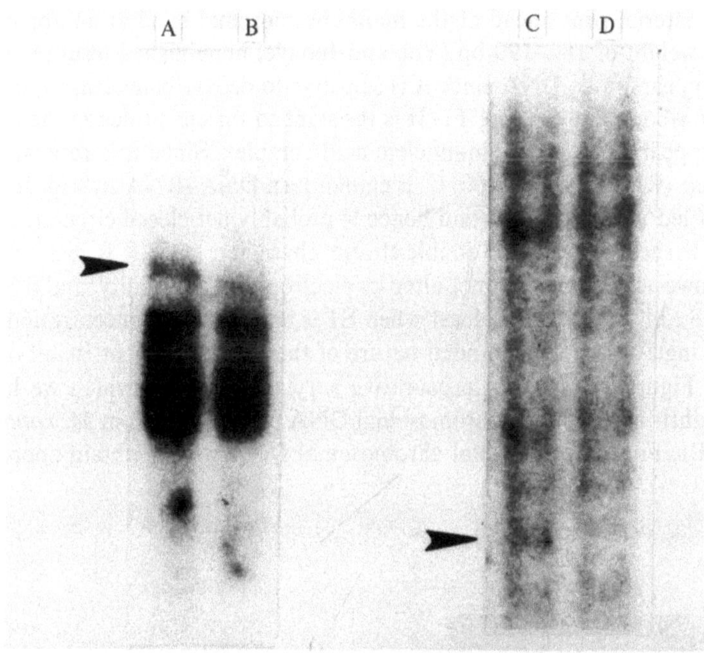


**Figure 3.4.** Hybridization analysis of yellow–tan heteroduplex DNA. The unique heteroduplex spots visible in Fig. 3.3(b) were cut out, electroeluted, nick translated, and used as a probe against nitrocellulose blots of DZ2 yellow (A,C) and tan (B,D) chromosomal DNA cut with *RSA*I (A,B) and *SAl*I (C,D). The arrows point out a 2.7 kb *Rsa*I fragment seen in lane A but not lane B, and a 3.9 kb *Sal*I fragment seen in lane C but not lane D.

sequences at multiple sites in the chromosomes is not clear. Nevertheless, clearly discernible differences are visible between the yellow and tan lanes, demonstrating that the heteroduplex spots do indeed result from DNA rearrangement.

We should emphasize that although we have detected a DNA rearrangement accompanying phase variation in *M. xanthus,* we have not yet proven any *causal* relationship between the rearrangement that we have observed and the phase transition. Nevertheless, the detection of this arrangement accompanying phase variation increases the likelihood that the rapid phenotypic variations observed in this organism are indeed a result of recombinational switching.

## Small Homogeneous DNA Fragment in Total Chromosomal DNA Preparations

When uncut, total chromosomal DNA preparations of *M. xanthus* are electrophoresed on 5% polyacrylamide gels, a satellite band of ethidium bromidestainable material runs ahead of the main chromosomal band at an apparent molecular weight of 180–190 bp (Yee and Inouye, unpublished results). This material appears to be DNA since it is sensitive to deoxyribonuclease I and is resistant to ribonuclease A and T1. It is resistant to various proteases, hence it does not appear to be a protein–nucleic acid complex. Since it is resistant to ribonuclease (RNase) even at 60°C, it cannot be a DNA–RNA hybrid. It can be end-labeled with T4 kinase and hence is probably not closed circular. Furthermore, it shows significant double-strand character, since it is nick-translatable. However, boiling does not alter its electrophoretic mobility, and it may be sensitive to S1 nuclease, at least when S1 is used in high concentration, so the exact single- or double-stranded nature of the DNA fragment is still open to debate. Figure 3.5 shows a preparative acrylamide gel on which we have loaded a lightly sheared total chromosomal DNA preparation from *M. xanthus* strain DZF1. From 1 mg of total chromosomal DNA we can obtain approxi-

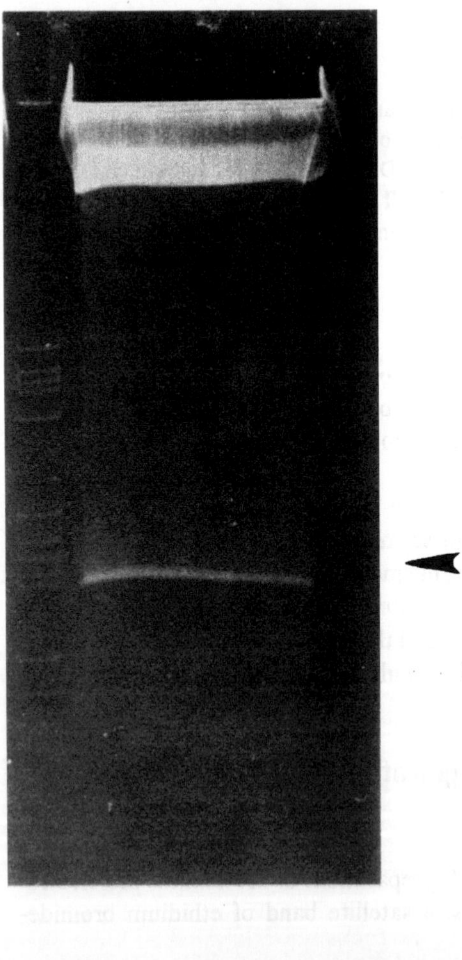

**Figure 3.5.** Polyacrylamide gel preparation of small fragment DNA. Total chromosomal DNA (150 μg) from *M. xanthus* strain DZF1 was mixed with glycerol and tracking dyes then loaded onto a 5% polyacrylamide gel prepared in 50 m*M* trisborate (pH 8.3), 1m*M* ethylenediaminetetrasicetate (EDTA). The molecular weight standard is pBR322 cut with *Hae*III. The arrow points out the small fragment.

mately 0.5 μg of fragment, indicating a copy number for this fragment of at least 15 per genomic equivalent.

When the end-labeled fragment was hybridized to total chromosomal digests, it was found to hybridize to unique *Eco*RI, *Bam*HI, *Hin*dIII and *Sal*I fragments; however, it hybridized to two *Pst*I fragments. The 18 kb *Sal*I fragment has been cloned into pBR322, from which subclones are being prepared suitable for sequencing. At present it is too early to offer more than extremely tentative speculations on the nature of this fragment. Might this fragment represent a viroid (Diener *et al.* 1982; Diener 1979)? If so, this would be the first example of a viroid composed of DNA rather than RNA, as well as the first viroid detected in a nonplant system. However, the viroid hypothesis is unattractive to us, primarily since this material is hybridizable to unique high molecular weight fragments in total chromosomal blots. Might this DNA represent extruded nascent strands from the origin of DNA replication, such as has been described for SV40 (Zannis-Hadjopoulos *et al.* 1981)? In the SV40 case, it was found that closed superhelical replicative intermediates were stable during incubation at elevated temperatures, but when these replicative intermediates were nicked or linearized, the nascent DNA strands were extruded. This would be a rather attractive hypothesis, except that the DNA fragment is present in too many copies per genomic equivalent to be associated with the chromosome origin. Even worse, the hybridization pattern of this fragment to total chromosomal blots does not seem to correlate with any of the restriction fragments that we in this laboratory have identified as being clustered about the origin (Komano, Yee, Inouye, and Inouye, unpublished results). Until we know the sequence of this fragment and have learned at least a few details concerning its synthesis and biologic effects, we should hold speculation to a minimum.

## Conclusion

One sees in vegetatively growing *M. xanthus* a pattern of DNA replication that is in many ways similar to the pattern of replication in slowly growing cultures of *E. coli;* in some ways, however, the pattern of replication is markedly different. Vegetatively growing *M. xanthus* exhibits a genuine DNA cycle with close phasing between chromosome replication and cell division. In some strains, *M. xanthus* exhibits an unusual pattern of sequential chromosome replication. Germinating myxospores exhibit a tight control between chromosome number, initiation of replication, and cell division. The characteristic high G + C content of myxobacteria is probably an adaptation to resist periodic exposures to high levels of ultraviolet radiation. The genome is quite large for a bacterium, and this mass may have some relationship with its complex life style. Exponentially growing cells maintain two chromosomes with nonrandom segregation of the DNA strands during replication. Intriguing changes in cytosine methylation are seen at specific DNA sites in developmental cells. DNA rearrangements have been observed which may be associated with phase variation in this organism. Total chromosomal preparations contain a bizarre, homogeneous short fragment of DNA.

It is not clear whether our opening questions to this chapter can ever really be answered. Teleological arguments applied to justify observed biologic adaptations are ever refutable. Clearly, however, study of the myxobacterial genome has uncovered an amazing number of unusual phenomena.

*Note Added in Pages.* We have completed sequencing the homogeneous short fragment DNA. It consists of a single linear strand 164 bases long showing extensive secondary structure and primed by a short RNA segment. An improved extraction procedure demonstrates that our initial copy number determination represented a gross underestimate: We now find that the fragment is present in 500–800 copies per genome equivalent, so that by weight it constitutes nearly 1% of the total DNA!

# 4

# Nutrition, Metabolism, and the Initiation of Development

Lawrence J. Shimkets*

Myxobacteria are aerobic chemoorganotrophs that are found throughout the world on plant material and in soil (Reichenbach and Dworkin 1981; McCurdy 1974). They secrete a variety of enzymes that hydrolyze proteins, nucleic acids, fatty acid esters, polysaccharides, and peptidoglycan. Myxobacteria may be divided into two metabolic groups based on the type of macromolecules they can hydrolyze and use as carbon and energy sources. Some species are able to use cellulose, and their nutritional requirements are relatively simple. Nitrate is the preferred nitrogen source, but ammonium and asparagine also support growth (Pronina 1962; Couche 1969). Cellulose and starch are used as effectively as monosaccharides by the cellulytic species (Couche 1969). However, most species of myxobacteria hydrolyze protein as a carbon and energy source instead of cellulose.

The cooperative use of secreted enzymes may have provided the selective pressure for the evolution of the multicellular behavior of myxobacteria (Dworkin 1972). According to this hypothesis, cells that live in a dense swarm accumulate a high local concentration of hydrolytic enzymes that support a faster growth rate by providing the cells with more nutrients. In support of this hypothesis, the growth rate of *Myxococcus xanthus* on casein increases with the cell density even though the rate of enzyme secretion remains constant (Rosenberg *et al.* 1977). This phenomenon might be considered a primitive form of communication in which the protease regulates cell density indirectly by providing nutrients.

More elaborate myxobacterial behaviors also depend on the use of secreted enzymes. For example, *M. xanthus* has occasional use for enzymes that hydrolyze peptidoglycan when it encounters prey bacteria that can serve as food

*Department of Microbiology, University of Georgia, Athens, Georgia 30602 U.S.A.

sources, as well as during development when cell lysis and differentiation occur. In either situation, if large numbers of cells are present, the cells will arrange themselves spatially in ridges that move in a rhythmic, pulsating manner (Kühlwein and Reichenbach 1968). These ridges are often referred to as ripples because they move across the agar like ripples moving across the surface of water. Many, if not all, of the members of the suborder *Cystobacterineae* ripple (Reichenbach and Dworkin 1981). In *M. xanthus,* rippling is induced by bacterial cells, peptidoglycan, and four components of peptidoglycan, *N*-acetylglucosamine, *N*-acetylmuramic acid, diaminopimelate, and D-alanine (Shimkets and Kaiser 1982a). *Myxococcus xanthus* secretes a number of different enzymes that hydrolyze peptidoglycan (Sudo and Dworkin 1972). Presumably, hydrolysis of peptidoglycan by extracellular enzymes produces peptidoglycan fragments that induce this multicellular behavior. Rippling may also involve the secretion of an intracellular signal, to synchronize the movement of cells (Shimkets and Kaiser 1982a).

The most complex multicellular behavior of myxobacteria is fruiting body formation, which is induced only when cells become nutritionally deprived (Dworkin 1963). There are probably two components of the system used to initiate development: (1) an intracellular signal(s) that senses growth-limiting nutritional conditions and causes the secretion of (2) an intercellular signal(s) that synchronizes the response of the population as a whole. This review focuses on the nutrition and metabolism of myxobacteria and the nature of the intracellular and intercellular stimuli for initiation of development.

## Nutrition

Since myxobacteria normally grow clumped, the isolation of dispersed growing mutants is necessary to obtain accurate measurements of cell growth (Dworkin 1962). One such strain, *M. xanthus* FB, has facilitated genetic and biochemical analysis of growth and development. Much of the research on myxobacteria has focused on *M. xanthus,* and, unless otherwise mentioned, the results described in this chapter were obtained with this species.

*Myxococcus xanthus* is able to grow on media containing purified casein (Rosenberg *et al.* 1977) or a mixture of amino acids (Dworkin 1963) as the only organic constituents. Some of the amino acids are required for growth, and others provide a source of carbon, energy, and nitrogen. Methionine and the three branched-chain amino acids, leucine, isoleucine, and valine, are required for growth (Dworkin 1963; Hemphill and Zahler 1968a; Witkin and Rosenberg 1970; Bretscher and Kaiser 1978). *Myxococcus xanthus* is not auxotrophic for any other organic compound. The most effective low molecular weight carbon and energy sources are amino acids, acetate, and pyruvate. Pyruvate is probably the most efficient carbon source (Bretscher and Kaiser 1978). Acetate and, to a lesser extent, succinate, fumarate, and oxaloacetate were shown to be oxidized to carbon dioxide by starved cells (Watson and

Dworkin 1968). Carbohydrates are not used as carbon and energy sources, nor are they incorporated very well into cells (Dworkin 1963; Watson and Dworkin 1968; Hemphill and Zahler 1968a; Bretscher and Kaiser 1978).

It is more difficult to assess the value of various nitrogen sources because of the presence of the required amino acids in the medium. If the concentration of required amino acids is kept to a minimum, ammonium sulfate and, to a lesser extent, asparagine and spermidine stimulate growth (Bretscher and Kaiser 1978). The composition of minimal medium A1 is listed in Appendix B. Vegetative cells can be propagated indefinitely in this medium with a doubling time of about 36 hr and will form colonies on A1 agar with greater than 75% efficiency of plating (Bretscher and Kaiser 1978). For comparison, the doubling time in a medium containing hydrolyzed casein is about 3.5 hr.

## Intermediary Metabolism

Studies of the intermediary metabolism of *M. xanthus* have been accomplished through the use of three techniques: growth of cells in minimal media, assay of specific enzymes in cell lysates, and use of radiolabeled compounds to follow the fate of a particular substrate. Together these techniques provide a general picture of the metabolic capabilities of this particular species. The cellulolytic species most certainly differ from the proteolytic species in many respects. Whether other proteolytic myxobacteria have an intermediary metabolism that is similar to that of *M. xanthus* is a question that beckons future investigation.

### Tricarboxylic Acid Cycle

The tricarboxylic acid (TCA) cycle plays a central role in the metabolism of *M. xanthus*. It is a major route for reducing nicotinamide adenine dinucleotide (NAD) for adenosine-5′-triphosphate (ATP) generation by oxidative phosphorylation. It also serves as a source of intermediates for amino acid biosynthesis. Although TCA cycle intermediates are not incorporated and oxidized very extensively in *M. xanthus*, carbon sources that enter the TCA cycle, such as acetate and pyruvate, are oxidized extremely well and provide much of the carbon for biosynthesis (Watson and Dworkin 1968; Bretscher and Kaiser 1978). For example, pyruvate is the major carbon and energy source of A1 minimal medium. Its carbon contributes 63% of the dry weight of cells, whereas about 90% of the pyruvate in the medium is used to generate energy (Bretscher and Kaiser 1978). All TCA cycle enzymes that have been examined have been observed in *M. xanthus* cell extracts. These include citrate synthase (EC 4.1.3.7), isocitrate dehydrogenase (EC 1.1.1.41), $\alpha$-ketoglutarate dehydrogenase (EC 1.2.4.2), succinic dehydrogenase (EC 1.3.99.1), and malate dehydrogenase (EC 1.1.1.37) (Watson and Dworkin 1968; Filer *et al.* 1977).

## Glyoxylate Cycle

The glyoxylate cycle is a bypass of some of the reactions of the TCA cycle that is used when a two carbon substrate, such as acetate, serves as a major carbon and energy source. This cycle is a means of providing four carbon compounds for biosynthesis and, at the same time, generating reduced NAD from operation of the TCA cycle. Two enzymes are uniquely required for a functional glyoxylate cycle, isocitrate lyase (EC 4.1.3.1) and malate synthase (EC 4.1.3.2). Both of these enzymes have been observed in extracts of *M. xanthus* (Bland *et al.* 1971; Orlowski *et al.* 1972). Isocitrate lyase cleaves isocitrate to succinate, which can reenter the TCA cycle, and glyoxylate. Malate synthase catalyzes the fusion of glyoxylate with the acetyl group of acetyl coenzyme A (CoA) to form malate. The glyoxylate pathway probably functions at a greatly reduced rate when *M. xanthus* is growing in an amino acid-rich medium. However, it may be very important during the developmental cycle when amino acids become scarce and there is a need to synthesize myxospore capsule polysaccharides. During sporulation the specific activity of the glyoxylate enzymes increases (Bland *et al.* 1971).

## Electron Transport

Like other aerobes, *M. xanthus* has a cyanide-sensitive respiratory chain composed of cytochromes (Dworkin and Niederpruem 1964). Cell extracts were shown to contain cyanide-sensitive reduced NAD (NADH) oxidase, NADH cytochrome *c* reductase, and cytochrome oxidase activities (Dworkin and Niederpruem 1964). The difference spectra wavelengths (nanometers) for the $\alpha$ bands were 602 (cytochrome *a*), 562 (cytochrome *b*), and 552 (cytochrome *c*) (Dworkin and Niederpreum 1964). A trough at 460 suggested the presence of flavoprotein. The $\beta$ band was at 524 nm and the soret band at 425 nm with a shoulder at 442 nm (Dowrkin and Niederpruem 1964).

## Glycolysis

In the Embden–Meyerhof pathway, glucose is converted to pyruvic acid. Three of the enzymes in the pathway—hexokinase, phosphofructokinase, and pyruvate kinase—catalyze reactions that are irreversible under physiologic conditions. The reverse of glycolysis, gluconeogenesis, uses many of the same enzymes but bypasses the irreversible glycolytic steps through the use of different enzymes. Several of the Embden–Meyerhof enzymes have been detected in crude extracts from *M. xanthus.* Although a hexokinase that uses ATP as the phosphate donor (EC 2.7.1.1) has not been found (Watson and Dworkin 1968), a hexokinase that uses pyrophosphate as the phosphate donor has been observed (Lubochinsky, unpublished). The other glycolytic enzymes that have been reported are phosphoglucoisomerase (EC 5.3.1.9), phosphofructokinase (EC 2.7.1.11), fructosediphosphate aldolase (EC 7.1.2.7), and glyceraldehyde-3-phosphate dehydrogenase (EC 1.2.1.12) (Watson and Dworkin 1968). The presence of four other enzymes, triosephosphate isomerase, phosphoglycerate kinase, phosphoglyceromutase, and enolase, has not been examined, but these enzymes are probably present because they are required for gluconeogenesis. Pyruvate kinase, which catalyzes the final step in glycolysis, was not detected in assays using ADP as the phosphate recipient (Watson and Dworkin 1968). Similar assays with pyrophosphate as the phosphate recipient have not been reported in *M. xanthus*. Were it not for the conspicuous absence of this enzyme, *M. xanthus* would have an intact glycolytic pathway.

It is unlikely that *M. xanthus* uses the Embden–Meyerhof pathway for conversion of glucose to pyruvate. Growth in minimal medium is not stimulated by a variety of carbohydrates (Bretscher and Kaiser 1978), and radiolabeled glucose, starch, glycogen, or ribose are incorporated very poorly into the cell (Watson and Dworkin 1968; Hemphill and Zahler 1968a). In addition, pyruvate kinase is apparently missing. These results suggest that the primary function of the Embden–Meyerhof enzymes in *M. xanthus* is the synthesis of glucose and other carbohydrates.

## Gluconeogenesis

Carbohydrates are important components of the cell wall, nucleic acids, lipopolysaccharide, and extracellular polysaccharide. Since *M. xanthus* incorporates sugars poorly, it must rely on gluconeogenesis for their synthesis. From the TCA cycle intermediate, oxaloacetate, phosphoenolpyruvate is synthesized using phosphoenolpyruvate carboxylase (EC 4.1.1.32) (Watson and Dworkin 1968). Driving the Embden–Meyerhof pathway in reverse results in the formation of fructose-1,6-diphosphate. The pathway from fructose-1,6-diphosphate to uridine-5′-diphosphate (UDP)-*N*-acetylglucosamine, a cell wall precursor, has been elucidated by enzymatic analysis of cell lysates. Fructose-1,6-diphosphatase (EC 3.1.3.11) converts fructose-1,6-diphosphate to fructose-6-phosphate (Watson and Dworkin 1968; Filer *et al.* 1977), which is then converted to D-glucosamine-6-phosphate by L-glutamine-fructose-6-phosphate transamidase (EC 5.3.1.19) (Filer *et al.* 1977). Glucosamine-6-phosphate acetyltransferase (EC 2.3.1.4) catalyzes the *N*-acetylation of the sugar, which is then attached to the nucleotide carrier with UDP-*N*-acetylglucosamine pyrophosphorylase (EC 2.7.7.23) (Filer *et al.* 1977). The resulting nucleotide hexosamine, UDP-*N*-acetylglucosamine, may be used in cell wall biosynthesis or be converted to UDP-*N*-acetylgalactosamine or other hexosamines by epimerases (Filer *et al.* 1977). It is clear that *M. xanthus* has the enzymes necessary for the synthesis of polysaccharide precursors from amino acids and that this synthesis can occur through the classic pathway. The concentration of these enzymes is elevated during myxospore formation when a polysaccharide capsule is being synthesized (Filer *et al.* 1977).

## Hexose Monophosphate Shunt

Ribose is not incorporated by *M. xanthus* unless it is attached to a purine or pyrimidine base (Hemphill and Zahler 1968a). Since pentoses are not required for growth in minimal medium (Bretscher and Kaiser 1978), *M. xanthus* must be able to generate ribose-5-phosphate from glucose-6-phosphate for use in nucleotide biosynthesis. The hexose monophosphate shunt, which is also called the pentose phosphate cycle, is the pathway by which *M. xanthus* accomplishes this task. There are four steps in the process, two of which involve the reduction of NADP and are easily assayed. The first step, the conversion of glucose-6-phosphate to glucono-δ-lactone-6-phosphate, is catalyzed by glucose-6-phosphate dehydrogenase (EC 1.1.1.49) (Watson and Dworkin, 1968). This is converted to phosphogluconate in the second step, which has not been examined in *M. xanthus.* Then 6-phosphogluconate dehydrogenase (EC 1.1.1.44) converts the phosphogluconate to a five-carbon sugar, ribulose-5-phosphate (Watson and Dworkin 1968). The final step, the conversion of ribulose-5-phosphate to ribose-5-phosphate, also has not been examined in *M. xanthus.* It is unlikely that the completion of the cycle, the conversion of ribose-5-phosphate back to glucose-6-phosphate is very significant in *M. xanthus* metabolism since ribose is not incorporated well. In fact, when ribose does enter the cell as part of the nucleoside uridine, it is primarily incorporated into nucleic acids (Hemphill and Zahler 1968a).

## Nucleotide Biosynthesis and Salvage

*Myxococcus xanthus* maintains rather small nucleotide pools in comparison with *Escherichia coli* (Hansen and Dworkin 1974). Adenosine-5′-triphosphate (ATP) is the principle constituent of these pools and is maintained at an energy charge of 0.81–0.87 (Hansen and Dworkin 1974; Smith and Dworkin 1980). Adenosine-3′,5′-cyclic monophosphate (cAMP) is also present in *M. xanthus* (Yajko and Zusman 1978; McCurdy *et al.* 1978) along with a cAMP phosphodiesterase that is not inhibited by the caffeine analogs theophylline or aminophylline (Yajko and Zusman 1978). A protein that binds 8-azido-cAMP, a photoaffinity analog of cAMP, has a molecular weight of about 12,500 daltons and appears to be located in the cytoplasm (Orlowski 1980). Guanosine-5′-diphosphate-3′-diphosphate (ppGpp) and guanosine-5′-triphosphate-3′-diphosphate (pppGpp) are minor constituents of the nucleotide pools, but their concentration increases as cells become depleted of nutrients (Manoil and Kaiser 1980a,b).

*Myxococcus xanthus* is capable of synthesizing all of its purine and pyrimidine nucleotides (Dworkin 1962). The first step in de novo purine biosynthesis, the formation of β-D-ribosylamine-5′-phosphate, can be accomplished in two different ways. The phosphoribosylpyrophosphate amidotransferase (EC 2.4.2.14) uses either L-glutamine or ammonium as the nitrogen donor and does not require ATP (Westby and Tsai 1974). It appears that this enzyme is feedback inhibited by adenosine-5′-monophosphate (AMP), guanosine-5′-monophosphate (GMP), and several other purines (Westby and Tsai 1974). The other enzyme, ribose-5′-phosphate aminotransferase, uses ammonium and requires ATP (Westby and Tsai 1974). It is feedback-inhibited by AMP but not GMP (Westby and Tsai 1974). Pyrimidine nucleotides have little effect on either enzyme (Westby and Tsai 1974). When estimates of the nucleotide pool size in *M. xanthus* (Hansen and Dworkin 1974) are compared with the feedback inhibition studies (Westby and Tsai 1974), it appears that little feedback inhibition of this step would occur in growing cells unless there is a synergistic enhancement of inhibition by several nucleotides.

Mutants defective in the synthesis of nucleotides may salvage their requirements from the media in a wide variety of forms. For example, the purine requirements of auxotrophic mutants can be fulfilled with purine nucleobases, nucleosides, nucleotides, deoxynucleosides, or nucleic acids (Hemphill and Zahler 1968a). Myxobacteria have a variety of enzymes for degrading nucleic acids (Hemphill and Zahler 1968a; Mayer and Reichenbach 1978), incorporating nucleotide derivatives (Tsai and Westby 1978), and metabolizing them. However, purines are probably not oxidized as an energy source by *M. xanthus* (Shimkets 1980).

## Amino Acid Biosynthesis and Metabolism

In spite of the fact that *M. xanthus* can derive all of its carbon, nitrogen, and energy from the metabolism of amino acids (Dworkin 1962; Hemphill and Zahler 1968a; Witkin and Rosenberg 1970), relatively little is known of the pathways involved. Somewhat more is known of the amino acid biosynthetic pathways.

The aspartate family is composed of aspartate, asparagine, methionine, threonine, isoleucine, lysine, and the cell wall component *meso*-diaminopimelate, all of which are derived from aspartate (Umbarger 1978). With the exception of isoleucine, all of these can be synthesized by *M. xanthus* (Bretscher and Kaiser 1978). However, synthesis of methionine requires exogenous vitamin $B_{12}$ as a coenzyme in the methyl transfer reaction (Bretscher and Kaiser 1978). Apparently *M. xanthus* lacks vitamin $B_{12}$-independent methyltransferase that many bacteria use in methionine biosynthesis (Umbarger 1978).

The formation of β-aspartylphosphate is catalyzed by two different aspartokinases (EC 2.7.2.4) in *M. xanthus* that have been partially purified by column chromatography (Shabtai *et al.* 1975). Aspartokinase I is repressed by L-lysine and feedback-inhibited by either *meso*-diaminopimelate or low concentrations of lysine. At higher lysine concentrations, the inhibition is not as severe and the concentration of *meso*-diaminopimelate determines the specific activity of the enzyme (Shabtai *et al.* 1975). This may ensure that cells do not become starved for *meso*-diaminopimelate in the presence of high lysine concentrations. Aspartokinase II is both repressed and feedback-inhibited by L-threonine. Aspartokinase I and, to a lesser extent, aspartokinase II are feedback-stimulated by L-isoleucine and L-methionine (Shabtai *et al.* 1975). Feedback stimulation may also assure synthesis of *meso*-diaminopimelate in a rich medium. A combination of L-threonine and low L-lysine concentrations virtually eliminate both aspartokinase activities unless methionine or isoleucine is present (Filer *et al.* 1973).

The branched-chain amino acids, leucine, isoleucine, and valine, are all required for growth of *M. xanthus* (Bretscher and Kaiser 1978). Although valine and isoleucine synthesis proceed through structurally different biochemical intermediates, four out of the five enzymes involved with their synthesis are shared in many bacteria (Umbarger 1978). The immediate precursor to valine, $\alpha$-ketoisovalerate, can substitute for valine and support growth of *M. xanthus* in A1 minimal medium, but it cannot replace leucine even though it is a precursor for it (Bretscher and Kaiser 1978). There are two methods for converting $\alpha$-ketoisovalerate to valine in *E. coli.* Transaminase B, which uses glutamate as the amino donor, is also used to convert $\alpha$-keto-$\beta$-methyl valerate to isoleucine, as well as $\alpha$-ketoisocaproate to leucine (Umbarger 1978). Transaminase C, which uses alanine or $\alpha$-aminobutyrate as the amino donor, is specific for valine synthesis (Umbarger 1978). A lack of transaminase B alone would not explain the lack of valine synthesis in *M. xanthus,* so it is likely that there are at least two missing steps in the branched amino acid biosynthetic pathway.

The aromatic amino acid family consists of phenylalanine, tyrosine, and tryptophan, none of which is required by *M. xanthus* (Bretscher and Kaiser 1978). However, growth in A1 minimal medium is greatly stimulated by phenylalanine, suggesting that it is synthesized at an extremely slow rate (Bretscher and Kaiser 1978). Under some conditions, phenylalanine can be oxidized to tyrosine, which is then secreted (Hemphill and Zahler 1968a).

Since amino acids can serve as the sole nitrogen source, it is not surprising that a variety of transaminase activities have been observed in cell extracts of *M. xanthus.* These include L-alanine:2-oxoglutarate aminotransferase (EC 2.6.1.2) (Hansen and Andreoli 1973), L-aspartate:2-oxoglutarate aminotransferase (EC 2.6.1.1) (Hansen and Andreoli 1973), L-glutamate:glyoxylate aminotransferase (EC 2.6.1.4) (Kottel *et al.* 1974). and L-alanine:glyoxylate aminotransferase (EC 2.6.1.12) (Kottel *et al.* 1974). Since these activities have not been purified, it is not possible to tell how many different enzymes are involved.

Ammonium sulfate enhances growth in A1 minimal medium, suggesting that ammonium may be assimilated by cells (Bretscher and Kaiser 1978). Glutamine synthetase (EC 6.3.1.2) is one enzyme that is used to assimilate ammonium in *M. xanthus* (Filer *et al.* 1977). Although asparagine synthetase has not been reported, it seems safe to assume it is also produced, since asparagine is not required for growth (Bretscher and Kaiser 1978). Dehydrogenases that catalyze reductive amination have also been observed in cell extracts, including glutamate deyhdrogenase (EC 1.4.1.2), alanine dehydrogenase (EC 1.4.1.1), and glycine dehydrogenase (EC 1.4.1.5) (Kottel *et al.* 1974). Since the $K_m$ of these enzymes is unknown, it is not clear whether their function is catabolic or biosynthetic. In general, NAD-dependent glutamate dehydrogenases, similar to the one reported in *M. xanthus,* serve a catabolic function unless ammonium is present at a very high concentration (Tyler 1978).

## Regulation of Nutritional Downshift in Other Bacteria

### Stringent Response in *Escherichia coli*

A sudden shift of *E. coli* cells to adverse nutritional conditions triggers a decrease in the rate of RNA synthesis and a series of changes in metabolism, collectively known as the stringent response (Gallant 1979). This shift down causes an increase in the concentration of two nucleotides, guanosine-5′-diphosphate-3′-diphosphate (ppGpp) and to a lesser extent guanosine-5′-triphosphate-3′-diphosphate (pppGpp). Many of the changes observed during the stringent response can be duplicated in vitro with ppGpp. This nucleotide can stimulate, or inhibit, in vitro transcription of specific genes in a manner that often correlates with transcription patterns observed during the stringent response (Gallant 1979). Even more impressive is the correlation between ppGpp regulation of enzyme activity in vitro and the metabolic adjustments observed during the stringent response (Gallant 1979). It is clear that ppGpp exerts a powerful regulatory effect on both transcription and enzyme activity.

There seem to be two pathways that lead to the accumulation of ppGpp. The most carefully characterized one occurs on the ribosome during amino acid starvation and requires the product of the *relA* gene, known as stringent factor. Stringent factor is stimulated by the codon-specific binding of uncharged transfer RNA (tRNA) to the ribosome A binding site to transfer the $\beta$–$\gamma$ pyrophosphoryl group of ATP to the 3′ hydroxyl group of GTP (Haseltine and Block 1973). The pppGpp is then dephosphorylated to ppGpp by the product of the *gpp* locus (Somerville and Ahmed 1979). The *relA* mutants are relaxed in that they fail to decrease their rate of RNA synthesis in repsonse to an amino acid shift down (Gallant 1979). Mutations in the *relB* and *relC* genes also cause the relaxed phenotype, probably by reducing stringent factor activity on the ribosome. The *relC* gene codes for the 50S ribosomal protein L11 (Parker *et al.* 1976) and the *relB* locus codes for a ribosome-associated protein (Mosteller 1978). However, *relA* mutants do respond to a shift down in carbon and energy sources by accumulating ppGpp, suggesting that there is an alternate pathway for ppGpp synthesis. The intracellular level of ppGpp is also controlled by the rate of its degradation. The *spoT* gene codes for a ppGpp-3′-nucleotide pyrophosphatase (Sy 1977).

## Initiation of Endospore Formation in *Bacillus subtilis*

When members of the genus *Bacillus* become nutritionally deprived, they differentiate dormant cells known as endospores. After a sudden shift from a medium that is rich in amino acids to one that lacks amino acids, *Bacillus subtilis* begins forming endospores. One of the earliest events after the downshift is a rapid but transient change in the size of the guanine nucleotide pools. The guanosine-5′-triphosphate (GTP) pool decreases in size by 80% with a concurrent increase in ppGpp and pppGpp (Lopez *et al.* 1981). The synthesis of ppGpp and pppGpp can account for about half of the decrease in the GTP pool. In addition, the activity of inosine monophosphate dehydrogenase, the penultimate enzyme in the biosynthesis of GMP, becomes inhibited (Lopez *et al.* 1981).

Mutations analagous to *relA* have been isolated in *B. subtilis* and studied to determine if the stringent response is involved in the initiation of endospore formation. A relaxed mutant, otherwise isogenic with the wild type, produces 100-fold fewer spores than the wild type in response to an amino acid downshift, suggesting that the stringent response can, under some circumstances, lead to the initiation of endospore formation (Lopez *et al.* 1981). The relaxed mutant fails to make adjustments in the guanine nucleotide pools that normally occur after amino acid step down. There is little synthesis of ppGpp, and there is only a slight decrease in the size of GTP pool. Three lines of evidence suggest that it is the sudden decrease in the GTP pool rather than the accumulation of ppGpp or pppGpp that is the intracellular trigger for the initiation of endospore formation. First, when the *relA* mutant is starved for amino acids in the presence of decoyanine, a specific inhibitor of GMP synthase, it yields spores nearly equal in number to that of the wild type. Under these conditions there is a rapid decrease in the size of both the GTP and ppGpp pools (Lopez *et al.* 1981). Second, when cells are grown in glucose–lactate and shifted to lactate, the endospore yield of the relaxed mutant is nearly equal to that of the wild type. During this carbon and energy source step down, both the wild type and mutant experience a rapid drop in GTP. The ppGpp and pppGpp pools rise dramatically in the wild type but do not increase in the mutant (Lopez *et al.* 1981). Third, cells can be induced to sporulate in the presence of excess ammonium, rapidly metabolizable carbon and energy sources, and orthophosphate, provided a partial inhibition of guanine nucleotide synthesis is imposed by either decoyanine or by leaky auxotrophic mutations. In fact, all nutritional conditions that induce endospore formation also result in a decrease of GTP (Freese 1981). Therefore ppGpp can trigger the initiation of endospore formation in *B. subtilis* by decreasing the pool of GTP. However, an increased level of ppGpp and pppGpp is not necessary for the initiation of endospore formation.

## Nutritional Control of Fruiting Body Formation

Two general features have emerged from studies of the nutritional control of fruiting body formation. First, development is a response to nutritional deprivation. In cellulolytic species, such as *Sorangium compositum,* development is initiated when a carbohydrate source becomes limiting (Couche 1969). In the proteolytic species, such as *M. xanthus,* amino acid starvation triggers development (Dworkin 1963). In both cases development does not occur on a rich or complex medium even when the cells enter a stationary phase. Second, gradual and partial starvation is more likely to result in rapid and synchronous development than is severe starvation (Wireman and Dworkin 1975; Manoil and Kaiser 1980a). Amino acids, carbon, and energy are required for the successful completion of development.

Nutritional control of development of *M. xanthus* has been investigated by using a defined medium that supports only vegetative growth. By deleting certain components of the medium, adding competitive inhibitors of cellular metabolites, or using auxotrophic mutants, one can identify specific nutrients whose limitations lead to development. Recently, A1 minimal medium has been used because it contains little more than the essential amino acids plus a separate carbon and energy source (see Appendix B for composition). Partial starvation for any of the required amino acids, isoleucine, leucine, and valine induces development (Manoil and Kaiser 1980b). Starvation for nonessential amino acids also initiates development. For example, tryptophan auxotrophs form fruiting bodies on A1 agar unless tryptophan is added to the agar (Manoil and Kaiser 1980b). Amino acid analogs such as serine hydroxamate and tyrosinol induce development but are prevented from doing so in the presence of serine and phenylalanine, respectively (Manoil and Kaiser 1980b). Development may also be initiated when pyruvate, which is the carbon and energy source, or inorganic phosphate are deleted from A1 (Manoil and Kaiser 1980b). Since the required amino acids provide a nitrogen source, it is not possible to determine if nitrogen starvation also induces development.

### Stringent Response in *Myxococcus xanthus*

Since development can be induced by amino acid starvation in *M. xanthus,* the guanosine polyphosphate pools were examined during the early stages of development. Cells growing exponentially in casitone (hydrolyzed casein) medium were concentrated in a small volume of medium and spotted on a nonnutrient agar surface. These conditions impose amino acid, carbon, and energy limitation and result in mature fruiting bodies by 72 hr. Within the first hour there was a 10–20-fold increase in ppGpp, a twofold increase in pppGpp, and a twofold decrease in GTP (Manoil and Kaiser 1980a,b). The ppGpp/GTP ratio increased from 0.004 to 0.160 and stabilized for at least 13 hr (Manoil and Kaiser 1980a). Although the increase in ppGpp was rapid, the overall yield was small in comparison with the size of the GTP pool, and a sharp decrease in GTP was not observed. In comparison, a similar shift down in *B. subtilis* that leads to endospore formation causes a rapid increase in ppGpp and pppGpp at the expense of the GTP pool and results in a ppGpp/GTP ratio of about 3.0 (Lopez *et al.* 1981). The increase in ppGpp in *M. xanthus* does not decrease the size of the GTP pool as dramatically as it does in *B. subtilis.*

A mutant was isolated that is resistant to thiostrepton. When compared with an otherwise isogenic wild type, the mutant showed little change in net RNA synthesis and ppGpp synthesis upon treatment with tyrosinol, as might be expected of a relaxed mutant (Gill and Kaiser unpublished). Yet the mutant formed fruiting bodies and myxospores in response to amino acid starvation (Gill and Kaiser unpublished). This suggests that the stringent response is not required for the initiation of development under at least some conditions.

Although the basic nutritional conditions that initiate development are similar in *B. subtilis* and *M. xanthus,* the manner in which development is initiated may differ. During the stringent response in *B. subtilis* the synthesis of ppGpp initiates development by draining the GTP pool. In contrast, the synthesis of ppGpp does not severely reduce the size of the GTP pool in *M. xanthus.* Nor is the presence of high levels of ppGpp sufficient to induce development. During growth in A1 minimal medium, the ppGpp pool is high and usually does not increase further in response to growth-inhibiting conditions (Manoil and Kaiser 1980b).

## Compounds that Induce Development of Nonstarved Cells

Many amino acids of the aspartate family exert a strong control over the initiation of development. Amino acids that modify the activity of aspartokinase (EC 2.7.2.4) in vitro affect fruiting body formation in a reciprocal manner. Methionine and isoleucine, which feedback-stimulate aspartokinase activity (Filer *et al.* 1973; Shabtai *et al.* 1975), inhibit the development of otherwise starved cells (Rosenberg *et al.* 1973; Campos and Zusman 1975). On the other hand, threonine and *meso*-diaminopimelate, which are strong inhibitors of aspartokinase acitivity (Shabtai *et al.* 1975), induce development of nonstarved cells (Rosenberg *et al.* 1973; Campos and Zusman 1975). Threonine-induced development can be prevented by supplementing the agar with either lysine or diaminopimelate (Manoil and Kaiser 1980b). This suggests that threonine induces the development by feedback-inhibiting aspartokinase causing the cells to starve for lysine (Manoil and Kaiser 1980b). It seems that nutritional imbalances caused by high concentrations (1–10 m*M*) of a particular compound can produce unusual cell responses due to their affect on allosteric enzymes.

Adenine-containing nucleotides also induce development when added to nonstarved cells (Campos and Zusman 1975). Most adenine-containing compounds induce fruiting body formation, and some are more effective than 3′,5′-cyclic adenosine monophosphate (cAMP). When adenosine-5′-diphosphate (ADP) and threonine are added together, there is a synergistic enhancement of the number of fruiting bodies produced due to the interaction of two different mechanisms (Campos and Zusman 1975). A mutant that is resistant to 2,6-diaminopurine and has lost the ability to be stimulated to develop in the presence of excess nutrients by adenine-containing compounds, retains the ability to be induced by threonine (Manoil and Kaiser 1980c). It is possible that adenine-containing compounds induce development indirectly by causing intracellular nutritional imbalances similar to that observed for threonine (Manoil and Kaiser 1980c). However, cells are nearly 5000-fold more sensitive to adenine 3.5 $\mu$M) than to threonine (17 m*M*). An alternate hypothesis is that adenine-containing compounds trigger development by mimicking an endogenous intercellular signal (see "Intercellular Signals Initiating Development of Myxobacteria").

## Intercellular Communication in Other Bacteria

Communication among bacterial cells of the same species is relatively rare. The few systems that have been investigated share three common features. First, the intercellular signal is secreted into the medium at a concentration proportional to the cell density. The sex pheromones of *Streptococcus faecalis* (Dunny *et al.* 1979) and the luciferase autoinducer of *Photobacterium fischeri* (Nealson *et al.* 1970) both reach their maximum extracellular concentration as the cells reach stationary phase. Second, the inducer is a small and stable molecule that acts at low concentrations. Third, the inducer is extremely specific in that only a small group of related strains respond to it.

*Streptococcus faecalis* strains that lack a conjugative plasmid secrete a substance that induces strains carrying the plasmid to become adherent (Clewell 1981). The pheromone-induced plasmid donor cells acquire a new cell surface antigen (Yagi *et al.* 1971) that binds to a substance located on the surface of the plasmidless recipient cells facilitating the formation of donor–recipient mating aggregates (Dunny *et al.* 1979). Once the recipient receives the plasmid, its production of the sex pheromone drops, and it becomes a conjugation donor that is sensitive to the pheromone concentration (Dunny *et al.* 1978). A strain lacking a conjugative plasmid produces at least three sex pheromones, each of which activates a donor carrying a specific conjugative plasmid (Clewell *et al.* 1982). Two of the pheromones seem to be small peptides with molecular weights of less than 1500 daltons (Clewell *et al.* 1982).

Luciferase in marine bacteria is an $\alpha$–$\beta$ protein dimer that catalyzes the oxidation of reduced flavin mononucleotide and a long chain aldehyde with the emission of light (Hastings and Nealson 1977). Production of luciferase is controlled by a small molecule that is secreted into the growth medium and increases proportionately with the cell density (Eberhard 1972). Dialysis of the growing culture against fresh medium prevents bioluminescence (Ulitzur and Hastings 1979). The inducer may regulate transcription of luciferase since rifampicin inhibits autoinduction (Nealson *et al.* 1970). Both *P. fischeri* (Eberhardt 1972) and *Beneckea harveyi* (Greenberg *et al.* 1979) produce an autoinducer of luciferase, but the autoinducer produced by one species is not active on the other species.

## Intercellular Signals Initiating Development of Myxobacteria

There are three basic requirements for the initiation of myxobacteria development: a solid surface, a limited nutrient supply, and a high cell density. In assays for intercellular signals that initiate development, one of these three requirements is adjusted so that development is not normally initiated. Developing cells, cell extracts, or partially purified extracts are then added to see if development is induced.

A solid surface is necessary for gliding motility, which is the means by which myxobacteria aggregate into fruiting bodies. Sporulation of *M. xanthus* also requires a solid surface (Wiremen and Dworkin 1977; Shimkets and Kaiser 1982b), with two notable exceptions. Myxospores may be chemically induced with glycerol in liquid culture (Dworkin and Gibson 1964). Glycerol induction bypasses all three requirements for the initiation of development, but glycerol-induced spores differ in many respects from fruiting body myxospores (see Chapter 9). Methionine starvation in liquid medium also induces sporulation of one strain of *M. xanthus* (Witkin and Rosenberg 1970), but the spores that are formed have a different ultrastructure from fruiting body myxospores (Witkin and Rosenberg 1970; Inouye *et al.* 1979b). Thus, there does appear to be a requirement for a solid surface and/or cell–cell contact for the production of spores resembling those produced in fruiting bodies. This requirement is fulfilled early in development before there is any change in cell shape (Wireman 1979). Soon after this period has elapsed, cells become committed to sporulate, even when transferred to a liquid growth medium (Wireman 1979). The events that establish this commitment to sporulation are not known but probably involve intercellular communication (Hagen *et al.* 1978).

There is evidence that myxobacteria secrete a substance that induces nearby cells to participate in development. An assay for such an endogenous factor involves plating cells at a high cell density on a solid surface but with enough nutrients to repress development. If thin pieces of agar containing nonstarved cells are placed on top of fruiting bodies, then the nonstarved cells will form fruiting bodies (Lev 1954). The stimulation appears to be at least genus specific since *Chondromyces exiguus* will not stimulate *M. xanthus* to initiate development (Lev 1954). A modified version of this technique that uses a cellophane membrane instead of an agar slab has also been used, and *M. xanthus* fruiting bodies above and below the membrane are often directly opposite each other (McVittie and Zahler 1962). An activity that induces nonstarved *Myxococcus virescens* cells to develop has been partially purified by paper chromatography but has not been identified (Jennings 1961). Although it does appear that *M.*

*xanthus* and other myxobacteria secrete signals that induce nonstarved cells of the same species to develop, this type of assay has not been used to identify the intercellular signal.

The morphogenesis of a single fruiting body requires about a million cells, and there appears to be a mechanism by which these cells measure their density to determine if enough cells are present to complete development. If *M. xanthus* is plated at low cell density on a nonnutrient agar surface and partially purified cell extracts are added, some fractions induce development without increasing the cell density. Two fractions have $R_f$ values similar to adenosine and AMP (Shimkets and Dworkin unpublished). Adenine and most adenine-containing compounds induce fruiting body formation at low cell densities without increasing the cell density (Shimkets and Dworkin 1981). If cells are plated at a high cell density on agar-containing hadacidin, a specific inhibitor of de novo AMP synthesis, aggregation is inhibited unless exogenous adenosine is added. Hadacidin, a structural analog of L-asparate, is a competitive inhibitor of adenylosuccinate synthase in *E. coli* (Shigeura and Gordon 1962a,b). It appears to inhibit the same enzyme in *M. xanthus* because growth in defined medium that is inhibited by hadacidin is restored by adenosine but not by guanosine (Shimkets and Dworkin 1981). Agar containing Norit, a substance that binds purines, also inhibits aggregation at high cell densities unless adenosine is added (Shimkets and Dworkin 1981). The two most abundant extracellular adenine-containing compounds bound by the Norit are AMP (at an extracellular concentration of 2 $\mu M$) and adenosine (6.7 $\mu M$). The extracellular concentration of cAMP is less than 10 n$M$. At these concentrations only adenosine will initiate development at low cell densities (Shimkets and Dworkin 1981). Reconstruction experiments show that secreted adenosine can account for about 30% of the aggregation centers established at high cell densities.

These results suggest that adenosine is secreted as a method of measuring cell density and that aggregation is initiated when the local adenosine concentration exceeds a threshold of about 1 $\mu M$. The source of the secreted adenosine appears to be newly synthesized AMP or purine nucleotides that are recycled through inosine 5′monophosphate (IMP) to AMP. Similar concentrations of adenosine also induce partially starved cells to initiate aggregation (Manoil and Kaiser 1980c). To avoid inducing development when nutrients are excessive, the secretion of adenosine, the perception of extracellular adenosine, or both must not be constitutive but instead be regulated by nutritional deprivation. In fact, high concentrations of glycine repress the ability of cells to respond to extracellular adenosine (Manoil and Kaiser 1980c).

Some myxobacteria, such as *Stigmatella aurantiaca,* are stimulated by light to form fruiting bodies if starved on a solid surface at low cell density (Qualls *et al.* 1978a,b). Blue light between 400–500 nm is the most effective, although other wavelengths work at high irradiances (White *et al.* 1980b). A low molecular weight lipidlike pheromone secreted by developing *Stigmatella aurantiaca* cells stimulates fruiting body formation in the absence of light (Stephens *et al.*

1982). However, such high pheromone levels are apparently achieved only at unusually high cell densities. At somewhat lower cell densities, light acts cooperatively with the pheromone to stimulate fruiting body formation, but light does not enhance production of the pheromone. Perhaps light sensitizes cells to their own pheromone (Stephens *et al.* 1982). Light also acts cooperatively with guanine-containing compounds to stimulate fruiting body formation in *Stigmatalle aurantiaca* (Stephens and White 1980). However, the *Stigmatella* pheromone (Stephens *et al.* 1982), guanosine (Shimkets and Dworkin 1981), and light (Shimkets unpublished) do not stimulate fruiting body formation of *Stigmatella aurantiaca* (Stephens and White 1980b). Thus it appears that the intercellular stimuli promoting fruiting body formation are species specific, and it seems unlikely that hybrid fruiting bodies of these two species would form.

## Conclusion

Biochemical, behavioral, and ecological studies of myxobacteria suggest a general scheme for their lifestyle in the environment. The acquisition and use of nutrients by the proteolytic myxobacteria in nature may begin when they prey on other bacteria. Myxobacteria secrete a variety of lytic enzymes and antibiotics that can lyse bacterial cells and hydrolyze the cell components. The protein fraction of the prey is particularly useful since amino acids serve as carbon and energy sources. These amino acids are degraded to acetate, pyruvate, and TCA cycle intermediates for use in generating energy or for synthesizing polysaccharides and lipids. Salvage of other components of the prey bacteria, such as purines and pyrimidines, may substantially reduce the biosynthetic burden.

Many of the predatory myxobacteria move in spatially and temporally synchronized waves known as ripples as they feed on other bacteria. The constituents of the prey bacteria that induce this behavior in *M. xanthus* are the peptidoglycan components *N*-acetylglucosamine, *N*-acetylmuramic acid, D-alanine, and diaminopimelate. Although the peptidoglycan layer of the prey is hydrolyzed very effectively by secreted enzymes, the carbohydrate portion of murein does not appear to serve as a substantial carbon and energy source. The mechanism by which these carbohydrates trigger oscillations in cell movement is unknown. Perhaps the presence of phosphofructokinase is a clue. This is an irreversible glycolytic enzyme whose function in *M. xanthus* is obscure since pyruvate kinase is apparently missing. Phosphofructokinase would of course regulate the level of fructose phosphate used in gluconeogenesis as well as the adenine nucleotide pools. In the yeast *Saccharomyces carlsbergensis,* this enzyme is allosteric and its activity oscillates to generate pulses of ADP, which in turn generate oscillations in other enzyme reactions dependent on ADP and ATP (Hess and Borteux 1971). Perhaps phosphofructokinase coordinates rippling of *M. xanthus* by synchronizing enzymes dependent on adenine nucleotides or by controlling the synthesis of polysaccharide-containing slime on which the myxobacteria move.

Another exciting area for future research is the initiation of the developmental cycle. What are the intracellular and intercellular signals that control initiation and how do they coordinate cell behavior and modify gene expression? The guanosine polyphosphates ppGpp and pppGpp are synthesized in response to rapid depletion of nutrients in *M. xanthus,* but they may represent only the first in a series of changes needed to trigger development. The level of ppGpp reaches a maximum under nutritional conditions in which development is not initiated. However, the level of pppGpp does increase further under conditions leading to development, and more work is required to determine if pppGpp is an intracellular signal for the initiation of development.

There may be two sets of intercellular signals that coordinate cell behavior in the initial stages of development, one involved with the aggregation of cells and the other involved with sporulation. Secreted adenosine appears to initiate aggregation, but reconstruction experiments indicate that other unknown signals are also involved. Although starving cells for adenosine does inhibit aggregation, it does not prevent sporulation, suggesting that sporulation is regulated somewhat independently of aggregation. Presently, four intercellular sporulation signals are envisioned based on experiments with conditional nonsporulating mutants (Hagen *et al.* 1978). These intercellular signals may be spatially localized to regulate the development of a three-dimensional fruiting body.

# 5

# Antibiotics and Lytic Enzymes

Eugene Rosenberg and Mazal Varon*

It has been known for many years that one of the characteristic features of myxobacteria is their ability to lyse other microorganisms. Only recently, however, have some of the factors responsible for this lytic behavior been isolated and studied. It is now clear that myxobacteria produce a wide variety of antibiotics, bacteriocins, cell wall lytic enzymes, lipases, nucleases, polysaccharidases, and proteases. Apparently, these agents play an important role in the predatory activities of myxobacteria. In addition, some of these agents may take part in developmental autolysis that preceeds fruiting body formation and sporulation. In this chapter, we shall survey the lytic phenomena as exhibited by myxobacteria, describe the lytic agents, and discuss the possible functions of these agents in growth and development of myxobacteria.

## Lytic Phenomena

### Lysis of Microorganisms by Myxobacteria

Following the initial discovery by Pinoy in 1913 (in Bender 1962) that *Chondromyces crocatus* is able to lyse other bacteria, several investigators have reported that a wide variety of myxobacteria are bacteriolytic (Snieszko *et al.* 1942; Oxford 1947; Singh 1947; Oetker 1953; Norén 1953, 1955b; Mathew and Dudani 1955; Margalith 1962; Kletter and Henis 1963; Gillespie and Cook 1965; Shilo 1970; Raverdy 1973; Singh and Yadava 1976; Roper and Marshall 1977; and others). The spectrum of microorganims susceptible to

*Department of Microbiology, George S. Wise Faculty of Life Sciences, Tel Aviv University, Ramat Aviv, Israel.

lysis by myxobacteria or their extracellular products is wide and includes many eubacteria, actinomycetes, cyanobacteria, yeasts, and other fungi. Even the lysis of nematodes by *Sorangium* sp. has been reported (Katznelson *et al.* 1964).

Table 5.1 summarizes some of the data related to lysis of eubacteria. With few exceptions, Gram-positive and Gram-negative, as well as acid-fast bacteria, are susceptible to one myxobacterium or another. Bacteria not shown in Table 5.1 that could be lysed by myxobacteria include *Aerobacter aerogenes, Proteus mirabilis, Klebsiella pneumoniae, Citrobacter freundii, Acinetobacter anitratum, Mycobacterium phlei, Micrococcus flavus,* and *Agrobacterium tumefaciens.*

Two bacterial species appear to be resistant to myxobacteria: *Bacillus cereus,* a spore former, and *Chromobacterium violaceum,* a pigmented bacterium. The resistance of the *Bacillus* has been attributed to the presence of spores (Singh and Yadava 1976; Raverdy 1973). Although the assays were performed by plating the two organisms together on the same nutrient agar medium and incubating them for several days, it is possible that *Bacillus cereus* grew more rapidly than the myxobacteria, reached stationary phase, and formed spores

**Table 5.1.** Lysis of Eubacteria by Myxobacteria[a]

| Substrate organism | Myxococcus | Chondrococcus | Angiococcus |
|---|---|---|---|
| Gram-positive | | | |
| *Bacillus cereus* | 0/7 | 0/5 | 0/3 |
| *Bacillus polymyxa* | 3/9 | 0/5 | 0/3 |
| *Bacillus subtilis* | 2/2 | — | — |
| *Corynebacterium faciens* | 3/5 | 5/5 | 3/3 |
| *Micrococcus luteus*[b] | 5/5 | 4/5 | 3/3 |
| *Sarcina lutea* | 5/5 | 5/5 | 0/3 |
| *Staphylococcus aureus* | 9/9 | 5/5 | 3/3 |
| *Streptococcus faecalis* | 3/7 | 1/5 | 2/3 |
| Gram-negative | | | |
| *Azotobacter indicus* | 1/5 | 2/5 | 2/3 |
| *Erwinia carotovora* | 3/5 | 5/5 | 3/3 |
| *Escherichia coli* | 8/8 | 5/5 | 3/3 |
| *Salmonella paratyphi* | 5/7 | — | — |
| *Salmonella typhimurium* | 9/9 | 5/5 | 3/3 |
| *Serratia marcescens* | 7/9 | 4/5 | 3/3 |
| *Vibrio cholerae* | 8/8 | — | — |
| *Pseudomonas aeruginosa* | 1/7 | 0/5 | 3/3 |
| *Pseudomonas fluorescens* | 2/3 | — | — |

[a]Data taken from Mathew and Dudani 1955; Kletter and Henis 1963; Raverdy 1973; Singh and Yadava 1976. The table indicates the fraction of myxobacteria examined that were active against the test substrate organism.

[b]Formerly *Micrococcus lysodeikticus.*

before the lytic activities of the myxobacteria could be expressed. Thus, the apparent resistance of *B. cereus* may be due to spores rather than vegetative cells. The role of pigment in protecting prey against myxobacteria is not clear. Raverdy (1973) demonstrated that a pigmented *Serratia* sp. was not lysed by *Myxococcus xanthus* or *Myxococcus virescens*, whereas a nonpigmented *Serratia* sp. was lysed by both myxococci. This correlation does not necessarily indicate cause and effect because the two *Serratia* strains may not have been isogenic.

A gliding bacterium that was reported to be a myxobacterium (but later reclassified as a *Lysobacter* by Christensen and Cook, 1978), lyses blue–green algae (Shilo 1970). This microorganism whose deoxyribonucleic acid (DNA) contained 70 mol % cytosine plus guanine (C + G) was found to lyze three unicellular strains (*Anacystis nidulans, Coccochloris penyocystis,* and *Synechococcus cedorum*) and five filamentous cyanobacteria (*Nostoc* sp., *Plectonema boryanum, Oscillatoria prolifera, Spirulina platensis,* and *Spiriulina tenuis*). Microscopic examination of the plaques on *Nostoc* lawns showed that all the vegetative cells in the algal filaments were completely lysed, whereas heterocysts remained intact (Fig. 5.1). Cyanobacteria were also shown to be lysed by a *M. xanthus* strain (Burnham *et al.* 1981). Lysis of *Phormidium luridum* accompanied by growth of the myxobacteria occurred in the absence of any additional organic components in the medium.

Yeasts and other fungi were susceptible to several myxobacteria strains, including *Myxococcus, Chondrococcus* and *Angiococcus* sp. (Singh and Yadava 1976). The most active species were *Myxococcus fulvus* and *Myxococcus stipitatus,* causing complete lysis of nine yeasts and yeastlike fungi out of 13 examined.

## Autolysis of Myxobacteria

Many myxobacteriologists have noted that their cultures undergo autolysis after they have reached stationary phase. The appearance of spheroplasts in the culture often preceeds lysis. Because this phenomenon is common in stationary phase cultures of many other proteolytic bacteria, it did not stimulate detailed examination. However, the reports by Wireman and Dworkin (1975, 1977) that autolysis is an integral part of fruiting body formation has brought this subject to the frontier of modern myxobacterial research. Wireman and Dworkin (1975, 1977) observed that fruiting body formation in *M. xanthus* was accompanied by lysis of 80–90% of the original vegetative population. The surviving cells appeared as myxospores in the mature fruiting bodies. Lysis began after the cells had aggregated into mounds and persisted throughout fruiting body and myxospore formation. Once the cells had reached a certain stage of "commitment" on the developmental plates, they continued to undergo lysis and myxospore formation even when suspended in buffer. Lysis of the committed cells in liquid media was independent of cell density, indicating an

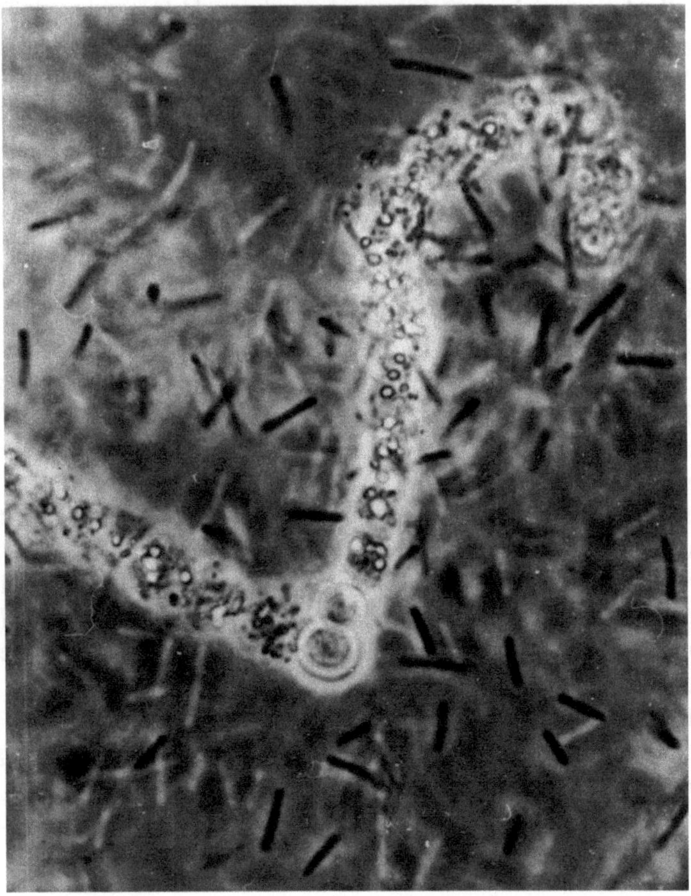

**Figure 5.1.** Lysis of *Nostoc* by myxobacter strain FP-1. A partly lysed filament showing intact heterocysts and gas vacuoles (courtesy of Shilo).

autolytic mechanism. Myxospore formation, on the other hand, was strongly dependent upon cell concentration. With $7.5 \times 10^8$ autolysing cells per milliliter, all of the survivors appeared to have sporulated; only 25% of the survivors sporulated with $7.5 \times 10^7$ autolysing cells per milliliter, whereas none of the survivors sporulated with $7.5 \times 10^6$ autolysing cells per milliliter. In all three experiments the total cell number was identical and only the volume (i.e., cell density) was varied. The authors concluded that the lysing cells provided products essential for the sporulation of the surviving cells. The specific nature of these products and the mechanism by which they function remain unclear. They could serve as a nutrient(s) for the sporulation process and/or specific chemical signals for myxospore induction. In any case, the concentration of those products is limiting in suspensions containing less than $10^8$ autolysing cells per milliliter.

Another indication for the involvement of lysis in sporulation comes from the recent findings of Shimkets and Kaiser (1982a). Certain mutants (Spo C) incapable of autolysis and sporulation were induced to carry out both functions by addition of peptidoglycan components such as *N*-acetylglucosamine, *N*-acetylmuramic acid, diaminopimelic acid, and D-alanine. Murein rescue of sporulation, such as fruiting body sporulation, required high cell density. It is not clear whether these murein components are direct inducers for sporulation or whether they affect sporulation indirectly by inducing developmental lysis that, in turn, results in the release of the inducer.

Studies in which myxosporulation was artifically induced by glycerol also indicated a correlation between autolytic activity and cellular morphogenesis (Kottel and White 1974; Dawson and Jones 1979). The autolytic activity was enhanced during the conversion of vegetative cells into myxospores. A mutant noninducible by glycerol showed no increased autolytic activity, whereas a late-converting mutant that formed myxospores after a 7 hr delay autolyzed after a similar delay. It was suggested that glycerol activated autolytic enzymes already present in the vegetative cells. These enzymes initiated the morphogenetic process, but then a subsequent synthesis of additional enzyme(s) was necessary for the shape change (rod to sphere) to occur. As discussed in Chapter 2, the envelope of the vegetative cells of *M. xanthus* consists of a discontinuous layer with patches of peptidoglycan separated by nonpeptidoglycan material. This structure is altered during morphogenesis, and the patches of peptidoglycan become connected to each other. Such a process would depend on the activity of autolytic enzymes.

## Lytic Agents Produced by Myxobacteria

### Antibiotics

Antibiotic production is frequently associated with soil fungi, actinomycetes, and, to a lesser degree, the genus *Bacillus*. Indirect evidence for antibiotic production by myxobacteria was reported by Oxford (1947), Norén (1953), and Norén and Raper (1962). However, these microbiologists failed to isolate the active component(s) from the culture fluid, and it is thus difficult to assess how much of the reported phenomena was due to antibiotic and how much to other factors such as bacteriolytic enzymes. Investigations carried out primarily during the last 10 yr have demonstrated that myxobacteria are, in fact, a rich source of antibiotics. Although relatively few myxobacterial antibiotics have been characterized, they exhibit a wide range of chemical structures, modes of action. and antimicrobial spectra.

## Myxin

The first clear and direct demonstration of antibiotic production by a bacterium originally reported to be a *Sorangium* sp. (later classified as a *Lysobacter*), was the isolation of myxin (Peterson *et al.* 1966). The chemical structure of myxin, 1-hydroxy 6-methoxyphenazine, 5,10-dioxide (Fig. 5.2), was established by Edwards and Gillespie (1966), Weigele and Leimgruber (1967), and Sigg and Toth (1967).

Myxin has an extremely broad spectrum of antimicrobial activity (Peterson *et al.* 1966; Grunberg *et al.* 1967). At 18 μg/disk, it was shown to be active against all bacteria and yeasts examined and 48 out of 51 test fungi. Studies on the mode of action of myxin on *Escherichia coli* (Lesley and Behki 1967, 1971; Behki and Lesley 1972) indicated that the primary action of the antibiotic is inhibition of DNA synthesis. This is followed rapidly by degradation of intracellular DNA, inhibition of ribonucleic acid (RNA) and protein synthesis, and death. The data suggest that myxin interacts directly with eukaryotic and prokaryotic DNA, causing extensive damage. This would explain its wide antimicrobial spectrum and may indicate that myxin is too toxic for general use [median lethal dose ($LD_{50}$) intraperitoreal (i.p.) in mice: 40 mg/kg].

## Antibiotic TA

In 1973, Rosenberg and co-workers reported the isolation of antibiotic TA from *M. xanthus* TA. Although the chemical structure of antibiotic TA is still not known, the compound has been crystallized and its chemical properties reported (Rosenberg *et al.* 1982). Antibiotic TA ($C_{34}H_{57}O_9N$, mol wt 623.8)

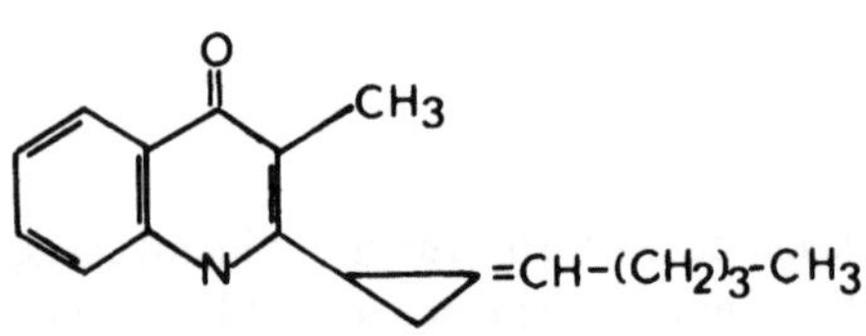


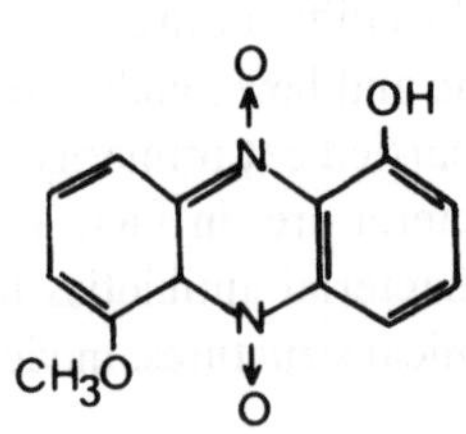


**Figure 5.2.** Myxin (Edwards and Gillespie 1966).

is a macrocyclic antibiotic containing the following functional groups: ketone, lactone, secondary amide, methoxy-substituted diene ($\lambda_{max}$ 239 nm), primary alcohol, and three secondary alcohols, two of which are *cis*-vicinal. The antibiotic activity is sensitive to mild alkaline hydrolysis and periodate oxidation.

Antibiotic TA is bactericidal for both Gram-positive and Gram-negative bacteria (Rosenberg *et al.* 1973; Vaks *et al.* 1974) and inactive against yeasts, fungi, and protozoa. The more sensitive bacteria (e.g., *Klebsiella pneumoniae*) are killed by 0.1–1.0 μg antibiotic TA/ml, whereas the more resistant strains (e.g., *Pseudomonas fluorescens*) are killed by 10–25 μg antibiotic TA/ml. *Myxococcus xanthus,* itself, is not sensitive to 100 μg TA/ml. Antibiotic TA interferes with peptidoglycan synthesis by inhibiting polymerization of the lipid–disaccharide–pentapeptide (Zafriri *et al.* 1981).

Production of antibiotic TA is repressed by vigorous aeration or high concentrations of peptone (Mizrahi *et al.* 1976). Glycine, alanine, and serine are precursors of the antibiotic and stimulate its synthesis (Fytlovitch *et al.* 1983). Both carbon atoms of glycine enter the antibiotic, whereas alanine and serine are converted to acetate before they are incorporated into antibiotic TA.

## Myxovirescin A

*Myxococcus virescens* strain Mx v48 has been reported to produce a family of at least 12 closely related antibiotics, referred to as myxovirescens (Gerth *et al.* 1982). Myxovirescin A, the main antibiotic in the mixture, appears to be very similar to antibiotic TA. The proposed structure of myxovirescin A is shown in Fig. 5.3 (Trowitzsch *et al.* 1982). Like antibiotic TA, myxovirescin A is a macrocyclic antibiotic containing a substituted diene, secondary amide, and three secondary alcohols, two of which are *cis*-vicinal.

Myxovirescin A is an efficient inhibitor of enterobacteria, as well as other Gram-negative bacteria. At higher concentrations, the antibiotic has also been found to inhibit certain Gram-positive bacteria. Calcium ions, bovine serum, and lecithin interfered with the bactericidal activity of myxovirescin A on *Escherichia coli.* Eukaryotes have been found to be completely resistant. The antibiotic appears to block cell wall synthesis.

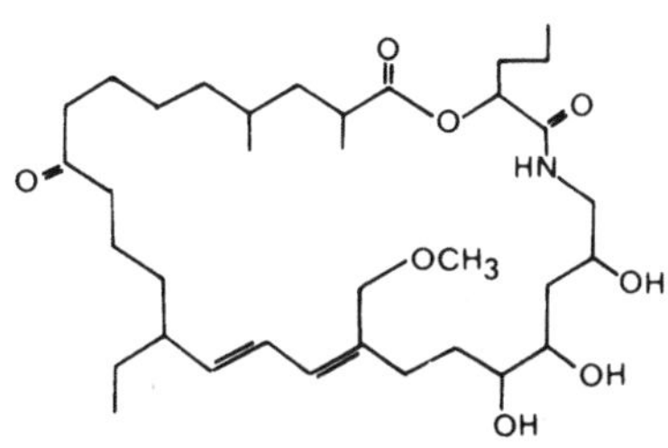


**Figure 5.3.** Myxovirescin A (Trowitzsch *et al.* 1982).

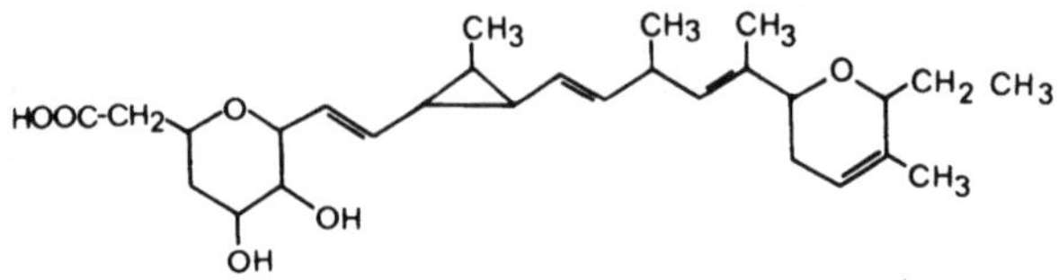


**Figure 5.4.** Ambruticin (Ringel *et al.* 1977).

## Ambruticin

*Polyangium cellulosum* ssp. *fulvum* was shown to produce a new antifungal antibiotic, ambruticin (Ringel *et al.* 1977). Ambruticin is a cyclopropyl–polyene–pyran acid (Fig. 5.4) and appears to represent a new class of antibiotics. This antibiotic is active against a variety of fungi, including important medical pathogens such as *Coccidioides imitis, Histoplasma capsulatum,* and *Blastomyces dermatitidis,* as well as the dermatophytic filamentous fungi. Ambruticin has low activity against Gram-positive and Gram-negative bacteria. The acute $LD_{50}$ values for the sodium salt of ambruticin in mice was 315 mg/kg.

## *Myxococcus coralloides* Antibiotic

An antibiotic active against Gram-positive bacteria has been isolated from *Myxococcus coralloides* (Arias *et al.* 1979). The antibiotic, purified by preparative thin-layer chromatography, had an absorption maximum at 300 nm and an apparent molecular ion of 260. The purified antibiotic was active against several Gram-positive strains, including *Micrococcus luteus, Streptococcus faecalis, Bacillus cereus, Staphylococcus aureus* and *Bacillus subtilis,* but was not active against Gram-negative bacteria or yeasts.

## Myxothiazol

Reichenbach and co-workers isolated from cells of *M. fulvus* strain Mx f16 a new antifungal antibiotic, myxothiazol (Gerth *et al.* 1980; Trowitzsch *et al.* 1980). The chemical structure of myxothiazol is shown in Fig. 5.5. Cysteine, threonine, leucine, and isoleucine appear to be biosynthetic precursors of the molecule.

The action of myxothiazol is essentially restructed to eucaryotic cells. At concentrations of 0.01–3.0 μg/ml it inhibits the growth of many yeasts and fungi by interfering with respiration (Thierbach and Reichenbach 1981a,b). Among the bacteria tested, only *Micrococcus leutus, Myxobacterium* sp., *Nocardia corallina,* and *Paracoccus denitrificans* showed any sensitivity. The

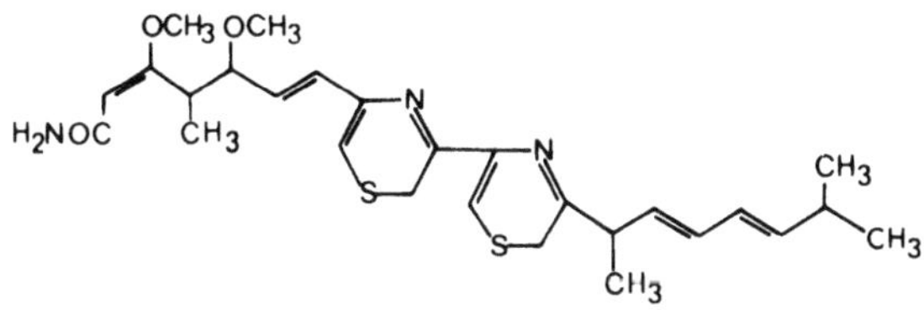


**Figure 5.5.** Myxothiazol (Trowitzsch *et al.* 1980).

effect of myxothiazol on membrane preparations from *Paracoccus denitrificans* was similar to its action on eukaryotic cells (Thierbach and Reichenbach 1983). Myxothiazol is highly toxic to mice ($LD_{50}$: 2 mg/kg) and chicken embryo fibroblasts.

## Althiomycin

Althiomycin (Fig. 5.6) can be isolated from the culture fluids of *Cystobacter fuscus, M. virescens,* and *M. xanthus* (Kunze *et al.* 1982). This is an example of an antibiotic produced by myxobacteria that had previously been discovered in a different organism, *Streptomyces* sp. Many other examples of antibiotics that are produced by totally different microorganisms have been described (Lechevalier 1975). Althiomycin (Sakakibara *et al.* 1974) is active against Gram-positive and some Gram-negative bacteria but is inactive against eucaryotes. The antibiotic blocks protein synthesis by interfering with transpeptidation on the 50 S subunit of the prokaryotic ribosome (Pestka and Brot 1971).

## Pyrrolnitrin

Recently, Reichenbach and co-workers (Gerth *et al.* 1982) have shown that *M. fulvus* strain Mx f147 produces the antifungal antibiotic pyrrolnitrin (Fig. 5.7). The same antibiotic was shown previously to be produced by several pseudomonads (Elander *et al.* 1968). The fungistatic properties of pyrrolnitrin have been described (Nishida *et al.* 1965). Pyrrolnitrin is also produced by other myxobacteria, including *Corallococcus exiguus* and *Cystobacter ferrugineus.*

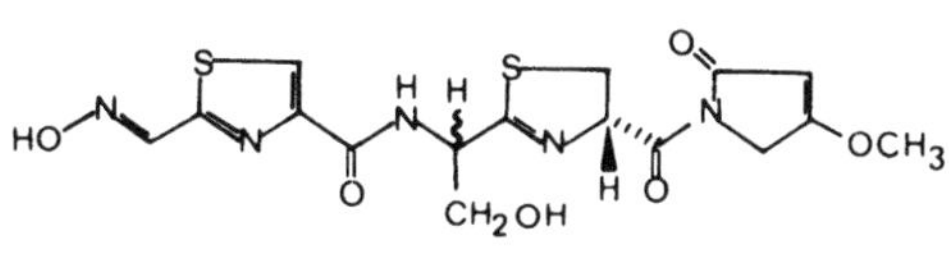


**Figure 5.6.** Althiomycin (Sakakibara *et al.* 1974).

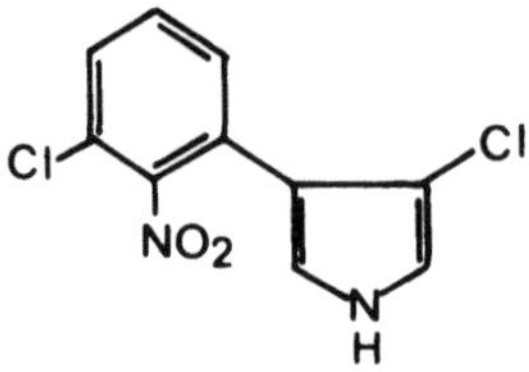

**Figure 5.7.** Pyrrolnitrin.

### Myxovalargins

*Myxococcus fulvus* strain Mx f65 produces a family of peptide antibiotics referred to as myxovalargins (Irschik *et al.* 1983). Amino acid analysis yielded valine, alanine, and arginine at molar ratios of 5:2:1. In addition, several unusual amino acids were reported, including agmatine at the carboxy end of the linear peptide. Myxovalargins were more active against Gram-positive bacteria than against Gram-negative ones. Yeasts and molds were insensitive. The myxovalargins appear to be protein synthesis inhibitors.

## Lytic Enzymes

Following the pioneer studies of Norén (1953) and others, a number of different lytic enzymes from myxobacteria have been purified and characterized. These include enzymes that cleave bacterial cell walls—proteases and peptidases, nucleases and lipases.

### Cell Wall Cleaving Enzymes

The three-dimensional network structure of peptidoglycan is responsible for the rigidity of the cell wall of most bacteria. The peptidoglycan consists of polysaccharide chains of alternating $\beta$-1,4 *N*-acetyl glucosamine and *N*-acetyl muramic acid units with polypeptide cross-links. The wall-cleaving enzymes can be divided into four general groups: (1) *muramidases* (e.g., lysozymes) that split the 1,4-bond between muramic acid and glucosamine, (2) *glucosaminidases* that split the 1,4-bond between glucosamine and muramic acid, (3) *amidases* that split the amide linkages between muramic acid and the alanyl residue of the *N*-terminus of the peptide chains, and (4) *peptidases* that hydrolyze peptide bonds in the polypeptide cross-links. Although the first two are highly specific, the latter are not necessarily so; they often can hydrolyze other polypeptides as well (Whitaker 1965). As seen in Table 5.2, one or more examples of each of these groups of enzymes have been isolated from a myxobacterium.

Harcke *et al.* (1971, 1972) isolated from the supernatant fluid of cultures of *Chondrococcus corraloides* an enzyme that hydrolyzes the $\beta$-1,4-bond between

**Table 5.2.** Cell Wall Cleaving Enzymes of Myxobacteria

| Enzyme | Source | Cell wall substrate |
|---|---|---|
| Muramidase | *C. corraloides*[a] | *E. coli* |
| Glucosaminidase | *M. xanthus*[b] | *B. subtilis, M. luteus* |
| | *M. virescens*[c] | *S. aureus, M. luteus* |
| Amidase | *M. xanthus*[d] | *B. subtilus, M. luteus* |
| | | *S. aureus* |
| | *Sorangium* sp[e] | *Arthrobacter globiformis, M. luteus* |
| Peptidase | *M. xanthus*[d] | *B. subtilis, M. luteus* |
| | *Myxobacter* AL-1[f] | *A. crystallopoietes* |
| | *M. virescens*[g] | *M. luteus* |
| | *Sorangium* sp.[e] | *A. globiformis, M. luteus* |

[a]Harcke *et al.* 1972.
[b]Hart and Zahler 1966; Sudo and Dworkin 1972.
[c]Haskå 1974.
[d]Sudo and Dworkin 1972.
[e]Whitaker 1965; Tsai *et al.* 1965. This species has been reclassified as a *Lysobacter* (Reichenbach unpublished).
[f]Ensign and Wolfe 1966; Tipper *et al.* 1967.
[g]Haskå 1974.

muramic acid and glucosamine, thus having the substrate specificity of a muramidase. Hart and Zahler (1966) isolated from the culture filtrate of *M. xanthus* an enzyme capable of lysing whole cells of *Micrococcus luteus* and a number of other (lysozyme-sensitive) bacteria. The purified enzyme catalyzed the release of *N*-acetyl amino sugars from cell walls. Enzymes of a similar description have been isolated by Sudo and Dworkin (1972) from *M. xanthus* (one enzyme) and by Haskå (1974) from *M. virescens* (two enzymes). The latter were described as endo-β-*N*-acetylglucosaminidases, with an activity similar but not identical to that of lysozyme. Several myxobacterial enzymes were described as amidases, splitting the wall backbone into intact glycan units and intact peptides (Katz and Strominger 1967; Tsai *et al.* 1965; Sudo and Dworkin 1972).

Peptidases having a variety of different specificities have been isolated from myxobacteria. An enzyme isolated from *Myxobacter* AL-1 (later classified as a *Lysobacter*) was found to hydrolyze the bond between the D-alanine terminus of the basal tetrapeptide and the cross-linking pentaglycine (in the case of *Staphylococcus aureus*) or L-alanine (in the case of *Arthrobacter crystallopoites*), which interlink the peptidoglycan strands (Tipper *et al.* 1967). The same enzyme also has amidase activity, that is, it can split the muramic acid–alanine bond (Katz and Strominger 1967). D-analyl-ϵ-*N*-lysine endopeptidases have been isolated from *Sorangium* (Tsai *et al.* 1965), from *M. xanthus* (Sudo and Dworkin 1972) and from *M. virescens* (Haskå 1974). The purified enzyme of *M. virescens* was an alanyl–lysine endopeptidase. It has no measurable amidase activity. The *M. xanthus* endopeptidase attacked alanyl–lysine and alanyl–diaminopimelic acid linkages with nearly equal facility.

## Proteolytic Enzymes

The majority of the myxobacteria (except for the cellulolytic species) use proteins as both carbon and energy sources and are unable to use sugars. Therefore, whether growing saprophytically or preying on other bacteria, they depend on proteolytic enzymes for their nutrition. Accordingly, it is common to detect proteolytic activity in myxobacterial cultures. Proteases have also been detected in cultures of a cellulose decomposer reported to be a *Sorangium* (Gillespie and Cook 1965; Whitaker 1965). The "bacteriolytic acitivty" described in several studies that used pretreated prey bacteria (Noren, 1960a,b; Harcke *et al.* 1971; Hütterman 1969) was probably due to proteolytic enzymes. Norén (1960b) showed that bacteria, previously damaged by heat, freezing, and thawing or merthiolate, were lysed by trypsin as well as by culture filtrates of the myxobacteria. Margalith (1962) demonstrated that the lytic activity of *M. fulvus* against autoclaved *Escherichia coli* coincided with the casein-hydrolyzing activity.

Several extracellular proteolytic enzymes have been purified and studied in detail. Gnosspelius (1978) purified one of at least three proteases produced by *M. virescens.* The purified enzyme hydrolyzed casein and hemoglobin and was specific for peptide bonds involving amino acids with nonpolar side chains. The molecular weight of the enzyme was estimated to be 26,000. Its optimal pH was between 7 and 10, and it was totally inhibited by diisopropylphosphorofluoridate. These properties, together with its substrate specificity, suggested that the enzyme was a serine protease, even though it was inhibited by metal chelating agents.

*Myxobacter* AL-1 *(Lysobacter)* produced two extracellular proteolytic enzymes (Ensign and Wolfe 1965, 1966; Jackson and Wolfe 1968; Wingard *et al.* 1972). AL-1 protease I (mol wt 8700) hydrolyzes cell walls (peptide cross-linkages) as well as proteins (approximately one-third of the peptide bonds present in albumin and casein). AL-1 protease II (mol wt 17,000) does not possess cell wall lytic activity and has a unique specificity for lysine residues. Both proteases have alkaline pH optima and are very stable enzymes.

Two different proteases have been purified from cultural fluids of *Sorangium* sp. (Whitaker 1965; Jurášek and Whitaker 1965). The molecular weight of both enzymes was estimated to be 19,000. The $\alpha$ enzyme had the greater activity toward casein, whereas the B enzyme was more active in lyzing suspensions of *Arthrobacter globiformis.* One of the proteolytic enzymes released by *Chondrococcus coralloides* has a high specificity for arginine–peptide bonds (Harcke *et al.* 1971).

## Nucleases

In a study of different isolates of *Sorangium (Lysobacter)*, Gillespie and Cook (1965) found that some of them released into the culture filtrate deoxyribonuclease (DNase) and ribonuclease (RNase). Two cell-associated restriction endonucleases were identified in cell-free extracts from *M. virescens* (Morris and Parish 1976). Nuclease I hydrolyzed the DNA of myxophage MX-1 at approximately 10 sites, whereas nuclease II hydrolyzed it at a large number of sites. Restriction endonuclease activity has also been found in certain strains of *M. fulvus, M. stipitatus, Archangium serpens,* and *Stigmatella erecta* (Mayer and Reichenbach 1978). Strains of *M. fulvus* also contained enzymes also to release the supercoiled state of closed circular DNA (Mayer and Reichenbach 1978). All these enxymes seemed to be located on the cell surface or in the periplasmic space. Because of their location it would be difficult to imagine they degrade the DNA of prey bacteria. It has been reported that myxobacteria do not use nucleic acid fractions for growth (e.g., Loebeck and Klein 1956).

## Lipases

A glycerol ester hydrolase was found by Sørhang (1974) in cultures of *M. xanthus*. This lipase released fatty acids from triglycerides, tricaprylin being the most readily hydrolyzed. The location of the enzyme was not clear. Although some of the activity could be recovered from the supernatant fluid and almost nonc from the cells, the recovery was incomplete.

## Regulation of Bacteriolytic Enzymes

Myxobacteria are not obligate predators; since they grow well on proteins, peptides, or amino acid mixtures as the only source of carbon and energy, it does not seem economic for them to make bacteriolytic enzymes constitutively. There has been, however, no systematic attempt to study the regulation of the synthesis of these enzymes. Mathew and Dudani (1955) examined the lytic activity of *M. fulvus* and *M. virescens* before and after subculturing on prey bacteria. The lytic activity was not affected by this procedure. On the other hand, the bacteriolytic activity of *M. virescens* was affected by the composition of the growth medium (Haskå and Stahl 1971). These latter authors also found that the appearance of bacteriolytic enzymes preceded that of the proteolytic enzymes, as would be expected if the former were required to make the cytoplasm of the prey accessible to the latter. It is tempting to assume that proteolytic enzymes are induced or activated after the release of their potential substrate by the "bacteriolytic" enzymes.

## "Helper" Mechanisms for Bacteriolytic Activity

Gram-positive bacteria are generally more sensitive than Gram-negative bacteria to bacteriolytic enzymes (e.g., to lysozyme). It is generally assumed that the outer layer of lipoprotein–lipopolysaccharide shields the peptidoglycan. Damaging the outer layer by (nonbiologic) treatment with organc solvents, ethyknediaminetetraacetate (EDTA), detergents, heat, or freezing and thawing sensitizes Gram-negative eubacteria to bacteriolytic enzymes. The need for prior damage to the outer layer of Gram-negative bacteria before myxobacterial enzymes can work on them has been reported (Oxford 1947; Singh 1947; Norén 1960a,b; Bender 1962; Gillespie and Cook 1965; Haskå *et al.* 1972).

Interestingly, the myxobacteria generate natural helper mechanisms for their bacteriolytic enzymes. Haskå *et al.* (1972) found that fatty acids (unbranched or isobranched) isolated from *M. xanthus* and added to eubacteria, otherwise resistant to lysis, sensitized them to lysis by myxobacterial enzymes. The authors suggested that the fatty acids changed the permeability properties of the cell wall. The location of the fatty acids on the surface of the myxobacteria might help explain the need for direct contact with the prey for effective lysis in certain cases.

Gnosspelius (1978) has isolated from the slime produced by *M. virescens* a protein–polysaccharide–lipid complex that bound proteins and affected their properties. The author suggested that this complex denatured native proteins, thus rendering them more suitable for proteolysis. In addition, the complex showed proteolytic activity that might enhance the denaturing capacity of the complex in situ.

## Bacteriocins

Bacteriocins have been found in at least two myxobacterial species: *M. xanthus* (McCurdy and McRae 1974) and *M. fulvus* (Hirsch 1977). Like other bacteriocins, the myxocins show a limited and species-specific antimicrobial host range. Three distinct myxocins were separated from the supernatant fractions of *M. fulvus* Mx f16, none of which was induced by mitomycin C. They differed from one another in their molecular weights and electrophoretic mobilities. The largest, referred to as fulvocin A, caused turbid zones in lawns of indicator bacteria, whereas the other two fulvocins (B and C) caused clear zones. Fulvocin C was purified and shown to be a small, stable protein of molecular weight 4672, composed of 45 amino acids (Hirsch *et al.* 1978; Tsai and Hirsch 1981). A bacteriocin isolated from *M. xanthus* FB after mitomycin C induction was an extremely stable, particulate fraction. Electron microscopy of the purified material showed circular bodies of varied size that resembled membrane fragments.

## Role of Lytic Enzymes and Antibiotics in Nutrition

Water-insoluble organic matter constitutes the major source of nutrients in soil. Of particular interest to the myxobacteria when it comes to feeding are other microorganisms, either dead or alive, fragments of cells, and denatured protein. For bacteria to use these insoluble substrates, they must either (i) make direct contact or (ii) produce extracellular, diffusable materials that solubilize the substrate. In the case of myxobacteria, evidence is available for both mechanisms.

Several authors have emphasized the need for direct contact between the myxobacterium and its prey (Burnham *et al* 1981; Kletter and Henis 1963; Margalith 1962; Roper and Marshall 1978; Shilo 1970). Lysis of the cyanobacterium *Phormidium luridum* by *M. xanthus* cells occurred within clumps formed by the myxobacteria, but did not occur when the two organisms were separated by a membrane of a 0.2 $\mu$m pore size (Burnham *et al.* 1981). Direct examination of lysis of *Nostoc* by *Myxobacter* FP-1 (*Lysobacter* Sp.) showed that the latter attaches to the prey at one pole for a certain period of time (10–20 min). During this period, the prey lyzes and the predator then moves on to a neighboring prey cell (Shilo 1970).

At present, little is known about the adherence mechanisms of myxobacteria. The frequent observation that freshly isolated myxobacteria adhere strongly to glass and plastic surfaces, suggests that hydrophobic interactions may play a role. The general problem of cell surface adhesions appears to be a particularly interesting area for future myxobacterial research.

One of the features of the cell-contact mechanism is that once the myxobacterium has attached to an insoluble substrate, it can commence growth independently of cell density. The intimate contact of the myxobacterium and its substrate ensures high local concentrations of lytic factors. In the absence of direct contact, high cell densities are required because of dilution of the lytic factors into the medium. In this regard, Kletter and Henis (1963) reported that direct contact of myxococci with heat-killed Gram-negative bacteria allowed lysis of the prey and growth of the myxococci long before lytic activity could be demonstrated in the culture fluid.

The solubilization mechanism is quantitatively important when bacteria are present at a high cell density, for example, the myxobacterial swarm. Although individual cells might not produce enough extracellular product to solubilize the substrate, the accumulated product from many closely spaced cells might reach a sufficiently high concentration to solubilize and/or hydrolyze the substrate and thus provide nutrients. In a relatively simple example, Rosenberg *et al.* (1977) showed that at cell densities less than $10^3$/ml, *M. xanthus* was incapable of growth on casein: From $10^4$ to $10^7$ cells per milliliter, growth rates increased dramatically. Cooperative growth was correlated with increased concentration of hydrolyzed casein in the medium. These data demonstrate the importance of the solubilization mechanism, at least with casein as the substrate, and suggest that the communal behavior of myxobacteria results in more efficient feeding.

When myxobacteria are feeding on complex substrates, such as intact prey bacteria, in which multiple "hits" would be necessary to solubilize the substrate, it would be expected that cell density dependence would be even more pronounced. In this regard, we have recently observed that *M. xanthus* fails to grow on *Escherichia coli* unless more than $10^7$ myxobacteria per milliliter were present (Gendelman and Rosenberg unpublished). When a mutant of *M. xanthus* unable to produce antibiotic TA was used in the latter experiment, growth of *M. xanthus* did not occur even at high initial cell densities. Thus, in addition to lytic enzymes and proteases, antibiotics, such as TA, may play a role in the growth of myxobacteria on microorganisms. The recent characterization of a number of different myxobacterial antibiotics with different spectra of activities makes this hypothesis amenable to experimental examination.

It should be emphasized that the cell-contact and solubilization mechanisms are extreme situations and that intermediate cases are more likely. For example, a slime layer around a single cell or small group of cells will impede diffusion of extracellular products. Thus, these cells will have a higher local concentration of extracellular enzymes and antibiotics than would be predicted from simple diffusion and mixing. Christison and Martin (1971) and Gnosspelius (1978) have reported that proteolytic activity appeared to be bound in a complex with polysaccharides of the slime layer.

## Development: Role of Lytic Enzymes and Antibiotics

As discussed previously, there is a strong correlation between *M. xanthus* sporulation and autolysis on development plates. To understand the relationship between these two phenomena, it would be useful to know (1) biochemical factor(s) responsible for the autolysis, (2) physiologic conditions for inducing the autolytic factor(s), and (3) the role of the lytic product(s) in development.

The term "autolysis" encompasses two different mechanisms: *cis*-autolysis, in which cell-associated factors are responsible, and *trans*-autolysis, in which the responsible factors are produced by other cells in the population. The result that lysis of induced cells (Wireman and Dworkin 1977) was independent of cell density and occurred several hours after transfer to buffer, suggests that developmental lysis is a *cis*-event. Thus, the lytic factor(s) is probably cell associated. In general, any factor that blocks cell wall synthesis can lead to lysis by the continuing action of enzymes involved in cell wall turnover and cell divsion (Rogers and Forsberg, 1971).

What induces autolysis and subsequent myxospore formation? Although the lytic process appears to be a *cis*-event, induction of lysis is cell density dependent, suggesting a cell-to-cell signal mechanism. Information on the nature of this hypothetical signal has come from studies on one group of nonsporulating mutants (group C, see Chapter 8). SpoC mutants do not undergo developmental lysis. However, these mutants can be induced to lyse and form spores by addition of peptidoglycan components (Shimkets and Kaiser 1982a) or carbohydrate-containing substances extracted from starving *M. xanthus* cells (Janssen and Dworkin 1980). It is not clear at this time how these materials induce lysis. Diaminopimelic acid will induce fruiting body formation (and presumably autolysis) in wild type cells (Campos and Zusman 1975). Since peptidoglycan components can induce lysis and fruiting body formation, it may be that developmental lysis proceeds in an autocatalytic manner.

Murein components cannot be the cell-density signal for normal development, because addition of these components to SpoC mutants induces lysis and sporulation *only* at high cell densities. The cell-density dependent step in development must, therefore, come after the murein-induced reaction. The biochemical mechanisms for inducing the release of murein components and causing lysis (directly or indirectly) are yet to be resolved.

Recently, we have extracted and partially purified from stationary phase *M. xanthus* cells several (at least five) anti-Myxo antibiotics. These chloroform–methanol soluble antibiotics are active against *M. xanthus,* but not against other bacteria. In general, the anti-Myxo antibiotics cause lysis of growing cultures of *M. xanthus.* In the case of at least two chromatographically pure fractions, the antibiotics also lysed nongrowing *M. xanthus* cells that had been suspended in buffer. Most interestingly, a SpoC mutant was significantly more resistant to AM V than its isogenic fruiting body producing wild type strain, DK1622. It will be interesting to see if one or more of these anti-Myxo antibiotics plays a role in development.

# 6

# Pigments of Myxobacteria

Hans Reichenbach* and Hans Kleinig**

Colors in nature arise either through absorption by certain molecules or atoms of selected wavelengths in the visible range, or through interference of light waves in material with a special physical structure. Structural colors are observed with some cytophagas: Their delicate swarm colonies may act like a thin film and show iridescence (Reichenbach and Dworkin 1981). The color of myxobacteria seems to be due exclusively to pigments with various chemical chromophores. The pigments are either localized within the cell or they are excreted. The extracellular pigments may be excreted as such, or they may arise in the environment through the action of enzymes liberated from the cells.

The myxobacteria are indeed a colorful assembly of organisms. The cell mass of most strains is more or less intensely yellow, orange, red, or purple. The fruiting bodies are almost always colored, and often very brightly so, in hues of yellow, orange, red, purple, light-to-dark brown, and even black, although rarely. The only myxobacterium that is often totally unpigmented is *Myxococcus stipitatus*. This organism, however, produces a substance of unknown chemical composition with a strong yellow fluorescence (Lampky and Brockman 1977) that seems to be excreted. It should be mentioned that there are some *M. stipitatus* strains with fawn-to-red brown swarms and fruiting bodies.

---

*GBF, Gesellschaft für Biotechnologische Forschung, Department of Microbiology, Mascheroder Weg 1, D-3300 Braunschweig, Federal Republic of Germany.

**Institut für Biologie II der Universität, Schänzlestr. 9-11, D-7800 Freiburg, Federal Republic of Germany.

## Chemical Structures and Biosynthesis

### Intracellular Pigments

It was suggested, although on quite insufficient evidence (Jahn 1924; Reichenbach 1965b; Burchard and Dworkin, 1966a), that the color of the cells of the myxobacteria is due to carotenoids. After some very preliminary attempts to investigate this assumption (Greene and Leadbetter 1962; Renwick and Peterson 1969), it was indeed corroborated. The intracellular pigments of a variety of myxobacteria were extracted, separated into pure compounds, and characterized by chemical and physical methods (Kleinig and Reichenbach 1969, 1973b; Kleinig *et al.* 1970, 1971; Reichenbach and Kleinig 1971, 1972). From these studies the following picture emerges.

Myxobacteria always contain mixtures of different carotenoids. Up to 60 different compounds have been obtained from one single strain, most of them, however, in minute quantities. It is thus obvious that spectra of crude pigment extracts cannot be very useful by themselves to characterize an organism, particularly as the quantitative relationships between the different pigments depend on the culture conditions, as well as on the specific strain. The myxobacterial carotenoids are of normal length ($C_{40}$) and either acyclic or monocyclic (Fig. 6.1). They show increasing degrees of desaturation, almost up to the theoretical endpoint (torulene, dehydrolycopene, aromatic carotenoids). Many contain hydroxyl and keto functions in various positions and combinations. In the hydrocarbon series, the complete Porter–Lincoln series (phytoene, phytofluene, ζ-carotene, neurosporene) can usually be demonstrated. The hydroxyl groups may be substituted with a sugar molecule, usually glucose, but rhamnose has also been found (Figs. 6.2, 6.3). Although unknown in eukaryotic organisms, carotenoid glycosides are not uncommon in bacteria (Isler 1971; Liaaen-Jensen and Andrewes 1972; Liaaen-Jensen 1973). In myxobacteria, as in most other bacteria, the sugar is attached to a tertiary carbon atom (tertiary glycosides: usually at a 1′-hydroxyl). The cyanobacteria, in contrast, produce secondary (2′-hydroxyl) glycosides. All myxobacterial glucosides, but not the rhamnosides, have been found to be esterified with a fatty acid at one of the hydroxyls of the sugar moiety. Although this fatty acid is variable, the fatty acid pattern of the carotenoid glucosides does not simply

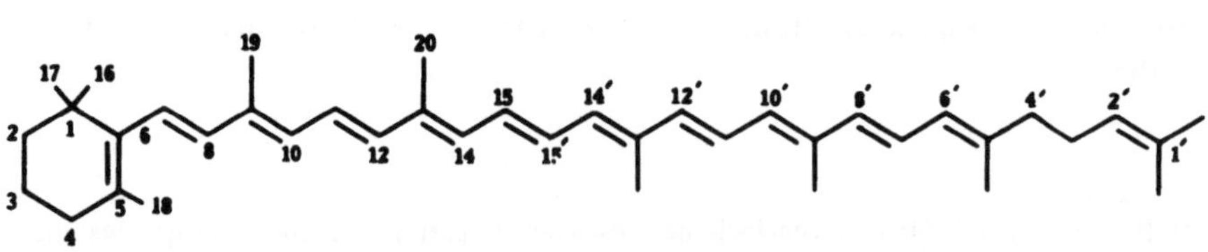


**Figure 6.1.** β,ψ-Carotene, the chromophore of many myxobacterial carotenoid pigments, showing the numbering of the carbon atoms.

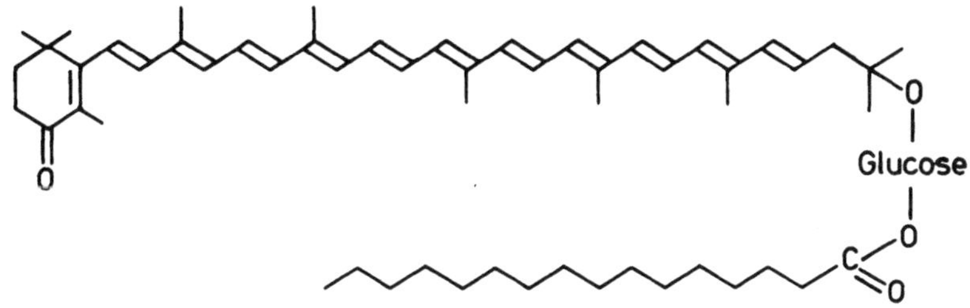


**Figure 6.2.** Myxobacton palmitate.

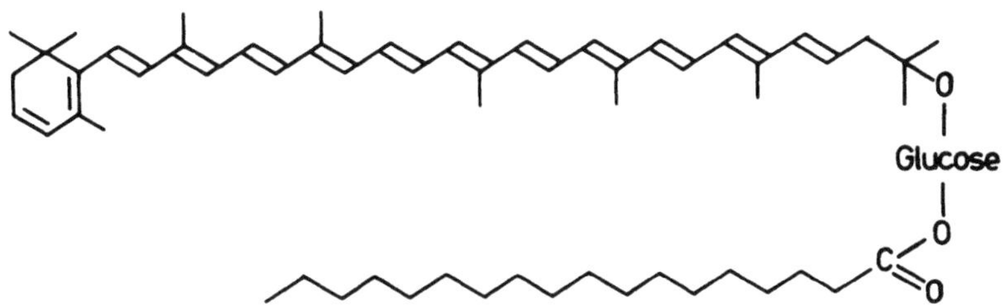


**Figure 6.3.** Myxobactin stearate.

reflect the overall fatty acid pattern of the cell: The latter is dominated by branched fatty acids, whereas the pigment glucosides contain mainly straight-chain fatty acids. Carotenoid glucoside fatty acid esters have been described in only one other organism: *Nocardia kirovani* (Vacheron *et al.* 1970). It is, however, questionable whether they are really so rare. In many laboratories, the crude pigment extracts are saponified before being chromatographed, in which case such esters would be hydrolyzed. *Myxococcus fulvus* produces another remarkable carotenoid derivative: a direct 2′-hydroxy fatty acid ester. In this case, too, the hydroxyl is esterified with a variety of straight-chain fatty acids. Although carotenoid fatty acid esters are quite common in higher plants, they have not been described previously in bacteria. However, again the possibility of hydrolysis during isolation has to be considered. Truly unique pigments are produced by *Nannocystis exedens*. Here, the main pigments are moncyclic aromatic carotenoids with highly substituted benzene rings (1,2,5- and 1,2,4,5-methylated and an additional hydroxyl group in 3-position; Fig. 6.4). Some of these pigments have a conjugated keto group in 2′-position, which shifts the electronic absorption spectrum to exceptionally long wavelengths (λ max 493 nm, ethanol) rarely reached by any carotenoid; these pigments have a deep blue–purple color (Kleinig *et al.*, to be published). Another unusual chromophore is that of myxobactin (see Fig. 6.3). This pigment is a monocyclic carotenoid glucoside with a 3,4-unsaturated β-carotene ring not known in any other natural pigment.

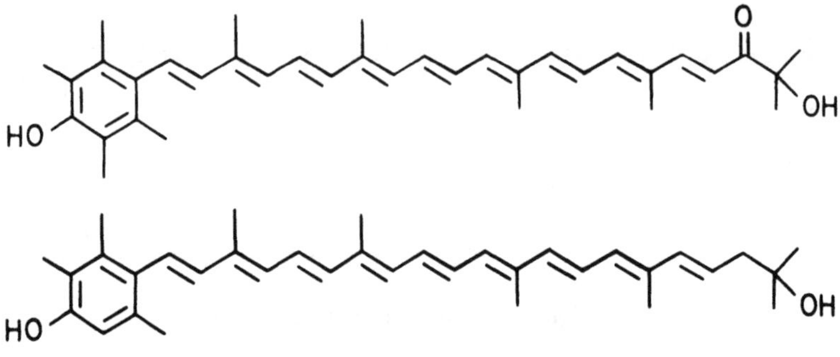

**Figure 6.4.** The main pigments of *Nannocystis exedens.*

Quantitatively, the main pigments of myxobacteria are almost always carotenoid glycosides, usually glucosides. Rhamnosides have so far been found only in *Sorangium compositum.* The chromophores of the glycosides differ characteristically in the two suborders of the *Myxobacterales* (see Chapter 1). In the *Cystobacterineae,* we find monocyclic carotenoids with a 4-keto (myxobacton) or a 3,4-unsaturated (myxobactin) β-carotene ring, and generally a dominance of dark-red-to-purple keto pigments. In the *Sorangineae,* the chromophores are acyclic or monocyclic, with an unsubstituted or a 3-hydroxylated β-carotene ring, and keto pigments are completely lacking. Consequently, the color of the cell mass is more in the red–orange range in the former case and yellow–orange in the latter. The phenotypic appearance of an organism depends on the carotenoid pattern, the quantitative relations of the individual pigments, and the occurrence of additional noncarotenoid pigments. Thus, *M. fulvus* and *Myxococcus xanthus* seem to have an identical carotenoid pattern, but in *M. xanthus* the dominating pigments absorb at shorter wavelengths. *Myxococcus xanthus* and *Myxococcus virescens* seem to contain the same carotenoids, but *M. virescens* in addition produces appreciable quantities of a greenish-yellow pigment of unknown structure that is also excreted into the environment.

*Nannocystis exedens* is exceptional among the myxobacteria in that it does not contain any carotenoid glycosides but contains instead the peculiar monocyclic aromatic carotenoids mentioned above, which are not found in any other myxobacterium.

As summarized in Table 6.1, the larger taxonomic groups of myxobacteria are characterized by different pigment types. Thus, the pigments are to a certain degree useful as chemosystematic markers. In this connection it is also interesting to compare the myxobacteria with other unicellular gliding bacteria. Carotenoids have been described in *Cytophaga* and *Flexibacter* strains, mainly the bicyclic $C_{40}$ carotenoids zeaxanthin (3,3′-dihydroxy-β,β-carotene) and its biosynthetic precursors, cryptoxanthin (3-hydroxy-β,β-carotene) and β,β-carotene. With many *Cytophaga*-like bacteria, including *Sporocytophaga,* the dominant pigments are, however, of the flexirubin type (Fig. 6.5), a novel

**Table 6.1.** The Occurrence of Various Types of Intracellular Pigments in Different Myxobacteria and *Cytophaga*-Like Bacteria

| Organism | Carotenoids[a] | | | | | | | | Flexirubin-type pigments |
|---|---|---|---|---|---|---|---|---|---|
| | | | | | | Glucosides | | | |
| | Di- | Mono- | Acyclic | Aromatic | With keto functions | Chromophore with keto functions | Chromophore with hydroxyl functions | Rhamnosides | |
| Myxobacterales | | | | | | | | | |
| Cystobacterineae | | | | | | | | | |
| *Myxococcus fulvus* | − | + | + | − | + | + | − | − | − |
| *M. xanthus* | − | + | + | − | + | + | − | − | − |
| *M. virescens* | − | + | + | − | + | + | − | − | − |
| *Cystobacter fuscus* | − | + | + | − | + | + | − | − | − |
| *C. ferrugineus* | − | + | + | − | + | + | − | − | − |
| *C. velatus* | − | + | + | − | + | + | − | − | − |
| *Stigmatella aurantiaca* | − | + | + | − | + | + | − | − | − |
| Myxobacterales | | | | | | | | | |
| Sorangineae | | | | | | | | | |
| *Chondromyces apiculatus* | − | + | + | − | − | − | + | − | − |
| *Sorangium compositum* | − | + | + | − | − | − | + | + | − |
| *S. cellulosum* | − | + | + | − | − | − | + | − | − |
| *Nannocystis exedens* | − | + | + | + | + | − | − | − | − |
| *Cytophaga*-like bacteria | | | | | | | | | |
| *Cytophaga* spp. | + | − | (+) | − | − | | | | + |
| *Flexibacter* spp. | + | − | (+) | − | − | | | | + |
| *Sporocytophaga* sp. | | | | | | | | | + |

[a]Symbols: +, occurrence demonstrated; (+) traces; −, excluded; open, not known.

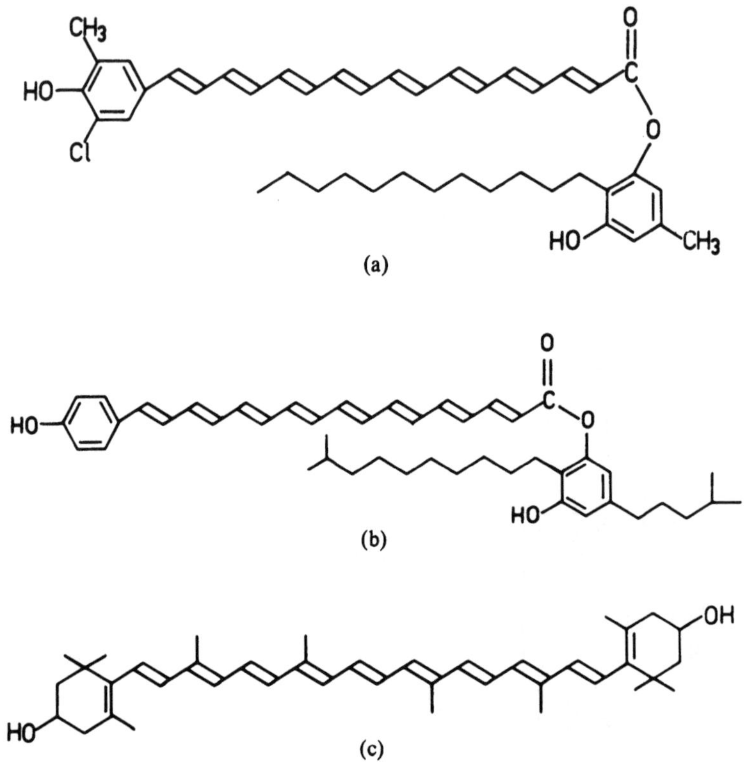


**Figure 6.5.** Typical pigments of *Cytophaga*-like bacteria: (a) Chloroflexirubin from *Flexibacter elegans;* (b) flexirubin-type pigments from *Cytophaga johnsonae:* (c) 3,3′-dihydroxy-β,β-carotene or zeaxanthin.

class of pigments discovered in *Flexibacter elegans* and so far found only in these bacteria and in certain flavobacteria (Achenbach *et al.* 1978; Reichenbach *et al.* 1981). Little is known about the pigments of *Lysobacter.* These organisms seem to produce mainly or exclusively extracellular pigments.

In addition to carotenoids other intracellular pigments may be found in certain strains of myxobacteria. *Myxococcus xanthus* FB produces a yellow non-carotenoid pigment with an absorption maximum at 379 nm that accumulates in stationary phase cells and becomes particularly conspicuous in cultures grown in the dark when carotenoid synthesis is much reduced. The same organism also contains protoporphyrin IX, although this pigment becomes visible only after extraction (Burchard and Dworkin 1966a). The pale yellow antibiotic myxalamid is synthesized by strains of several myxobacteria (*M. xanthus, M. virescens, M. stipitatus,* and *Stigmatella aurantiaca*) in sufficient quantity to become visible in the culture. Although it can be extracted from the cell mass, it is not absolutely certain that it is really intracellular (Gerth *et al.* 1983). Other intracellular antibiotics with a chromophore, such as yellow myxovirescin (Gerth *et al.* 1982), are present, as a rule, in minute amounts and thus do not ordinarily contribute to the color of the culture.

## Extracellular Pigments

Little is known about the chemical nature of the extracellular pigments of myxobacteria. *Myxococcus virescens* excretes a characteristic greenish-yellow water-soluble pigment into the medium. While this pigment seems to be excreted in its finished form, this is not the case with the dark brown or black pigments appearing in cultures of several myxobacteria (some *M. fulvus, Cystobacter* and *Stigmatella* strains). For example, cultures of *Stigmatella aurantiaca* in peptone liquid medium may become pitch black within 1 or 2 hr after having reached stationary phase (Reichenbach and Dworkin 1969). Very probably all these pigments are melaninlike and arise through the action of phenol oxidases that are either excreted or set free when the cells lyse.

Certain *Cystobacter* strains, either physiologic variants of *Cystobacter velatus* or a species of its own (*Cystobacter violaceus:* Kühlwein and Gallwitz 1958), excrete a deep violet pigment into the medium. The pigment is mainly produced in old and degenerating cultures on rich media in presence of oxygen. Its physiocochemical properties suggest that it is a melaninlike material, although the relevant enzymes have not been demonstrated (Reichenbach 1965b).

The specific fruiting body pigments also have to be regarded as extracellular. They are located within the sporangiole walls, in the slime matrix, and in the slime envelopes. They often are very bright and intense in color and seem to be very resistant to degradation. Nothing is known about their chemistry. The dark and red–brown pigments are perhaps also melanins. The fruiting body pigments are formed relatively late during morphogenesis (see Chapter 1, Fig. 1.10a).

## Biosynthesis

It appears that carotenoids are synthesized by myxobacteria from mevalonic acid via "lycopersene" and phytoene in the classic way by a series of symmetrical dehydrogenation steps with or without later cyclization of one end to form a $\beta$-carotene ring. It is not known why in all myxobacteria ring closure occurs on only one end. The biosynthetic scheme is supported by efficient incorporation of radioactive mevalonic acid into myxobacterial carotenoids (Kleinig and Reichenbach 1973a) and by the regular occurrence of the Porter–Lincoln series in myxobacteria. For a better understanding of the reaction sequence beyond the level of phytoene, studies with certain metabolic inhibitors were undertaken (Kleinig and Reichenbach 1973a; Kleinig 1974, 1975). In *M. fulvus,* D- or L-nicotine at concentrations below 1 m*M* specifically inhibited cyclization of the newly synthesized carotenoids. Thus, instead of the usual monocyclic myxobacton (Fig. 6.2) a new, acyclic carotenoid glucoside was formed. When the nicotine was then removed and the culture incubated anaerobically, the acyclic glucoside was transformed into another new glucoside with a 3′,4′-didehydro-$\beta,\psi$-carotene chromphore, that is, cyclization but not oxydation took place. If the culture was then aerated, the keto function at the 4-position of the $\beta$-carotene ring was introduced to give myxobacton. Incubation under aerobic conditions, after removal of nicotine, led directly to myxobacton. In presence of higher concentrations of nicotine (1–4 m*M*), a second effect was observed: Introduction of the tertiary hydroxyl in 1′-position was blocked, so that glucosides could not be synthesized any more, and finally over 95% of the total pigment was lycopene. During all these experiments, the specific pigment content of the cultures (i.e., amount of pigment per cell) remained constant (about 0.42 nmol/mg dry cell mass, culture under blue light, 2500 lux). The same effects as with nicotine were also observed with the inhibitor 2-(4-chlorophenylthio) triethylamine (CPTA) hydrochloride.

The lycopene that accumulated in presence of the inhibitors was not used for the synthesis of other carotenoids when the inhibitor was removed. Indeed, lycopene, which is a highly oxydized structure, is not a likely precursor for the cyclization step or other enzymatic reactions. However, if after removal of nicotine another inhibitor, the herbicide Sandoz (San) 6706, was added, at least some of the lycopene was reintroduced into the biosynthetic pathway and converted into myxobacton. San 6706, 4-chloro-5-(dimethylamino)-2-$\alpha,\alpha,\alpha$ (trifluoro-m-tolyl)-3(2H)-pyridazinone, specifically prevents the desaturation of phytoene to the colored carotenoids, as does diphenylamine (DPA). While DPA is incorporated into the cytoplasmic membrane and cannot quantitatively be removed again, San 6706 has the advantage that it can easily be washed out.

The DPA at very low concentrations (0.001–0.01 m*M*) led to a considerable increase in the specific pigment content that was due, however, almost entirely to phytoene. A new acyclic 7,8-desaturated carotenoid glucoside also appeared (6% of the total pigment at a DPA concentration of 0.002 m*M*). When phytoene desaturation was blocked with San 6706, phytoene also accumulated, but in this case the specific pigment content remained constant. As soon as the inhibitor was removed, a fast synthesis of myxobacton and 4-keto-torulene, the main carotenoids of the strain, began, and the concentrations of the two pigments rose steeply to normal levels. Curiously, the phytoene pool remained high, although some conversion of radioactive phytoene into colored carotenoids was demonstrated. Obviously, phytoene synthesis remained high initially to supply the urgently needed precursor molecules; there was, however, only limited exchange between the molecule populations, and the biosynthetic pathway leading to the colored compounds was preferentially fed with newly synthesized phytoene. Experiments with chloramphenicol showed that this efficient regulation of carotenoid biosynthesis took place with preexisting enzymes.

From these studies we may cautiously deduce that in *M. fulvus*, as in other organisms, a sequence of closely linked enzyme reactions is involved in carotenoid biosynthesis (phytoene to myxobacton or to lycopene) but that this sequence is not so rigid that use of unusual precursors (lycopene, acyclic glucosides) is excluded. It is not yet clear at what stage the glycosylation step normally takes place. It may happen relatively early in pigment synthesis, as is suggested by the occurrence of partly desaturated acyclic carotenoid glycosides. In any case, carotenoid glycosides are still available for biochemical modification of the chromophore (cyclization, introduction of keto function). There may indeed be a certain flexibility in the pathway.

The inhibitor experiments allow certain conclusions to be drawn with respect to carotenoid turnover. The absolute concentrations of the highly desaturated pigments, myxobacton and 4-keto-torulene, which were no longer synthesized in inhibited but growing cultures of *M. fulvus*, remained completely constant over 22 hr or four generations, even under (mild) illumination of the bacteria (2500 lux, blue light). As mentioned above, accumulated lycopene also remained unchanged under normal culture conditions, and when the inhibitor San 6706 was added it was only transformed into other pigments and was not decomposed. Thus, carotenoids once synthesized are not easily catabolized.

## Physiology

The carotenoids of myxobacteria are localized in the cytoplasmic membrane of the cell (Kleinig 1972). Their synthesis is stimulated in many cases, but not always, by light; in illuminated cultures, the carotenoid concentration may rise by a factor of 10–20 (Reichenbach 1965b; Burchard and Dworkin 1966a; Grimm 1967; Reichenbach 1971; Kleinig and Reichenbach 1973b). The mechanism of this induction is not clear. It is sufficient to illuminate the cultures for a short time and then to incubate them in the dark, even at low temperature (6°C), to elicit the full response. In *M. xanthus* FB, the action spectrum of induction shows maxima at 410 nm and, to a lesser degree, at 512, 533, 548, 585, and 635 nm (Burchard and Hendricks 1969). As these maxima correspond to the absorption maxima of a porphyrin and protoporphyrin IX is known to be a photosensitizing pigment in FB (Burchard and Dworkin 1966), it was proposed that protoporphyrin IX may be responsible for carotenoid induction.

In growing cultures, the specific carotenoid content is constant and tightly regulated (Kleinig 1975) albeit, of course, dependent on environmental factors such as light. Carotenoid content increases, however, in late logarithmic and stationary phase (Burchard and Dworkin 1966a); obviously, the machinery for pigment synthesis remains fully active, or nearly so, for some time after growth. In *M. fulvus* Mx f2, the carotenoid content of acetone extracted dry cells was 0.003% when grown exponentially in the dark, 0.03% when grown exponentially in the light, and 0.06% when grown in late log phase (Reichenbach and Kleinig 1971). The carotenoid content also depends on the strain. By treatment with the mutagen *N*-methyl-*N*′-nitro-*N*-nitroso-guanidine, it was possible to obtain mutant strains from Mx f2 with either decreased or increased pigment content, although at the same time the pigment pattern, too, was changed.

At least one of the functions of the carotenoids in myxobacteria has been established. In *M. xanthus* FB, the carotenoids prevent photolysis, which is mediated by a sensitizer, protoporphyrin IX, produced by the same organism (Burchard and Dworkin 1966a; Burchard *et al.* 1966). Only when the cells are grown in the light, however, do they contain sufficient carotenoid to be resistant to photolysis; when they are grown in the dark, the carotenoid levels become so low that they are no longer protective, and the cells quickly lyse when exposed to light in the stationary phase. Photolysis is also observed when cultures are grown in the light in presence of $10^{-4}$ *M* diphenylamine, which prevents the synthesis of colored carotenoids. On the other hand, mutants that efficiently synthesize carotenoids in the dark do not become sensitive to light when grown in the dark.

In *M. fulvus* Mx f2, cultures can be obtained that have a low content of colored carotenoids or are completely carotenoid free, either by using unpigmented mutants or by growing the wild strain in presence of inhibitors such as diphenylamine or the herbicide, San 6706 (see above). Under aerobic conditions such cultures become very sensitive to illumination with daylight intensities (about 40,000 lux); the cells are killed within minutes. Wild type cells, in contrast, survive such intense illumination for a long time (Reichenbach, unpublished data). Thus, there is no question that the carotenoids have a photoprotective effect in myxobacteria, as is known from other bacteria, too.

The peculiar carotenoid–glucoside–fatty acid esters with their polar head and hydrophobic tails obviously fit particularly well into the membrane and might serve as structural elements and lipase-resistant stabilizers for the membrane in addition to performing their photoprotective function.

Pigmentation in myxobacteria is not always a stable character. When *M. xanthus* strain FB is plated, two types of colonies appear: one with a *tan* and another with a *yellow* color (Burchard and Dworkin 1966a). The frequency of the *yellow* type varies between 5 and 20%. Although most *yellow* colonies give *yellow* subclones only, the *tan* colonies usually shed a certain proportion of *yellow* clones upon subcultivation; stable *tan* strains can be obtained, however. It appears that both variants contain an identical set of carotenoid pigments but that the *yellow* variant produces in addition a noncarotenoid pigment, P 379, which in methanol has an absorption maximum at 379 nm. This pigment seems to be identical with the greenish-yellow pigment produced by *M. virescens* (Burchard *et al.* 1977b). In fact, the organisms classified as *M. virescens* may be merely stable *yellow* variants of *M. xanthus.* On the other hand, the appearance of the *tan* phenotype in *M. virescens* strains is also not unusual. Incidently, the expression of the *yellow* phenotype is sensitive to temperature:

At 36.5°C, the temperature maximum for growth of FB, P 379 is not synthesized any longer (Burchard *et al.* 1977b). The proportion between *tan* and *yellow* cells can be influenced by a number of environmental factors (Sudo and Dworkin 1969; Burchard *et al.* 1977). An increase in the proportion of the *yellow* type is observed in cultures with a low cell density (below $10^6$/ml). Ultraviolet (UV) irradiation leads to an increase of the *yellow* type among the survivors. The reason may be a higher resistance of *yellow* cells to UV dammage. The UV inactivation curve of *yellow* cultures shows a shoulder not observed in the inactivation curve of *tan* cells. This seems to indicate a multiple-hit inactivation in the case of the *yellow* type. However, a stimulation of the conversion from *tan* to *yellow* by UV is also not excluded. An increase of the *yellow* type is further observed upon addition of mitomycin C, nalidixic acid, or phenethyl alcohol to the culture but not with acridine orange or ethidium bromide. A mechanistic explanation of the phenomenon is not yet available. Perhaps a genetic switch is at work, as is known for the phase variations of other bacteria. When the switch is thrown and the cells grow on plates, the system obviously remains stable for several generations, perhaps under the influence of the high cell density in colonies, for only rarely are colonies with an intermediate color or with color sectors seen.

In *Archangium (Cystobacter) violaceum* production of the characteristic deep-violet pigment was found to be variable and connected with a certain colony type (Reichenbach 1965b; Grimm 1967; Grimm and Kühlwein 1973a,b). Again, the conversion between slow spreading, soft slimy, intensely colored K colonies and fast spreading, tough slimy, poorly pigmented G or S colonies is controlled by environmental factors. Each colony type may become more or less stable under certain conditions, and variable under others. Thus, in liquid culture the conversion goes exclusively S to K, whereas on plates it goes into the opposite direction. The mechanism underlying this variability is not known.

# 7

# Gliding Motility and Taxes

Robert P. Burchard*

In 1892 Thaxter described the myxobacteria as "possessing a power of slow locomotion . . . a distinctly visible movement characterizes the active rods and consists in a sliding locomotion in conjunction with lateral bending." Almost one century later there are still conflicting ideas about the mechanism of gliding. At a genomic level it is evident that *Myxococcus* gliding is a complex phenomenon (Chapter 8). Elucidation of the enigma of gliding motility has been much slower than that for swimming motility of flagellated bacteria, since myxobacteria (and other gliders) have no obvious organelles of motility protruding from their cell envelopes. Indeed, gliding has recently been described as a "backwater" in the motility field (Allen 1981).

## Description of Movements

### Gliding of Individual Cells

Among the best narrative and pictorial descriptions of myxobacterial gliding and related behaviors are those in the writings and films of Reichenbach. Individual rod-shaped cells move on surfaces or at an air–water interface (Reichenbach 1965b) in a direction parallel to their long axes (Fig. 7.1). Reversal of direction may occur frequently by reversal of the anterior and posterior poles of the cell. Or a cell may trace a curvilinear path to achieve a 180 deg turn (Kühlwein 1957). Unlike some other gliding bacteria (Burchard 1981), there

*Department of Biological Sciences, University of Maryland Baltimore County, Catonsville, Maryland 21228 U.S.A.

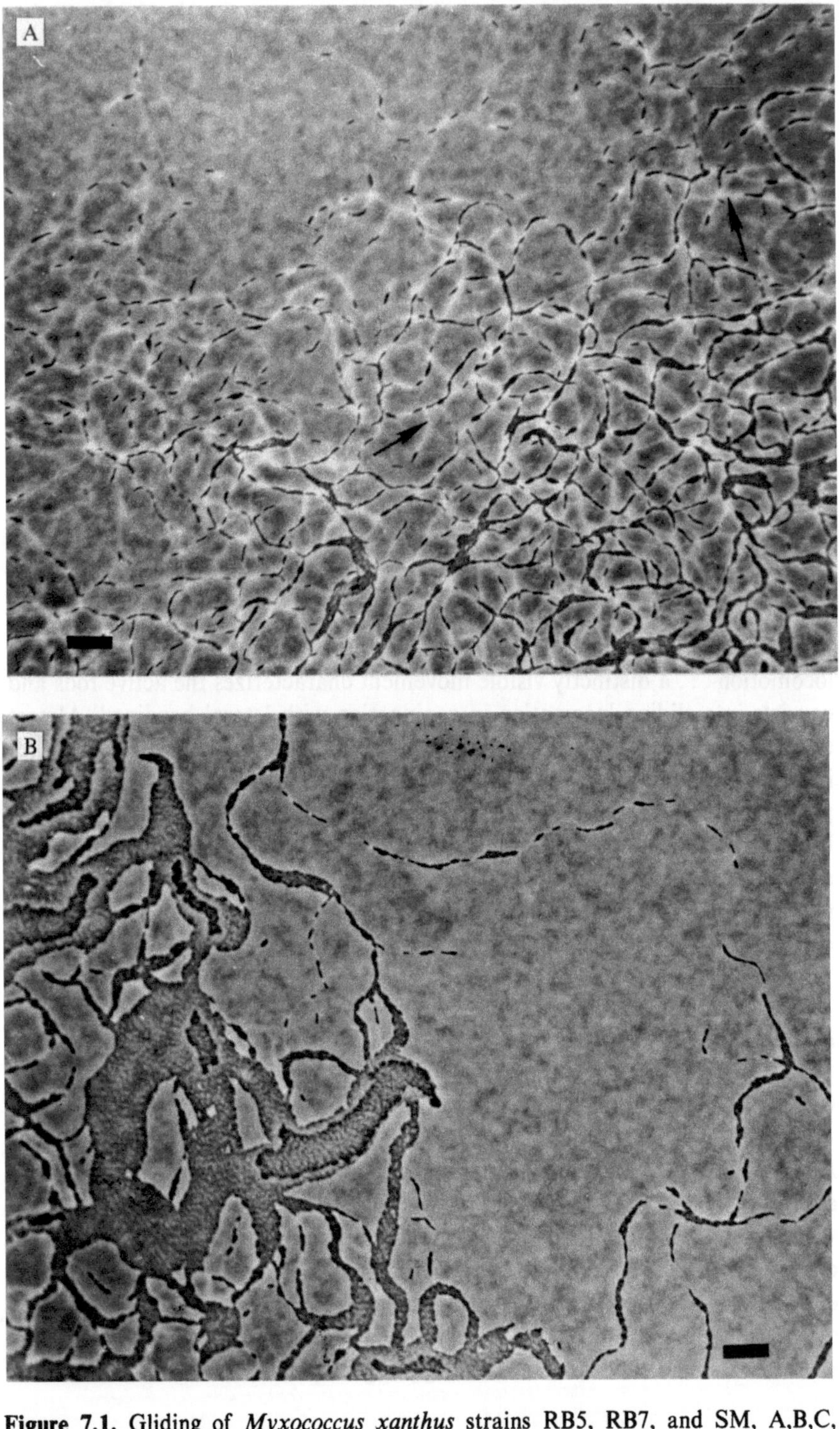


**Figure 7.1.** Gliding of *Myxococcus xanthus* strains RB5, RB7, and SM, A,B,C, respectively, on Gelrite (Kelco) gels. Slime trails (arrows) are particularly evident in A. Bar represents 25 $\mu$m.

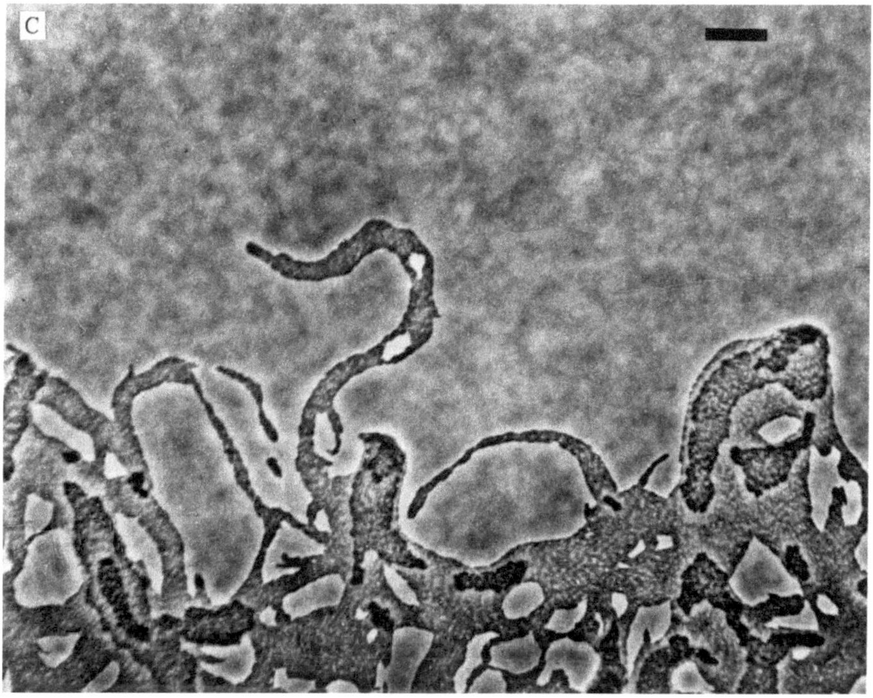


**Figure 7.1** (Continued.)

is no evidence for rotation of the gliding myxobacterium around its long axis during translocation (Meyer-Pietschman 1951; Reichenbach 1965b).

Some species demonstrate considerable cellular flexibility during movement, which has been attributed to wall elasticity by Thaxter (1892). This is normally limited to bending or flexing. However, there have been reports of rather extreme deformations suggestive of amoeboid cells (Kühlwein 1957; Kühlwein and Reichenbach 1968; Reichenbach 1962). Such cellular flexibility may be the result of the relatively high length-to-diameter ratio (~10) of certain myxobacterial cells as suggested by Dworkin (1966) and/or the unusual patchwork nature of the peptidoglycan layer described for *Myxococcus* (White *et al.* 1968; Chapter 2). In contrast, other species, for example, *Chondromyces apiculatus,* consist of rather rigid cells that are relatively short and wide (4–8 × 1.2 μm).

Compared to the velocity of swimming of flagellated bacteria and, for that matter, other gliding bacteria (Castenholz 1973; Henrichsen 1972), myxobacterial gliding is rather slow. *Archangium violaceum* has a maximum velocity of 60 μm/min (Kühlwein 1957). *Myxococcus* cells have been observed to move as slowly as 1 μm/min or as fast as 13 μm/min (Kaiser *et al.* 1979; Reichenbach 1966). The rate of gliding is not constant. Cells accelerate, decelerate, stop, and reverse direction in no regular fashion. Gliders have momentum adequate to force themselves through dense clusters of cells and to tunnel into an agar gel (Reichenbach 1965b).

The myxobactera undoubtedly encounter and glide on a variety of substrata in their natural habitats. In the laboratory their gliding is typically observed on glass slides or on agar. Substratum-dependent differences in gliding behavior have been described. For example, *Archangium* glides smoothly and steadily on agar, whereas on glass, translocation is uncertain and shaky or vibrating (Reichenbach 1965b), possibly reflecting poorer adhesion to the glass. Although some gliding bacteria in the order *Cytophagales* require polyanionic polysaccharides in their substrata, at least two myxobacteria are able to glide on agarose that is lacking or low in such polymers (Arlauskas and Burchard 1982). Gelrite (Kelco) also provides an excellent substratum for gliding of these and other bacteria (Fig. 7.1).

## Coordinated Gliding

Although myxobacteria have the capacity to glide as individual cells, their gliding behavior is characteristically communal. They tend to glide in dense swarms (or rafts) of closely associated cells (Fig. 7.1). The biological advantages of maintaining high population density have been reviewed (Kaiser *et al.* 1979; Rosenberg *et al.* 1977). That the associations among cells in the swarm are temporary is obvious from the shifting of cell positions relative to one another and from the often observed breaking away and rejoining of single or small groups of cells, described in the films of Reichenbach (1966, 1968). Entire swarms may merge to form larger ones or may split into smaller ones. In spite of this dynamic behavior, the general maintenance of such cell–cell associations may be viewed as a form of coordinated gliding.

Kaiser and Crosby (1983) have methodically characterized swarm behavior in several *Myxococcus xanthus* strains. They have found that the rate of expansion of a swarm over an agar surface is highly cell density dependent. Swarm expansion rate is directly proportional to cell density over a range in which mean cell–cell separations are of the order of one cell length. It is apparent that cell–cell interactions increase swarm movement. Once the cell center-to-center distance is $\leq 1\ \mu m$, no additional increase in swarm expansion rate occurs with further increase in cell density.

Another remarkable and complex type of coordinated movement characteristic of at least several genera of myxobacteria (Dworkin 1973; Reichenbach 1965a, 1966) has been described as patterns of ripples or as rhythmic, pulsating waves. These were initially observed to arise spontaneously during fruiting of and predation by swarms of myxobacteria. The phenomenon has been partially characterized by Shimkets and Kaiser (1982b) who define ripples as ridge-shaped accumulations of oriented cells moving in a synchronous and pulsating manner. A field of ripples, which may initiate at one point and spread and continue rippling for 24 hr or more, is characterized by an interripple wavelength of approximately 45 μm (approximately 10 cell lengths). The ripples move at approximately 2 μm/min, comparable to the gliding velocity of single cells. Ripples in adjacent tracks move in opposite directions. Since there is no significant net movement of the swarm, the authors speculate that cells may move from a ripple in one track to another in the adjacent track (moving in opposite direction) via organized crossover zones. This rippling system may serve to prevent cells from wandering off, thus maintaining the high population density required for fruiting and for feeding on prey bacteria. Both of these phenomena are characterized by the release of peptidoglycan components from cells walls: by autolysis in fruiting (Wireman and Dworkin 1977); by production of wall lytic enzymes in feeding (Sudo and Dworkin 1972). Shimkets and Kaiser (1982b) have clearly demonstrated that exogenous peptidoglycan or a combination of its subunits will induce rippling in an otherwise quiescent population.

Another form of coordinated gliding in the myxobacteria occurs during fruiting body formation. The morphogenetic events that result in such complex multicellular structures have been discussed in Chapter 1. The actual movements of individual and groups of cells during the process of building a fruiting body have not been methodically followed. Based on its structural complexity and the arrangement of cells within the fruiting body (Grilione and Pangborn 1975; Stephens and White 1980a), it is likely that coordinated gliding is a requisite for such morphogenesis.

## Other Movements

In addition to gliding motility, the myxobacteria demonstrate nontranslocational movements that also require attachment of a part of the cell to a substratum. Thaxter (1892) described a lateral bending in any plane, occurring in conjunction with gliding. This bending was sometimes so "extreme that the rod may form a loop with its ends approximated, after which the normal straight position may be assumed with considerable rapidity." Such apparently active flexing (and subsequent release of tension) has been observed in *Chondromyces* (Bonner 1952), *Archangium* (Reichenbach 1965b), and *Myxococcus* (Reichenbach 1966). The basis for such flexing may be contractile, as it is in *Flexibacter* FS-1 (Burchard 1982a).

Myxobacteria have also been observed to attach to a surface (e.g., in a microscopic wet mount) by one pole and undergo swinging and nutational movements (Kühlwein and Reichenbach 1968; Reichenbach 1965b). In *Myxococcus* such movements have been described as *Suchbewegung* (searching movements) (Reichenbach 1966). Jerky movements have also been noted (Fluegel 1963a; Reichenbach 1965b, 1966).

Other gliding bacteria have been observed to propel latex microspheres actively along their surfaces (Pate and Chang 1979; Lapidus and Berg 1982;

reviewed by Burchard 1981). *Myxococcus* has been observed not to move microparticles in this fashion by Dworkin *et al.* (1983) who attribute absence of such activity to the lack of moving cell surface components (e.g., rotary assemblies or adhesive tracks; see below). Lack of such propulsive activity may also be a reflection of the fact that *Myxococcus* glides at a velocity from one to two orders of magnitude more slowly than do *Cytophaga* and the other gliders that demonstrate this behavior.

## Biological, Physical and Chemical Conditions that Affect Gliding

Gliding motility stops during the process of cell division (Meyer-Pietschmann 1951; Reichenbach 1965b, 1966, 1974b. Once the daughter cells have separated, movement begins almost immediately. This point in the cell cycle is characterized by marked changes in macromolecular synthesis patterns (e.g., Zusman *et al.* 1971).

The velocity of gliding of groups of *A. violaceum* cells increases almost linearly with increasing temperature between 20 and 40°C (Grimm *et al.* 1971); the $Q_{10}$ is approximately 2 between 25 and 35°. At $\leq$ 20° no gliding was measurable; at 45° gliding halted.

Kühlwein (1953) noted a relationship between nutrition and gliding motility, and Dworkin (1973) suggested a regulatory couple between the rate of growth and the velocity of gliding. This relationship is supported by measurements of colony expansion, most of which is accounted for by gliding, not cell division. The rate of expansion increased with nutrient concentration up to 1 or 2% protein hydrolyzate (Burchard 1974a), above which it decreased. A contrasting picture was presented by Reichenbach (1965b) for *A. violaceum,* which demonstrated increasing velocity of swarm spread with decreasing nutrient concentration until a rather low nutrient concentration was reached, at which point the rate declined. Lonski *et al.* (1977) have reported that an extract from cells undergoing fruiting induced a fourfold increase in rate of *Myxococcus* gliding.

Subjection of *Myxococcus* cells to shear forces or to ultrasound adequate to destroy 50% of the cells did not markedly affect gliding of the survivors, supporting the observation that there are no shear-sensitive structures required for motility (Burchard 1974b). Motility of *M. xanthus* is inhibited by treatment with proteolytic enzymes, indicating that cell surface proteins are required for gliding (Burchard 1974b). Reversible inhibition of gliding by osmotic shock indicates that periplasmic proteins are required for motility. Furthermore, antibodies raised against *Myxococcus* periplasmic proteins blocked gliding of untreated cells (Burchard 1974b).

Continuous protein synthesis is not required for gliding since movement (short-term) occurs on media containing growth-inhibitory levels of chloramphenicol.

Cytochalasin B and colchicine, agents that inhibit eukaryotic cell motility phenomena involving microfilaments and microtubules, respectively, were reported to be inhibitors of *Cystobacter* gliding (Heumann and Kühlwein

1979), but the site(s) of drug action is not known. The former drug had no demonstrable effect on *Myxococcus* (Burchard unpublished). Cytokinins, multifunctional plant hormones, have been reported to suppress swarming of *Myxococcus* by a yet to be determined mode of action; effect on single cell gliding was not reported (Jones 1978).

Removal of $Ca^{2+}$ by treatment of *Cystobacter* and *Myxococcus* with the chelator ethyleneglycoltetraacetic acid (EGTA) inhibited gliding (Heumann and Kühlwein 1979; Burchard and Burchard unpublished). Motility recovered in *Myxococcus* with addition of $Ca^{2+}$ to treated cells (Fig. 7.2). The $Ca^{2+}$ ionophore A23187 also inhibited movement of this bacterium.

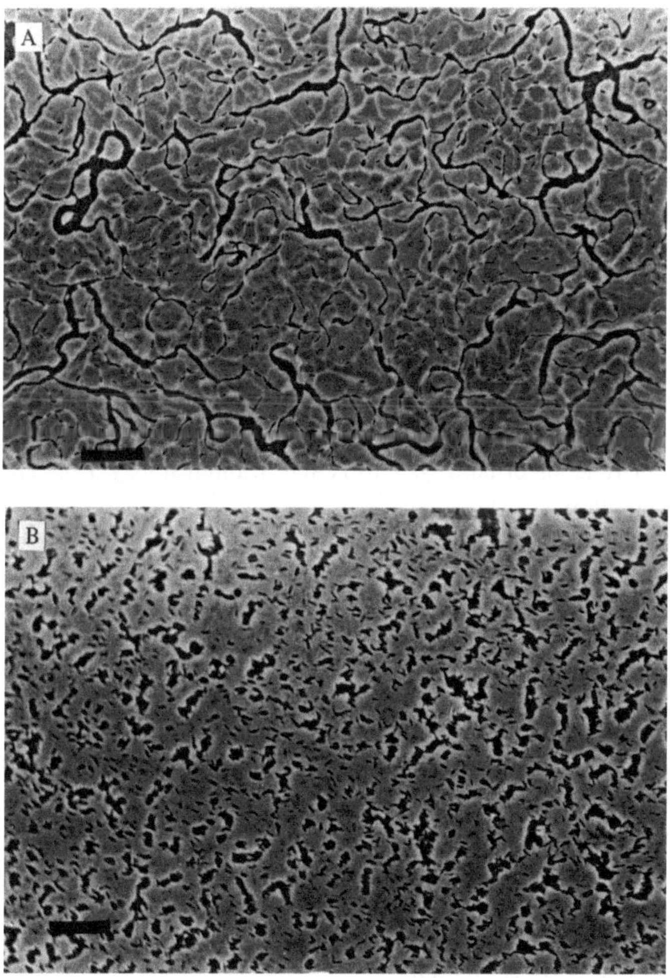


**Figure 7.2.** $Ca^{2+}$-dependent gliding of *M. xanthus*. EGTA-treated bacteria were spotted on phosphate-buffered agarose containing $MgCl_2$ ($10^{-3}$ *M*) with (A) and without (B) $CaCl_2$ ($10^{-3}$ *M*). Photographed after 3 hr. Bar represents 50 $\mu M$.

There is good evidence that gliding motility of *Flexibacter polymorphus* is energized by a proton motive force (Ridgway 1977). As in *Flexibacter,* the proton conducting uncouplers 2,4-dinitrophenol (DNP) and carbonylcyanide m-chlorophenylhydrazone (CCCP) inhibit motility of *Myxococcus.* However, unlike *Flexibacter,* arsenate inhibits *Myxococcus* gliding. The inhibitory effect of all of these agents on *Myxococcus* gliding, an obligately aerobic process, is reversed in the presence of pyruvate, the significance of which is not presently clear (Burchard unpublished).

## Gliding Motility Mutants

The initial report of two distinct *Myxococcus* motility mutants, one able to glide only when cells are in apposition to one another (semimotile or SM; Fig. 7.1C) and one unable to glide at all (nonmotile or NM, Burchard 1970), has been elucidated by the isolation, characterization and mapping of many gliding motility mutants (Hodgkin and Kaiser 1979a, 1979b; Chapter 8). Briefly, Hodgkin and Kaiser have presented evidence for the existence of two motility systems, each with multiple genetic loci. The adventurous (A) gene system controls single cell gliding; the social (S) gene system controls movements of groups of cells or swarms. $A^-$ mutants are still able to glide (via the S system) if two or more cells are in apposition or close association (see below). $S^-$ mutants demonstrate predominantly single cell gliding; such mutants fruit poorly, if at all, and do not swarm. The mean gliding velocities of individual $A^+S^+$, $A^-S^+$, and $A^+S^-$ cells are similar (Kaiser and Crosby 1983). $A^-S^-$ mutants are nonmotile, as are mutants in one other motility locus, mgl, required for function of both systems A and S. Interestingly, rippling requires both gene systems.

Approximately 10% of the motility mutants isolated by Hodgkin and Kaiser (1977) are conditional in that mutant cells making direct contact with wild-type cells or those of another mutant class may be stimulated to glide for a short period. This has been demonstrated for both motility systems. The donor of the stimulus need not be viable, since formaldehyde-killed cells are effective in stimulating three classes of recipients. The stimulation can also be provided by outer membrane vesicles of donor cells (Helman personal communication).

In comparing motile and nonmotile strains of *M. xanthus,* MacRae and McCurdy (1976) observed that polar fimbriae were present only on the former, leading them to suggest a function for these proteinous organelles in gliding motility, possibly involving retraction. Subsequently, Dobson *et al.* (1979) described motile strains that themselves lacked fimbriae. Kaiser (1979) clarified the question by demonstrating a correlation between the presence of fimbriae (pili) and S system motility.

Fimbriae are hypothesized to mediate the cell–cell associations required for S system gliding (Kaiser 1979). Recently, Kaiser and Crosby (1983) have examined the behavior of $A^-S^+$ cells at a density low enough to isolate most cells. Analysis of the positions of neighboring cells indicated that in some pairs $\leq 0.88$ cell lengths apart, one of the two cells moved substantially, presumably by gliding; the second cell did not move. It is likely that the cells in these pairs were in contact via polar fimbriae that are long enough to bridge the gap between them. Such cell–cell interactions partially account for the cell density dependence of swarm expansion rate in *M. xanthus.*

Antiserum directed against isolated fimbriae inhibited colony spreading by swarming; neither single cell gliding nor growth was affected (Dobson *et al.* 1979). These fimbriae are also responsible for the adhesiveness characteristic of $S^+$ strains, which results in cell clumping and adhesion to the walls of culture flasks.

It is noteworthy that bacteriophage Mx1-resistant mutants of *M. xanthus* frequently demonstrate aberrant aggregation behavior but are apparently not affected in their ability to glide (Murillo and Kaiser personal communication), as might be expected if alterations in cell surface components serving as phage receptors affected the mechanism of gliding or the interactions of the cell envelope with the substratum. This contrasts with a variety of bacteriophage-resistant mutants of *Cytophaga johnsonae,* nearly all of which are nonmotile (Pate *et al.* 1979).

## Mechanistic Theories of Myxobacterial Gliding

Early studies of gliding motility were most frequently performed with trichome-forming cyanobacteria. A number of the theories proposed to account for gliding of myxobacteria has been derived from those proposed for these larger, faster gliders. Although the idea of a unified theory of bacterial gliding motility is appealing, in view of the taxonomic diversity of these organisms (Reichenbach 1981), there may well be multiple mechanisms of gliding motility. However, it is reasonable to expect that among the myxobacteria there be only one mechanism. Furthermore, it would appear reasonable that this mechanism should also drive the other movements demonstrated by myxobacteria.

Why is the mechanism of gliding motility still in dispute? In contrast to swimming bacteria, there are no obvious organelles of movement protruding from the cell envelope of the myxobacteria. Furthermore, that envelope through or in which the mechanism of gliding must operate is at least morphologically similar to the envelopes of other nongliding, Gram-negative bacteria (Fig. 7.3).

### Slime and Gliding

It has long been recognized that the production of extracellular slime or sheath material is characteristic of most if not all gliding bacteria. The slime of *Myxococcus virescens* has been analyzed by Gnosspelius (1978a) who reported that

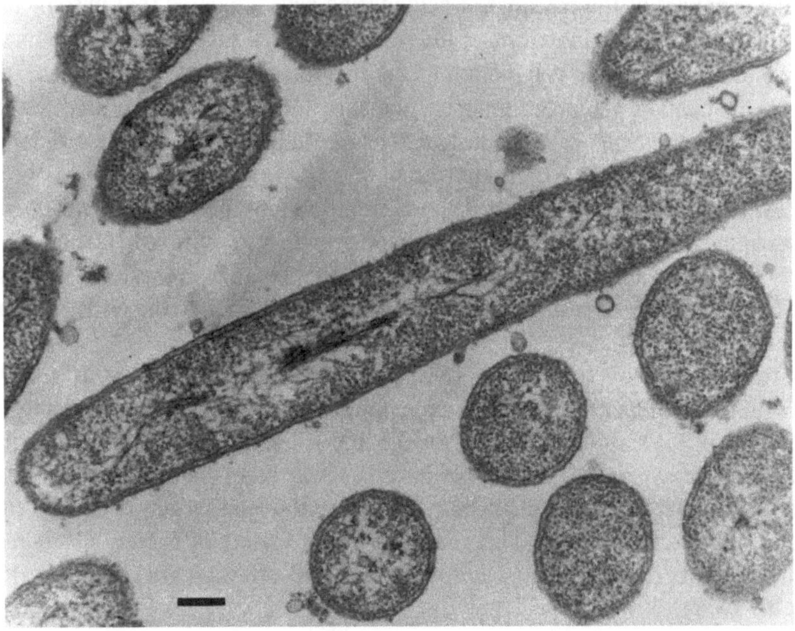

**Figure 7.3.** Thin section of *M. xanthus* cells demonstrating typical Gram-negative envelope form. Bar represents 200 nm.

it consisted of a protein–polysaccharide–lipid complex. The polysaccharide component of the slime of several myxococci has been characterized by Sutherland and Thomson (1975).

Production of this material must be energetically expensive for the myxobacterium. It is likely that extracellular slime mediates adhesion of the cell to the substratum on which it is gliding. Additionally, in light of evidence for inter-polysaccharide associations giving rise to three-dimensional networks (e.g., gels) (Dea *et al.* 1977), the polysaccharide component(s) of myxobacterial slime may associate with polysaccharides in substrata (agar, plant materials, soil) to produce a matrix on or in which gliding occurs. The absence of such interpolysaccharide associations may explain the "unsteady and vibrating" gliding on glass slides observed by Reichenbach (1965b).

Directed release of slime has been suggested to provide a propulsive force for translocation (Jahn 1924; Kühlwein 1953). Such a role for slime production is conceivable in light of the report that *Acetobacter* cells can be propelled along a surface by the directed polymerization and crystallization of extracellular cellulose microfibrils along the longitudinal axes of the cells (Brown *et al.* 1976). However, there is no evidence that myxobacterial slime production is directed posteriorly nor that the slime has ribbon-like organization. There is presently no evidence that slime production is the driving force of gliding. An ancillary function for slime in one proposed mechanism of gliding will be considered below.

It has also been suggested that production of excess slime may inhibit the gliding of the K variants of *Archangium violaceum* (Grimm and Kühlwein 1973b).

## Surface Tension as the Driving Force for Gliding

Recognition that a requisite for gliding is association of a cell with a surface probably accounts for the origin of the theory that surface tension forces drive gliding. The concept was originally presented to explain the movement of cyanobacteria (Burkholder 1934; Coupin 1923). Recently, Keller *et al.* (1983) have revived this hypothesis with a theoretically and experimentally based model for *Myxococcus* gliding driven by a cell-generated surfactant gradient. Their analysis indicates that it is metabolically feasible to produce enough extracellular surfactant from a localized position on the cell to propel it at the velocities observed for these gliders. Furthermore, the model predicts that gradients generated by two cells will permit them to pass one another in close apposition and in opposite directions. The model also predicts that isolated cells would be able to join a swarm (as they do), rather than being repelled by it, as might be expected in response to a localized concentration of surfactant.

Some experimental evidence supports the theory (Dworkin *et al.* 1983). (1) *Myxococcus* does produce surfactant, and nonmotile mutants produce less than the wild type. (2) Addition of exogenous surfactant (e.g., bovine serum albumin) at a level that would be expected to swamp out any organism-produced gradient inhibits gliding with no detectable effect on growth. (3) Gliding occurs at an air-agar interface. Elimination of the interface by overlaying the agar with an aqueous layer should eliminate the interfacial surface tension and thereby would be expected to block motility. In fact, Meyer-Pietschmann (1951) observed that gliding cells overlaid with a drop of water straightened out and stopped gliding. Dworkin *et al.* (1983) have made a similar observation, having carefully examined several strains under comparable conditions. Even when some movement was detected, it was at least one order of magnitude slower than was gliding at an interface (Dworkin personal communication).

In contrast, Carpenter and Burchard (unpublished results) have found that cells of strains that adhere tenaciously to agar [*M. xanthus* A2 and 5A from I. Sutherland and RB5 (Burchard 1975)] are able to glide on an 0.6% gel under an aqueous overlay. Average velocities as high as 8$\mu$m/min were measured for randomly selected cells. Since Dworkin (personal communication) also used strain RB5 in his rate measurements, the discrepancy between the two sets of observations is yet to be resolved.

Dworkin (personal communication) has observed underwater gliding on glass and plastic surfaces. Thus, it is not surprising that fruiting body formation, the aggregative phase of which requires gliding motility, can occur on a glass surface under water (Fluegel 1964; Kuner and Kaiser 1982).

As predicted by the surfactant theory, *Myxococcus* colonies were observed not to spread significantly when growing within soft agar (Dworkin *et al.* 1983). Reichenbach (1966), however, has observed that *M. virescens* cells are capable of boring into and moving long distances within agar, particularly in gels of low concentration (0.5–0.75%). Other myxobacteria, *Chondromyces crocatus, Archangium violaceum* and *Nannocystis exedens* (McCurdy 1969b; Reichenbach 1965b, 1970) also can move through agar. In the case of *Nannocystis,* such movement may be facilitated by the ability of the bacterium to hydrolyze the gel. The interpretation of these observations hinges on the nature of the "channels" made by the gliding cells. Are they simply pathways through a gelled, aqueous environment or are they tunnels with slime "walls" that could provide surfaces? The latter idea is supported by McCurdy's (1969) description of *Chondrococcus* penetration in that "cells are located within slime tubules which ramify through the agar."

At this point in its development, the surfactant theory does not account for the other, active movements of myxobacteria, that is, flexing and pendulum-like or nutational movements.

**Contractile Mechanisms**

Again, the origin of the idea that there may be a contractile basis for gliding dates back some years (Meyer-Pietschmann 1951; Reichenbach 1965b; reviewed by Burchard 1981). Observations of what were described as amoeboid-like changes in *Polyangium* cell form led Kühlwein (1957) to endorse a contractile mechanism of gliding. This theory was revived several years ago with the description of a membrane-associated organelle in the cytoplasm of *M. xanthus,* comprised of 4–5 nm diameter filaments that are arrayed longitudinally (Fig. 7.4). The filaments appear to demonstrate some helicity and have a herringbone-like periodic cross structure (∽12 nm spacing; Burchard *et al.* 1977a). Filaments of comparable size have been observed in association with envelope fragments of French pressure cell-disrupted cells (Burchard, unpublished results). That this structure is not an artifact of fixation is suggested by its appearance in cells fixed and stained by three procedures. Additionally, the structure was altered (lacking periodic striations) in one nongliding mutant. Recently, rows of "particles" spaced at intervals comparable to the periodic substructure and positioned similarly to the filaments have been observed in freeze-fractured cells (Bayer and Burchard unpublished).

Filaments of slightly larger diameter have been observed in lysates of motile (but not in nonmotile) *Myxococcus* strains (MacRae and McCurdy 1976). Also, Schmidt-Lorenz and Kühlwein (1968) described filaments (or tubules)

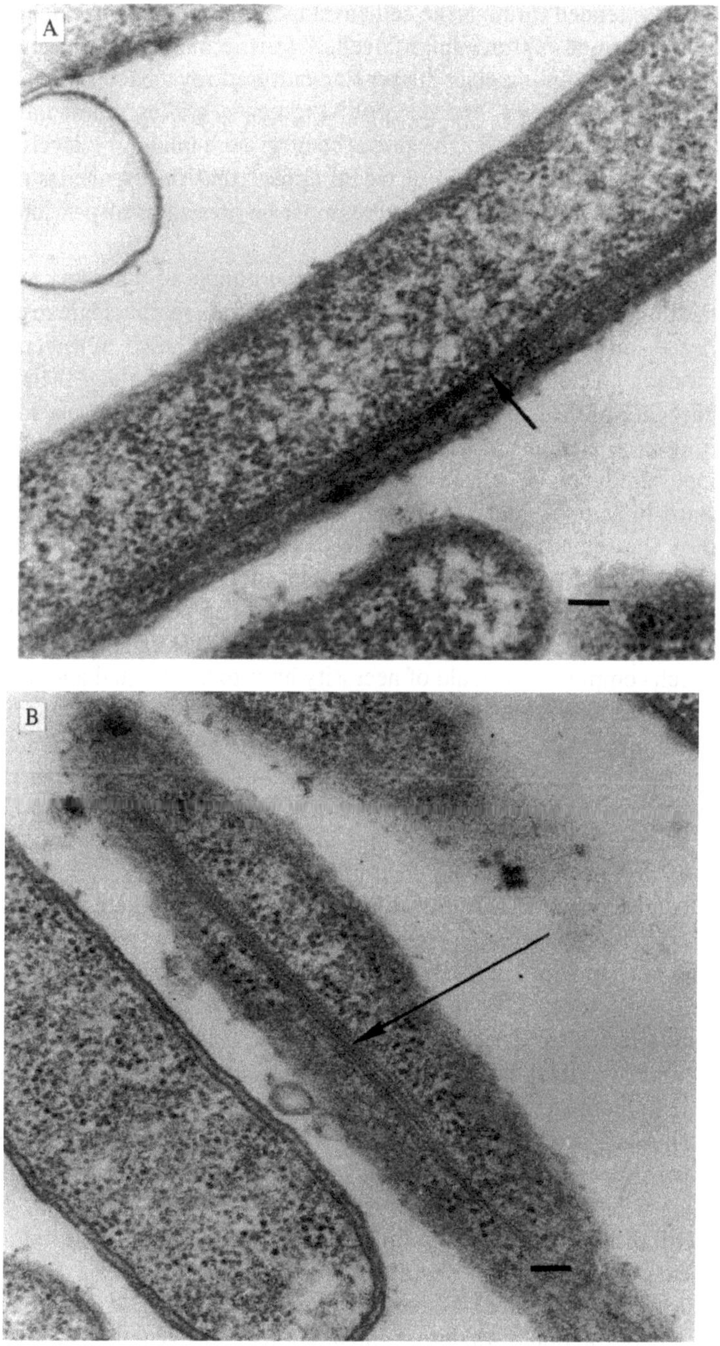


**Figure 7.4.** Thin sections of *M. xanthus* cells demonstrating a bundle of envelope-associated filaments (arrows), hypothesized to be a contractile organelle. Bar represents 100 nm.

in 12 of 14 species of myxobacteria; however, it was later reported that these structures extended through the cell envelope, emerging as what Schmidt-Lorenz (1969) termed "extracellular flagella." Furthermore, these structures were only observed in gliding cells; suspension-cultured myxobacteria lacked them. Since suspension-grown *Myxococcus* commence gliding immediately after placement on agar, even in the presence of growth-inhibitory levels of chloramphenicol (Burchard 1974b), it would appear that the organelles or mechanism of gliding, filaments or otherwise, must be present in suspension-cultured cells.

Burchard *et al.* (1977a) proposed that the bundle of filaments might be a contractile organelle of gliding, since it bears a crude morphologic resemblance to striated muscle. If the filaments were, in fact, anchored to the envelope at or near each pole of a gliding cell and if they were linked along their length to the inner face of the cell envelope, it would be possible to envision a contraction initiating at a terminus of the structure that could, through its linkage to the envelope, produce a localized envelope compression or pucker (Hughes and Burchard to be published). If the contraction and its coupled compression initiated at the temporary posterior pole of the cell and if it were propagated anteriorly, it would be dissipated at the anterior pole and be expressed as a small, forward extension of the cell (Fig. 7.5). The posterior end would have been pulled forward by the amount determined by the extent of the compression. Such compressions would of necessity have to be initiated and propagated continually as long as the cells were gliding in one direction; reversal of direction of gliding would be preceeded by reversal of the pole at which compressions started. It was initially proposed (Burchard *et al.* 1977a) that the organelle may propel gliding in an inchwork-like fashion. This appears to be ruled out by reports of no significant change in cell length during gliding (Reichenbach 1965b; Burchard and Bradley, unpublished results).

Such a mechanistic basis of translocation assumes that there be no backward slippage of the cell as compressed portions of it move forward. Such slippage could be prevented by an underlying layer of slime. Hughes and Burchard (to be published) propose that this slime has the properties of a plastic. It is adhesive until stressed by shearing forces that cause it to "give" and behave like a viscous (nonadhesive) fluid (Fig. 7.5). The shearing force is provided by the traveling compressions on the cell envelope. Once the shear stress has moved anteriorly, the slime reanneals to regain its adhesive character, thus preventing backward slippage. Slime would thus have a vital but ancillary role in gliding motility.

A contractile organelle of gliding motility could also explain other, nontranslocational movements characteristic of the myxobacteria (Hughes and Burchard, to be published). For example, flexing could result from a unilateral and rapid contraction of the organelle along its entire length. The compressed cell envelope might subsequently act as an extensor, resulting in relaxation of the flex. The recent report that flexing in *Flexibacter* FS-1 is a contractile event (Burchard 1982a) provides support for the concept of contractile phenomena

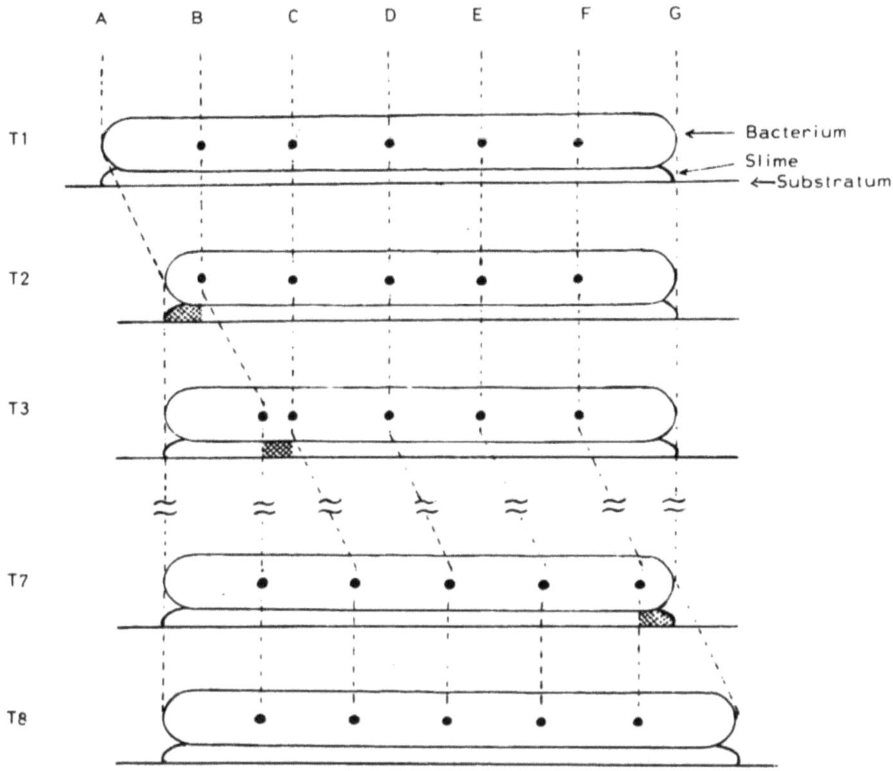


**Figure 7.5.** A model for gliding motility of myxobacteria. Submicroscopic contractions, expressed as localized cell envelope compressions, initiate at the cell's posterior pole (T1–T2) shearing the adjacent, extracellular, plastic slime until it yields (crosshatched region). The compressions are propagated forward (T3–T7) until released at the anterior pole as forward extensions of the cell. Once each compression has passed forward, the underlying slime reanneals and regains its adhesive properties, thus preventing backward slippage of the cell (Hughes and Burchard, to be published).

among the enveloped gliding bacteria. Previously, it had been proposed that contractions in flexibacters might be generated by a "directional sequence of 'make-and-break' interactions between the flexible outer membrane and the more rigid peptidoglycan layer" (Humphrey *et al.* 1979). Also, gliding mycoplasmas undergo configurational changes suggestive of contractile events (Bredt *et al.* 1973).

There are, however, a number of unanswered questions regarding the presence and function of the putative contractile organelle. It has been observed in a relatively small proportion of cells examined and, where observed, only one end could be traced to a point of association with the cell envelope. The other terminus of the bundle of filaments appeared to be disconnected and to splay out in the cytoplasm (Fig. 7.4). This might be explained by an inability to resolve the structure when it is attached (and presumably functional) at both ends. Only when one or both points of envelope association were broken and the filaments contracted might it be resolvable. What it is that triggers these hypothetical breaks is not known.

If the bundle of filaments is contractile and if, in fact, it is tightly linked to the cell envelope along its length, the question arises as to how its contractions are translated into compressions in that part of the envelope, presumably the outer membrane, that interfaces with the adjacent slime. Expression of the contractions on the cell surface might be facilitated by the unusual patchwork nature of the peptidoglycan layer (White *et al.* 1968) that may confer adequate flexibility to make this layer compressible. Furthermore, the adhesions described by Bayer (1968) in *Escherichia coli* (if they exist in myxobacteria) may serve to link the cytoplasmic membrane, to which the organelle is linked, with the cell wall.

What is the molecular basis for the putative contractility in *Myxococcus?* Actin is known to be an integral component of eukaryotic cell motility systems (*e.g.*, Clarke and Spudich 1977). Recently, a prokaryotic, actin-like protein has been reported in two gliding cyanobacteria, as well as in *E. coli* and a methanogen (Gilbert and Fulton 1982). There has also been a report of actin-like filaments in a gliding mycoplasma (Meng and Pfister 1980). However, several investigators have failed to find actin in *Myxococcus*. Recently, Burchard and Wu (unpublished results) have sought, with no success, a *Myxococcus* nucleotide sequence homologous to an eukaryotic actin gene probe (Fyrberg *et al.* 1980).

If there are contractile proteins in *Myxococcus* and other gliding bacteria, they may have properties distinct from those of eukaryotic cell actin and its associated proteins.

## Other Mechanistic Models

Two recent additions to the list of mechanistic theories of gliding motility invoke moving cell surface components. These theories were developed as models for gliding of members of the *Cytophagaceae;* neither is necessarily valid for the myxobacteria. However, Pate and Chang (1979) have suggested a unified theory of bacterial motility, founded on rotary motors. They described ring-like structures associated with the envelopes of *Cytophaga johnsonae* and *Flexibacter columnaris*, apparent after treatment of the cells with potassium iodide. These "rotary assemblies" bear some similarity in size and form to the basal bodies of flagella of swimming bacteria, leading to the proposal that gliding might be driven by spinning of these rotors against the substratum. Similarly, these assemblies could propel microspheres on the cell surface and might power the spinning of polar-attached cells. One nongliding mutant of *F. columnaris* lacked rotary assemblies.

Applying the isolation procedures of Pate and Change (1979) to *M. xanthus* and *Cytophaga* U67, Burchard (unpublished results) has found structures that may be comparable to rotary assemblies. But, these were heterogeneous in size and irregular in form. Lapidus and Berg (1982) applied procedures for the extraction of flagellar basal bodies from *E. coli* to *Cytophaga* U67. No structures resembling basal bodies were found.

Based on observations of gliding and nontranslocational movements of *Cytophaga* U67, as well as on the movements of latex microspheres on its surface, Lapidus and Berg (1982) have proposed a second, novel model for gliding, invoking a dynamic cell surface. The model states that "absorption sites within the cell's outer membrane move along tracks fixed to the rigid framework of the cell wall." These absorption sites are proposed to adhere to the substratum over which the cell is gliding or to microspheres that make contact with the cell surface. What cellular machinery might drive the movement of these sites in the outer membrane is unknown. In contemplating application of this model to the myxobacteria, it should be noted that there is no evidence for microsphere propulsion by these bacteria (Dworkin *et al.* 1983) nor have myxobacteria been observed to glide continuously on a fixed sphere, to spin continuously, or to flip over, all of which are characteristic movements of *Cytophaga* U67.

One generally overlooked theory of gliding motility is an ionic counterpart to the structural track system just described. Mitchell (1956 and personal communication) proposed an electrokinetic theory of gliding, involving a stream of ions passing longitudinally within the bacterium in one direction and over the surface in the other, an ionic caterpillar track. A variation on this theme has been proposed as a driving force for the "make-and-break" gliding theory (Duxbury *et al.* 1980).

## Tactic Responses

### Chemotaxis

In their natural habitats, the feeding of myxobacteria either on soluble nutrients or on prey bacteria (which might signal their presence by release of diffusible components), would be facilitated by their ability to respond to gradients of such materials. Time-lapse microcinematographic studies have, in fact, depicted movement of myxobacteria toward food bacteria (Kühlwein and Reichenbach 1968). Daft and Stewart (1973) have reported that myxobacterium CP-1 glides toward cyanobacteria on which it preys. These observers proposed that oxygen evolved by the prey bacteria may "help in attracting" the myxobacteria. Dworkin (1983) recently demonstrated that swarms (flares) of *M. xanthus* move toward clumps of *Micrococcus luteus* on agar. It is not clear why, under similar experimental conditions, *Myxococcus* did not move toward *E. coli* (Dworkin and Eide 1983). However, such swarms will readily glide toward 5 $\mu$m-diameter polystyrene latex beads and 8 $\mu$m glass beads, suggesting that the response is a physical one, directed at a particulate target, either animate or inanimate. How the myxobacterium perceives such a target is not certain. Dworkin (1983) has suggested an elasticotactic response as one possible explanation for his observations.

Reichenbach (1965a) observed that individual *Chondromyces* cells glide faster than a swarm of this species expands. He suggested that maintenance of swarms, which might otherwise disband, may be achieved by chemoattraction among the cells. When two such swarms were within a few millimeters of each other, they migrated toward one another with increasing velocity, until they fused (Reichenbach 1974b). It was suggested that such swarms could "smell" one another. From a critical analysis of the movements of *Chondromyces* swarms, as depicted in Reichenbach's (1974b) film, Lauffenburger and Keller (personal communication) have formulated a mode' to account for swarm cohesiveness and the movements of proximal swarms, based on the production of a diffusible chemoattractant by the bacteria. The model accounts for maintenance of the integrity of the isolated swarm by balancing random motility with gliding directed by the chemoattractant. "The presence of another nearby swarm upsets this balance by altering the chemical gradient, allowing cell migration in directions of the diminished gradient," with subsequent swarm merger (Lauffenburger and Keller, personal communication).

Aggregation, culminating in fruiting body formation, may also be triggered by chemoattractants, as is the case for the cellular slime molds. Bonner (1952) separated an aggregating center of *Chondromyces crocatus* from cells moving toward that center and cut a groove in the agar between the resultant two populations. The aggregating cells moved around the groove to join the displaced center, leading to the suggestion that a factor governed such morphogenetic movements. However, it may be that the physical disturbance of the agar, resulting from separation of the cells and cutting, could have accounted for this behavior by setting up stress lines to which cells might have responded elasticotactically (Lev 1954).

Several investigators have provided evidence for diffusible factors that stimulate myxobacterial fruiting body formation (Fleugel 1963b; Lev 1954; McVittie *et al.* 1962). However, there is no indisputable evidence that such factors are chemoattractants. Evidence against the existence of gradients of morphogenetic chemoattractants is provided from observations of Fluegel (1964) and Kuner and Kaiser (1982) that myxobacteria can aggregate to form fruiting bodies on glass under an aqueous overlay; it is unlikely that gradients could be maintained in such an aqueous environment (Kuner and Kaiser 1982).

As early as 1962, Reichenbach had concluded that chemotactic behavior among the myxobacteria was unlikely. He suggested that an alternate basis for aggregation could be a combination of random movements, slime trail following, elasticotaxis, and the effects of cell–cell adhesion. More recently, White

(1981) has hypothesized that aggregation could be explained solely by slime trail following (see below).

Recently, others have tested specific chemicals, putative morphogenetic signals, and nutrients as chemoattractants. Using two devices for generating relatively stable concentration gradients, it has been reported that adenosine 5′-monophosphate (AMP) and cyclic guanosine 5′-monophosphate (cGMP) are attractants for *M. xanthus* (Ho and McCurdy 1979; Shimkets *et al.* 1979). Dworkin and Eide (1983) subsequently were unable to reproduce these initial positive results with either apparatus. The latter investigators have used three other approaches to examine the response of *M. xanthus* to concentration gradients of amino acids, peptide mixtures, cell extracts, and various nucleotides. No evidence for chemotaxis was found for any of the putative attractants.

Chemotaxis studies on the relatively fast moving, flagellated, enteric bacteria indicate the presence of a temporal sensing mechanism or "memory" (Brown and Berg 1974; MacNab and Koshland 1972). But, as Dworkin and Eide (1983) have suggested, temporal discrimination of chemical gradients by myxobacteria would appear to be impossible in light of the fact that the velocity of gliding is from two to three orders of magnitude slower than that of swimming of flagellated bacteria; gradient "molecules are moving past the cell faster than the cell itself is moving."

However, it is noteworthy that an analysis by Berg and Purcell (1977) indicates that in principle a strictly spatial sensing mechanism for swimming bacteria is feasible. Similarly, such a mechanism should not be ruled out for the myxobacteria, particularly in view of their relatively great length.

## Phototaxis

Swarms of two of five unspeciated isolates of myxobacteria from pond mud demonstrated negative phototaxis, gliding in a direction away from a light source; single cells of these strains did not respond to light (Aschner and Chorin-Kirsh 1970).

Fruiting body formation of *Stigmatella aurantiaca* is markedly stimulated by visible light, but whether the effect is due to enhancement of gliding is not known (Qualls *et al.* 1978b).

## Elasticotaxis

Over 40 years ago, Stanier (1942) noted the regular alignment of *Myxococcus* fruiting bodies in parallel array, perpendicular to the line of inoculation. In exploring the basis of this behavior he found that myxobacteria glided parallel to stress lines in the agar, a phenomenon he termed *elasticotaxis* (Fig. 7.6). Presumably, direction of movement was being guided by the ultrastructural

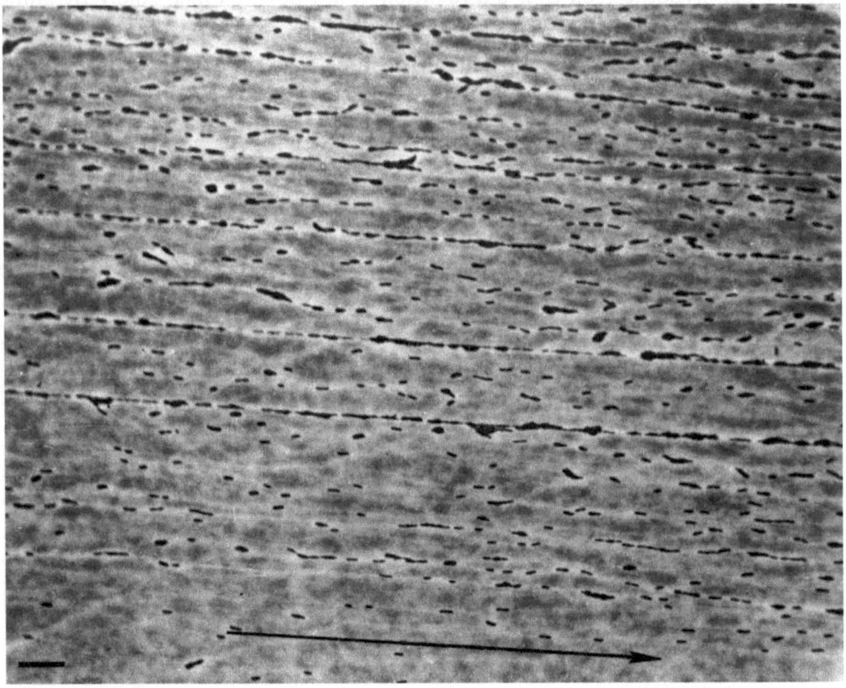

**Figure 7.6.** *Myxococcus xanthus* RB5 gliding on a nutrient gel (Gelrite) scored (direction indicated by arrow) with 3 $\mu$m aluminum oxide abrasive paper. Bar represents 25 $\mu$m.

organization of the substratum. *Archangium* (and presumably other myxobacteria) also glide in a direction dictated by the substratum (Reichenbach 1965b). Finally, Dworkin's (1983) observations of attraction of *Myxoccus* to *Micrococcus* cells and to inanimate particles may be explained by elasticotactic guidance.

## Slime Trail Following

Slime trails play an important part in the guidance of the motion of other myxobacterial cells (Bonner 1952 and others). These trails, which mark the paths traversed by gliders, are easily visualized by phase contrast microscopy by virtue of refractive index differences on relatively high concentration (2–3%) agar gels or on Gelrite (Kelco) (Fig. 7.1). Although definitive evidence is lacking, it is likely that trails are, in fact, depositions of extracellular slime (Bonner 1952; Reichenbach 1965a, 1970; Kühlwein 1957). When a myxobacterium gliding on an untrailed area encounters a trail deposited by another cell(s) at a relatively shallow angle, it will typically glide onto it, along it, and sometimes off it again. Once on the trail, gliding velocity increases (Reichenbach 1974b). Kaiser and Crosby (1983) have suggested that the increase in likelihood of a cell encountering a slime trail and thus gliding faster may account for their observation that swarms of an $A^+S^-$ strain of *M. xanthus* demonstrate a rate of spreading that is cell density-dependent. This is so even though $A^+S^-$ cells do not require cell–cell contact for motility, as do $A^-S^+$ cells.

Myxobacterial aggregation may be explained by invoking slime trail following. "Aggregation could begin when cells that are moving in straight lines in the slime trails begin to move in tight circles. Other cells in the connecting slime trails would accumulate in the areas of circular motility as a result of following the preexisting, circular slime trails. As cells accumulate, they would begin to move in a circular pattern on top of each other, as well as beside each other. In this way, small centers of aggregation could form without chemoattractants" (White 1981).

Since extracellular slime has a fibrillar substructure (Kühlwein 1957; Burchard and Loch unpublished results; Fig. 7.7), trail following may be a

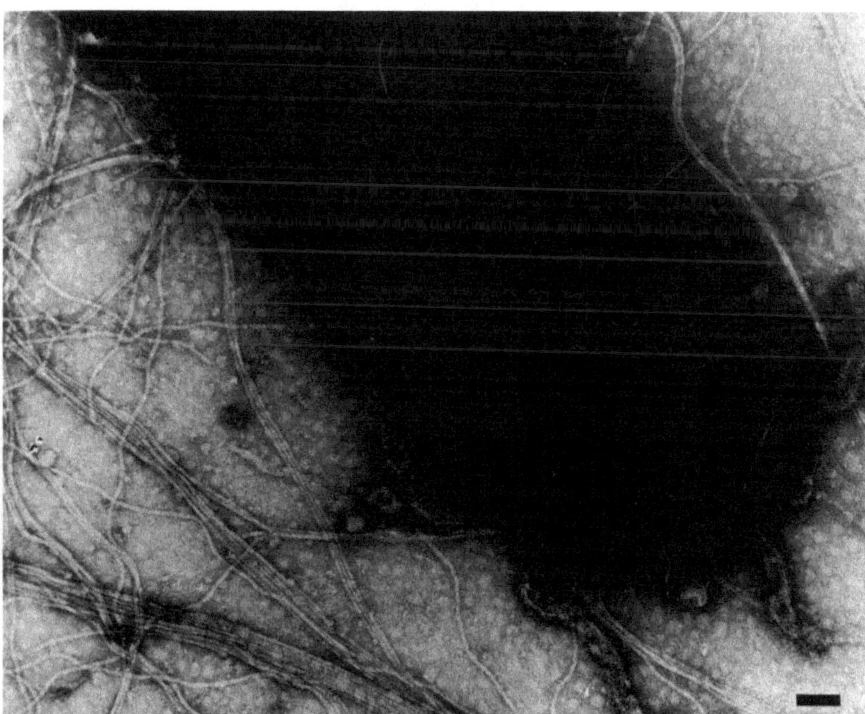

**Figure 7.7.** Slime fibrils on agar over which *M. xanthus* cells had been gliding. An electron microscope grid was pressed onto the agar surface, lifted off, and negatively stained with phosphotungstic acid. Dark area is a cell. Bar represents 100 nm.

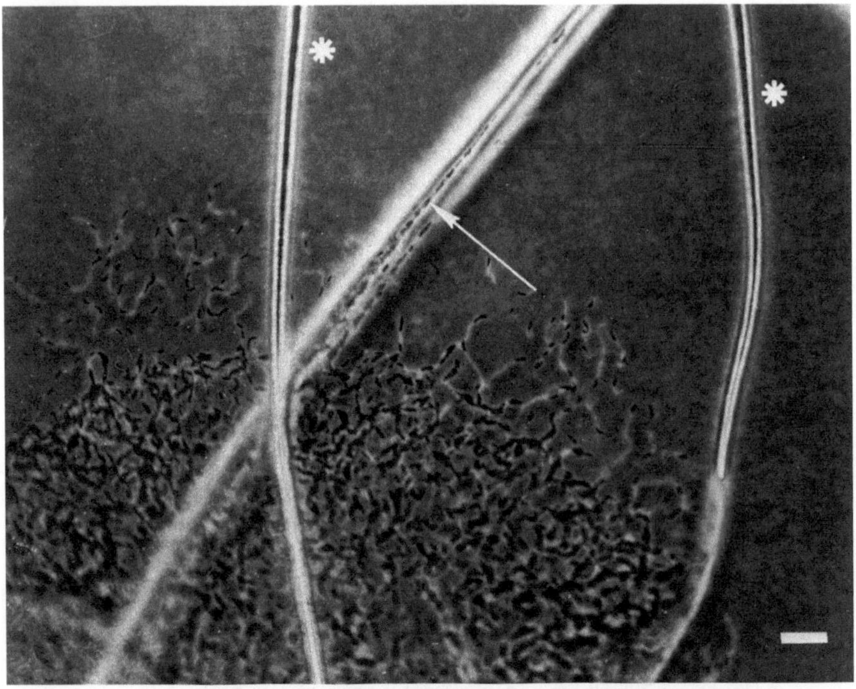

**Figure 7.8.** *Myxococcus xanthus* RB7 spotted onto and gliding along (arrow) a trail produced by an *Oscillatoria animalis* trichome (*) on Gelrite. Bar represents 25 $\mu$m.

mechanical phenomenon, a manifestation of elasticotaxis (Bonner 1952; Burchard 1982b). It has also been suggested that the trails reflect depressions in the agar over which cells have glided (Meyer-Pietschmann 1951; Reichenbach 1965b). Such grooves could also provide elasticotactic guidance. However, the existence of such grooves has been questioned; they may represent optical artifacts (Burchard 1982b).

Slime trails, albeit chemically distinct among the species of gliders (Burchard 1981), apparently provide no species-specific information, as manifested in guidance of gliding. Bacteria will glide on trails deposited by unrelated gliders as well as those of their own species (Burchard 1982b; Fig. 7.8). This behavior may be of value to those myxobacteria that prey on other gliders (Daft and Stewart 1973), allowing them to track their prey.

## Conclusion

The mechanism of gliding motility of the myxobacteria remains enigmatic. Results from fine structural, experimental, and theoretical approaches to the phenomenon have lead to distinctly different models, as have extensions of studies of other gliders that may have distinct motility mechanisms. What links the "gliding bacteria" may, after all, be only their ability to translocate without flagella on surfaces. Ongoing explorations of myxobacterial fine structure and genetics and the use of biochemical, biophysical, and molecular biologic tools should permit us to distinguish among the models discussed in this chapter. It is reasonable to propose that one motility mechanism powers not only gliding but the other surface-associated movements characteristic of these bacteria: flexing, swinging, and nutation.

Although gliding appears not to be directed by chemotaxis and only limited evidence exists for phototactic guidance, the movement of the myxobacteria should not be viewed as random wanderings of individual cells. They display elasticotactic behavior and a variety of coordinated movements—swarming, rhythmic rippling, and aggregation and fruiting body formation—all of which are important to the lifestyle of these complex prokaryotes. In addition to the guidance provided by such cell–cell interactions, individual pioneer cells provide long-distance guidance to follower cells via the slime trails they deposit. The slime itself may form complex associations with polymers in the surfaces on which the myxobacteria glide, thus providing the actual substratum on or in which gliding occurs. Slime may also account for cell adhesion to surfaces and may have plastic properties that have been postulated to be required for the expression of the mechanism of gliding motility.

## Acknowledgments

I gratefully acknowledge the contributions of my students and technicians, the translations of the German literature by my father, the manuscripts made available by Martin Dworkin and Dale Kaiser, and the support of the National Science Foundation.

# 8

# Genetics of Myxobacteria

Dale Kaiser*

Myxobacteria offer the opportunity to investigate the multicellular state. For many years myxobacteria were mistaken for fungi because their fruiting bodies have a quite complex morphology (Thaxter 1892). Like other bacteria, myxobacteria can grow and divide as independent cells; they also behave as a primitive multicellular organism in which cells feed and move in multicellular units and form multicellular fruiting bodies. The prokaryotic structure of myxobacteria renders the communal properties of cells readily accessible to experimental study.

The multicellular nature of myxobacteria is clearly illustrated in the phenomena of rippling and swarming and in the formation of fruiting bodies. Fruiting bodies arise in response to starvation. Many thousands of cells move to their chosen aggregation center where they arrange themselves in heaps and bundles in ways that are unique to each myxobacterial genus (see Chapter 1). Finally, some cells differentiate into dormant, nonmotile, thick-walled myxospores, whereas other cells lyse.

Rippling was first observed in time-lapse microphotographic studies of fruiting body development in *Myxococcus* (Reichenbach 1966) and was later observed in *Stigmatella* and *Archangium* (Reichenbach and Dworkin 1981b). Ripples (Fig. 8.1) consist of discrete alternating bands of oriented cells separated by bands depleted of cells (Shimkets and Kaiser 1982). They move rhythmically with a period of about 15 min, progressing in trains having a wavelength of 50–100 $\mu$m, or about 10 cell lengths.

Myxobacterial cells also move coordinately in multicellular units known as swarms. Swarming appears as a thin zone of cells at the edge of a colony (Fig. 8.2). This zone expands progressively outward by establishing a lacy network

*Department of Biochemistry, Stanford University, Stanford, California 94305 U.S.A.

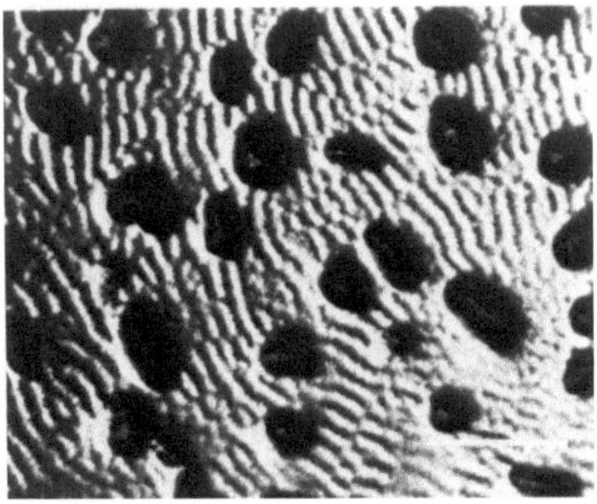

**Figure 8.1.** Ripples formed spontaneously in the early stages of fruiting body development. The ripples are evident as equispaced ridges among the immature fruiting bodies (dark masses). The scale bar is 400 μm long. From Shimkets and Kaiser (1983a).

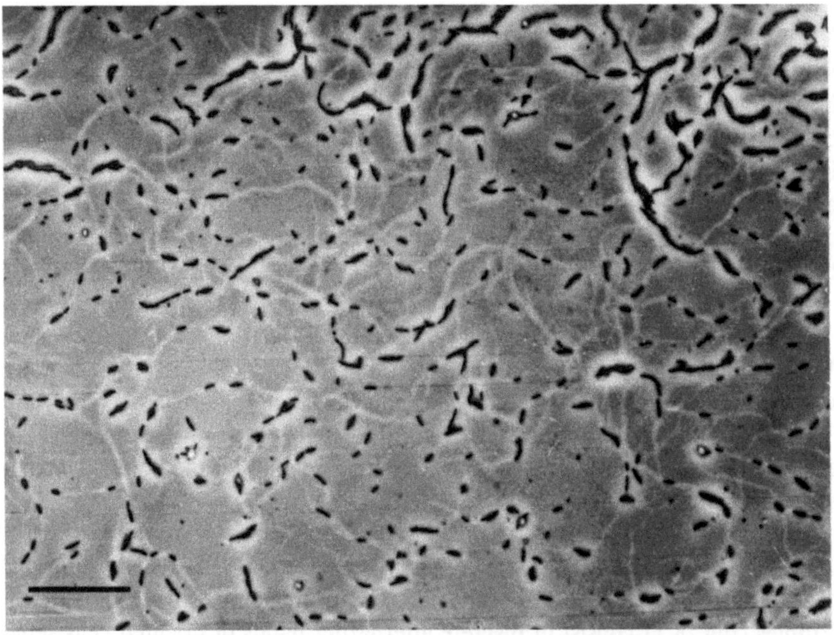

**Figure 8.2.** Swarming at the edge of a colony on agar. A colony located above the photographic frame is expanding outward (downward in the photograph). Individual cells and clusters of moving cells are evident. The bright lines are trails of slime deposited on the agar by moving cells. Cells tend to follow slime trails. The scale bar is 35 μm long. From Kaiser, D., Epoel D: Regulation of cell activity at fertilization by intracellular $Ca^{+2}$ and intracellular pH. In Lerner RA, Bergsma D (eds): "The Molecular Basis of Cell-Cell Interaction." New York: Alan R. Liss for the National Foundation-March of Dimes, BD:OAS 14(2): 377–388, 1978. Photograph by J. Hodgkin.

of cells, usually one cell layer thick, then later filling in the gaps (Kaiser and Crosby 1983). In swarms formed by the germination of spores from a cyst of *Chondromyces apiculatus,* Reichenbach (1966) observed that when a single cell emerged from the main mass of cells, it usually rejoined the rest after a few minutes. What keeps myxobacterial cells together and how are the movements of individual cells coordinated in ripples and swarms?

The process of fruiting body development raises other questions of coordination and control in multicellular units. How is fruiting body development simultaneously started in many cells? Is there a precise sequence of biochemical events underlying the morphologic changes during fruiting? If so, how is that series regulated in time and in space? How is cell movement directed into and within the nascent fruiting body? What changes in that sequence account for the morphologic differences between the fruiting processes in the several genera of myxobacteria? How is sporulation deferred until the cell movements necessary for fruiting body morphogenesis have occurred?

It seems likely that a combination of biochemical, cellular, and genetic experiments will be needed to answer these questions. Fortunately, the simple structure of myxobacterial cells, the modest size of their 5700-kb genome, and an extensive knowledge of the biochemistry and genetics of other bacteria possibly applicable to myxobacteria should facilitate the necessary experiments.

Several genetic tools are prerequisite to such an experimental program. Standard reference strains with known developmental or other multicellular behaviors are required. It must be possible to isolate mutants and to move the new mutations into the standard strains for careful analyses of their phenotypes. Transfer to a new strain allows a separation of mutations of interest away from adventitious mutations that might confuse the phenotype, and an efficient mode of gene transfer would be most advantageous. In order to saturate genes controlling particular functions with mutations, we must have effective mutagenic agents and must be able to handle many colonies. Complementation tests that divide mutations into units of genetic function require a means to form partial diploids. Tests of epistasis, important in the analysis of pathways and of regulatory interactions, necessitate the construction of multiple mutants without prior knowledge of the phenotype of the multiple mutant (Jarvik and Botstein 1973; Hartwell *et al.* 1974). Methods for cloning myxobacterial genes in *Escherichia coli* and for returning them to myxobacteria are needed for a molecular analysis of the genes and their products.

Although the genetics of myxobacteria had been underdeveloped previously compared to the enteric bacteria, the required genetic tools listed above are now available. Advances in transposon genetics (Kleckner *et al.* 1977; Botstein and Maurer 1982), as well as more classic genetic methods (Hayes 1968), provide essentially all the necessary techniques. To the extent that myxobacteria are typical of interesting but genetically underdeveloped microbes, the set of genetic tools described here may also be relevant to the investigation of other noncoliform bacteria.

**Table 8.1.** Transducing Phages

| Phage prototype | Head diameter (nm) | DNA (kb) | Guanine plus cytosine (mol %) | References |
|---|---|---|---|---|
| Mx4 | 67 | 62[a] | 70 | Campos *et al.* 1978; Geisselsoder *et al.* 1978 |
| Mx8 | 58 | 56 | 70 | Martin *et al.* 1978 |
| Mx9 | 60 | 60[b] | | Martin *et al.* 1978 |
| P1 | 85×115 | 98 | 46 | Ikeda and Tomizawa 1965 |

[a]Revised estimate based on published DNA length ratio of Mx4 to PM2 DNA of 6.02 (Geisselsoder *et al.* 1978), length ratio of PM2/λ of 0.213 (Kriegstein and Hogness 1974) and 48.5, the length of λ DNA (Sanger *et al.* 1982).
[b]Estimate based on head size (Martin *et al.* 1978).

## Transducing Phages and Transduction

Three major groups of transducing myxophages and one coliphage that transduces myxobacteria are known. The prototypes of these phage are described in Table 8.1.

### Phage Mx4

Mx4 was the first transducing myxophage isolated and came from a mixture of soil and manure (Campos *et al.* 1978). Mx41 and Mx43, which are structurally and serologically closely related to Mx4, are also transducing phages, and were discovered in carrier strains of *Myxococcus* (Martin *et al.* 1978). By analogy with Mx8 (below), it is possible that the Mx4 family of phages is temperate, but no lysogenic bacteria have yet been isolated from infected cells (Campos *et al.* 1978; Geisselsoder *et al.* 1978). The natural host range of Mx4, Mx41, and Mx43 appears to be limited to a set of nonfruiting mutants of *Myxococcus xanthus,* but the limitation is at a stage beyond adsorption (Campos *et al.* 1978) and can be overcome by a phage mutation, hrm-1, isolated by Wolfner (unpublished). The capacity of Mx4 for general transduction was shown by transduction of 11 different auxotrophic mutants to prototrophy, by transduction of three different antibiotic resistances (Campos *et al.* 1978) and by transduction of many different insertions of Tn*5* as described below.

Mx4 deoxyribonucleic acid (DNA) is terminally redundant. Limited digestion of the DNA with exonuclease III followed by annealing gave unit length circular DNA molecules with single-stranded tails, implying that the same DNA sequence resides at both ends (Geisselsoder *et al.* 1978). The mean length of redundant DNA is 7% of a genomic unit. Cyclic permutation of Mx4 DNA is suggested by the presence of a few circular molecules in denatured and annealed DNA preparations from Mx4 phage not exposed to exonuclease

III, although permutation was not favored by the authors in their description of the circles. A mutation to high frequency transduction has been combined with another mutation to temperature sensitivity that reduces the killing of transductants to give a transducing strain of high efficiency (Geisselsoder *et al.* 1978).

**Phage Mx8**

Mx8, Mx81, and Mx82 were isolated from carrier strains of *Myxococcus* sp. (Martin *et al.* 1978), and particles of similar morphology have been seen occasionally in cultures of *M. xanthus* FB (McCurdy and MacRae 1974). The host range of these phages includes both fruiting and nonfruiting strains of *M. xanthus* (Martin *et al.* 1978). The capacity of Mx8 for generalized transduction in *M. xanthus* has been demonstrated by transduction of 30 different motility loci (Hodgkin and Kaiser 1979a,b), two different fruiting loci, and many different insertions of Tn*5* into *M. xanthus,* all at frequencies of $10^{-5}$–$10^{-7}$ per particle. Plaque-forming and transducing particles of Mx8 have the same buoyant densities, and since the density of DNA depends on its guanine + cytosine (GC) content and Mx8 DNA and host DNA have the same mole percent GC, this suggests that the two types of particles contain equal amounts of DNA (Martin *et al.* 1978). The DNA molecules isolated from purified phage particles have a length corresponding to 56 kb (Martin *et al.* 1978). By digestion of Mx8 DNA with restriction endonucleases, Orndorff *et al.* (1983) have shown that it is circularly permuted. Mx8 is a temperate phage, and comparison of the restriction map of the prophage with that of DNA from phage particles has demonstrated the absence of unique end fragments, the circular permutation of phage DNA, and unique prophage attachment sites in Mx8 and *M. xanthus* (Orndorff *et al.* 1983).

A mechanism for general transduction by Mx8 can be suggested on the basis of the structure of Mx8 DNA and by analogy with coliphage P1 (Ozeki and Ikeda 1968) and *Salmonella* phage P22 (Susskind and Botstein 1978). The fact that Mx8 DNA is circularly permuted but has uniform length implies that DNA is frequently selected for packaging into phage particles without reference to its base sequence and that the amount of DNA packaged is determined by the size of the Mx8 head (Streisinger *et al.* 1967). In a lytic infection of *M. xanthus* by Mx8, a headful of host DNA would occasionally be packaged in place of phage DNA, and a transducing particle having the same DNA content as a plaque-forming particle would be formed. This scheme is consistent with the finding that plaque-forming and transducing particles have the same buoyant density (Martin *et al.* 1978). The resulting phage stock would contain two kinds of particles: a majority type, containing only phage DNA, and a minority type (the transducing particles), containing only host DNA. When recipient bacteria are infected with the phage stock, the transducing particles would inject their piece of (donor) DNA. Donor DNA pairs with the homologous DNA segment in the recipient and donor genes are incorporated by crossing over within the region of pairing. The fact that all genes so far tested have proven to be transducible by Mx8 (and by Mx4), albeit at different frequencies, suggests that any (or almost any) segment of host DNA can be carried in a transducing particle.

No reports of abortive transduction by Mx8 or any other myxophage have appeared.

### Phage Mx9

Myxophage Mx9 is serologically and morphologically distinct from Mx4 and Mx8 and is also a transducing phage (Martin *et al.* 1978). It was isolated from carrier strains of *Myxococcus* and may also be a temperate phage by analogy with Mx8. Its head diameter, 60 nm, is similar to that of Mx8 and Mx4. By analogy with the composition of Mx4 and Mx8, a DNA content for Mx9 in the range of 50–65 kb seems likely.

### Interspecific Transduction by Coliphage P1

Coliphage P1 can absorb to *M. xanthus* and cause its lysis or inhibit its growth (Kaiser and Dworkin 1975). Lipopolysaccharide is the adsorption receptor for P1 in *E. coli* (Lindberg 1973), and *Myxococcus fulvus* possesses lipopolysaccharide resembling that of other Gram-negative bacteria (Rosenfelder *et al.* 1974). Although P1 does not multiply in *M. xanthus*, it does inject its DNA, as shown by the fact that infection by P1::Tn*5* yields kanamycin-resistant strains of *M. xanthus* (Kuner and Kaiser 1981). No infectious P1 particles issue from these cells, and no lysogenic bacteria are formed. The kanamycin-resistant transductants contain Tn*5* DNA sequences but no P1 DNA detectable by the Southern blot technique (Kuner and Kaiser 1981). Specialized transduction thus occurs in the absence of P1 multiplication.

Generalized transduction of *Myxococcus* by P1 has been observed in the transfer of plasmids from *E. coli* to *M. xanthus* (O'Connor and Zusman 1983). In sum, it is possible to introduce DNA in general and transposons in particular from *E. coli*—and possibly from other P1 hosts (Murooka and Harada 1979)—into *M. xanthus* via phage P1. Because P1 is unable to replicate its DNA or to produce phage particles in *M. xanthus* when *M. xanthus* is infected with a stock of specialized transducing P1::Tn*5*, only transposed sequences are reproduced. These conditions facilitate the use of transposon techniques (Kleckner *et al.* 1977) to study *M. xanthus* genetics.

## Transposon Tn*5* in *Myxococcus xanthus*

Tn*5* is a 5.7-kb transposable element originally isolated from *E. coli* (Berg *et al.* 1975). It carries a gene for kanamycin phosphotransferase, whose product catalyzes inactivation of kanamycin to which wild-type *M. xanthus* is normally sensitive. Cells that have Tn*5* inserted in their chromosome are consequently highly resistant to kanamycin. The Tn*5* contains a central segment 2.6 kb long flanked by two end sequences of 1.5 kb each. The first 0.8 kb of the central segment encode for aminoglycoside 3′-phosphotransferase II (Berg *et al.* 1975; Beck *et al.* 1982) and the remaining 1.8 kb of the central segment contains open reading frames, but no protein products have been found (Beck *et al.* 1982). The terminal sequences of 1534 base pairs each are arranged in reverse orientation and have almost identical sequences. These termini identified as IS50L for left end and IS50R for right end are necessary for transposition of the entire element (Berg *et al.* 1981). The IS50L differs from IS50R by a single base pair at position 1442, and this difference generates a rightward promoter in IS50L at which transcription of the aminoglycoside 3′-phosphotransferase II gene is initiated (Rothstein *et al.* 1980). When Tn*5* transposes in *E. coli,* nine bases of host sequence are duplicated in direct orientation (Auerswald *et al.* 1981). When *M. xanthus* is infected with P1 carrying the transposon, Tn*5* can transpose to the *Myxococcus* chromosome, generating kanamycin-resistant strains (Kuner and Kaiser 1981). These kanamycin-resistant transductants are stable in the absence of kanamycin and, as described above, contain DNA sequences homologous to Tn*5*, demonstrated by restriction enzyme analysis and DNA–DNA hybridization; but they have no sequences that are homologous to P1.

The Tn*5* transposes from P1 to many different chromosomal sites in *M. xanthus* (Kuner and Kaiser 1981). In each of the first nine independent kanamycin-resistant transductants examined, the transposon was located in a different fragment generated by cleavage with restriction endonuclease *Eco*RI. Tests of genetic linkage also gave evidence for an approximately random distribution of insertion sites. A method for isolating Tn*5* insertions linked to a mutation of interest has been developed. The method, illustrated in Fig. 8.3, has been used successfully by many workers (Torti and Zusman 1981; Zusman 1982; LaRossa *et al.* 1983; Sodergren and Kaiser 1983). The frequency with which insertions of Tn*5* were found near to any particular locus—about 1%—corresponded to the ratio between the amount of DNA in an Mx8 phage particle (50 kb) and the amount of DNA in the genome of *M. xanthus,* 5700 kb (Yee and Inouye 1981,1982). It is as if at this level of resolution, Tn*5* is able to insert anywhere in the *M. xanthus* chromosome.

Once inserted, Tn*5* remains at the original site of insertion. Initially, 20 strains were tested by restriction analysis during vegetative growth and five during transduction from one *M. xanthus* strain to another (Kuner and Kaiser 1981). Subsequent examination of many other Tn*5* insertions has confirmed their general stability. Tn*5* at one site in *M. xanthus* generates a second copy at another site (intrachromosomal transposition) at a frequency of less than 1/1000 cells (Sodergren and Kaiser 1983; Avery and Kaiser 1983a).

The structure of Tn*5* lends itself to modification by replacement of its unique central segment with other antibiotic resistance genes. These modifications preserve the overall Tn*5* structure but change the encoded antibiotic resistance.

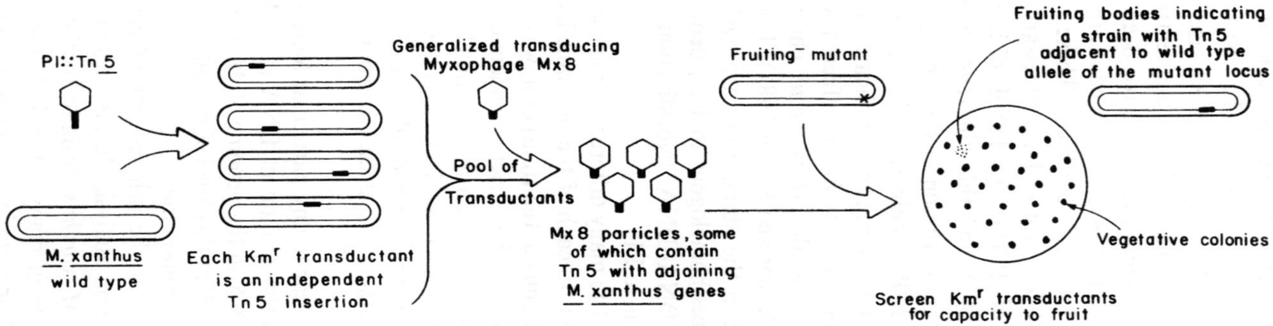


**Figure 8.3.** A method for isolating a Tn*5* insertion linked to a particular mutant locus, illustrated in the figure by a fruiting defective mutant of *M. xanthus*. P1::Tn*5* is bacteriophage P1 into which Tn*5* has transposed. $Km^r$ indicates resistance to kanamycin. From Kuner and Kaiser (1981).

One modified Tn*5*, called Tn*5*-132, encodes resistance to tetracycline and retains almost all of both IS50 units intact (Rothstein *et al.* 1981). Avery has developed a technique for the replacement of Tn*5* by Tn*5*-132 that keeps the original site of insertion (Avery and Kaiser 1983a). In the Avery technique, Tn*5*-132 is introduced into *M. xanthus* by P1::Tn*5*-132. Homologous crossovers occur in each IS50 unit, as these units are the only homologous segments between phage and cell genomes, to generate a replacement (see Fig. 8.4).

If Tn*5*-132 were to transpose to a new site in *M. xanthus* instead of replacing an existing Tn*5*-wild type (wt), the resulting tetracycline-resistant transductant would remain resistant to kanamycin, now having one copy of Tn*5*-wt at the original site and one copy of Tn*5*-132 at a new site. A replacement strain, by contrast, has only one copy of the transposon, a Tn*5*-132, and has by virtue of the loss of Tn*5*-wt become sensitive to kanamycin (Avery and Kaiser 1983a). The Tn*5*-132 now occupies the same site that Tn*5*-wt did originally, which is shown in three ways. First, Tn*5*-wt in the original strain and Tn*5*-132 in the replacement strain behave as genetic alleles (Avery and Kaiser 1983a; Sodergren *et al.* 1983). Second, Tn*5*-132 is found linked to the same mutations in *M. xanthus* as Tn*5*-wt was originally (Avery and Kaiser 1983a). Finally, the restriction pattern of *M. xanthus* DNA in the vicinity of the ends of Tn*5*-132 in the replacement strain is shown to be the same as it was in the vicinity of the ends of Tn*5*-wt in the original strain (Avery and Kaiser 1983a).

The presence of a versatile transposon such as Tn*5* in *M. xanthus* opens many possibilities for genetic manipulation and analysis. Some of these possibilities have already been realized and are discussed in the following sections. Others, such as the construction of specialized transducing myxophages, saturation transposon mutagenesis of particular genes, construction of deletions and translocations, and mapping around the entire chromosome, remain to be carried out. Combinations of in vivo genetic engineering with transposons (Kleckner *et al.* 1977) and in vitro constructions using restriction enzymes are beginning to be explored and are also discussed.

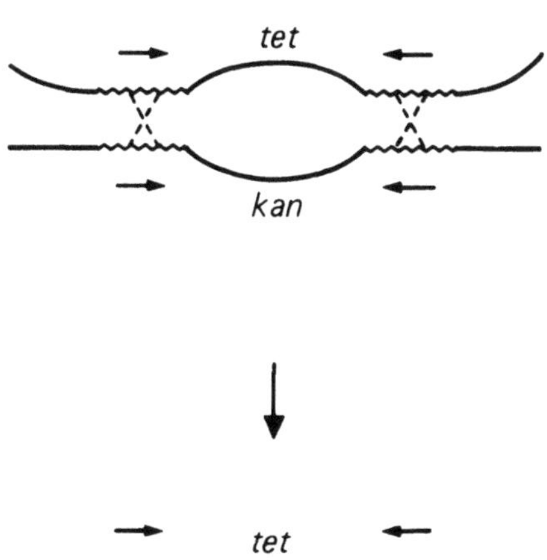


**Figure 8.4.** Replacement of Tn*5* (encloding kanamycin resistance) with Tn*5*-132 (encoding tetracycline resistance) by homologous recombination (dashed crosses) in the IS50 units represented by wavy lines. Short arrows indicate the orientation of IS50 units at the ends of Tn*5*. *Myxococcus xanthus* containing Tn*5* is infected with P1::Tn*5*-132, and tetracycline-resistant, kanamycin-sensitive transductants are selected. From Kuner *et al.* (1981).

## Gene Mapping by Cotransduction

A chromosome map can, in principle, be constructed entirely from results of cotransduction experiments. When a donor and recipient strain for generalized transduction differ by two or more genetic markers, the simultaneous transduction of any pair of donor markers depends on the distance that separates them. Only markers that are separated by less than one headfull unit of DNA can be included in the same transducing particle and be transduced together. Transducing particles of Mx8, for example, carry 56 kb of DNA, therefore the cotransduction of a pair of markers by Mx8 implies that they are closer than 56 kb, or about 1% of the bacterial chromosome.

Encapsulation of two or more different DNA segments in the same bacteriophage particle, which would lead to spurious indications of genetic linkage, is generally not found. No cotransduction was observed among 13 different insertions of Tn*5* and the motility loci *mgl-1, tgl-3, cglB2, cglC2, cglD1, cglE1,* and *cglF1* (Sodergren and Kaiser 1983). Similarly, linkage between a group of 35 loosely linked frizzy aggregation-defective mutants and an outlying Tn*5*, Ω3080, has been tested by Zusman (1982). Between each pair of unlinked markers less than 1% cotransduction was observed. Cotransduction was also sought among seven different unlinked insertions of Tn*5*, but from a total of 5246 transductants only one potential joint transductant was found (Cheng, unpublished observations). The frequency of spurious cotransduction due to incorporation of two molecules thus appears to be 0.02%, or less.

Because Tn*5* can transpose to many different sites in *M. xanthus*, the transposon can be used to focus genetic mapping on any interesting region of the chromosome, as diagrammed in Fig. 8.3. With one or more Tn*5* insertions linked to the same region, cotransduction between Tn*5* and other mutations can be tested. Application of this procedure has demonstrated the existence of clusters of independent mutations and thus defined genetic loci for *mgl, cglB, cglC, cglD* (Sodergren and Kaiser 1983), for "rough" nonfruiting mutants, (Torti and Zusman 1981), for "frizzy" nonfruiting mutants (Zusman 1982), and for *spoC* and *spoD* mutants (LaRossa *et al.* 1983).

Cotransduction could in principle be used to map the entire chromosome of *M. xanthus*. A set of about 60 appropriately chosen insertions of Tn*5* could cover the whole chromosome in cotransducible segments.[1] Once insertions have

[1]If transductional crosses sensitive enough to detect cotransduction frequencies as low as 1% were performed with Mx8, then each Tn*5* insertion could be cotransduced with any genetic locus within 45 kb on either side of the site of insertion [See discussion of equation (1) below.] Thus each insertion would cotransduce with loci within a particular 90-kb segment, and a minimum of 63 different Tn*5* insertions would be required. This value is derived by dividing the total genome size of 5700 kb (Yee and Inouye 1982) by the 90 kb size of each segment. In practice, an equal spacing of Tn*5* insertions would be difficult to obtain, and for a random set of Tn*5* insertions, at least 375 strains would be necessary (Avery 1983).

been placed in order of their chromosomal position, transducing phage stocks grown on each of the insertion strains could be used to map a new mutation. A new mutant strain would be transduced by appropriately constructed pools of transducing stocks until, by successive approximation, the segment that included the new site would be identified.

The cotransduction of a pair of genetic markers shows that they are in the same chromosome segment. Beyond simply grouping markers into segments, the frequency of cotransduction provides an estimate of the physical distance between the two members of a marker pair. Two cotransducible markers have three independent chances of being separated from each other in a cotransduction experiment: one chance when a DNA segment from the donor is packaged into the transducing phage particle and two more chances when two crossovers incorporate a fragment of DNA from the transducing particle into the recipient chromosome. Because the probability of genetic separation should increase with the physical distance for each of the three steps in transduction, the frequency of cotransduction should be very sensitive to distance.

Assuming that the probability of separating two markers at each of the three steps in transduction is directly proportional to the physical distance ($d$) between them, Wu (1966) showed that the cotransduction frequency ($C$) should be

$$C = (1 - d)^3 \tag{1}$$

Sodergren *et al.* (1983) measured the cotransduction frequency for two different pairs of Tn*5* insertions whose physical separation was known from mapping with restriction enzymes. Cotransduction frequencies that were calculated using Wu's equation on the measured distances of 16.5 and 10.8 kb agreed with the measured cotransduction frequencies within experimental error (Sodergren *et al.* 1983). The agreement suggests that where physical distances are not available, two point cotransduction frequencies can be used to estimate them. The estimation should be taken with caution because it does not include the effects of finite marker size or gene conversion (Shimkets *et al.* 1983).

Three factor crosses in which Tn*5* is one of the markers have been used to place a series of mutations that have related phenotypes in map order. Torti and Zusman (1981) and Zusman (1982) mapped "rough" and "frizzy" aggregation-defective developmental mutants, finding in both cases that the mutations lay in clusters and that the distance between the most distant members was the order of 10 kb. Sodergren and Kaiser (1983) used three-point tests to order *cglB, rif,* and *tgl;* this order agrees with that derived from frequencies of segregation from tandem duplications (Avery and Kaiser 1983b). However, the reliability of three-point crosses as tests of order appears to break down when the markers are closely linked. Stephens (unpublished experiments) found that *mgl* mutations that behaved in three-factor crosses as if they were on opposite sides of Tn*5* at Ω1901, actually all mapped on a DNA segment cloned from one side of the Tn*5*.

## Tandem Duplications

Tandem genetic duplications arise spontaneously in *M. xanthus* as they do in other organisms (Anderson and Roth 1977). Figure 8.5 shows that a tandem duplication contains one new DNA sequence not present in the parental strains, called the "novel joint." Although duplications arise frequently during growth of a culture, they also decay to the original single copy state by homologous recombination between duplicated parts, for example, between b and b′ or between c and c′ in Fig. 8.5. Therefore, unless duplications are selected in some way they will be lost. Selection for duplications may be natural as well as experimental and, in fact, wild-type *M. xanthus* has evolved a duplication of the structural gene for protein S (Inouye *et al.* 1983).

Tandem duplications can be used for tests of gene dominance and for complementation tests, which define genetic units of function. In *M. xanthus,* where F′ strains are not available, tandem duplications can serve many of the same purposes (Bachmann and Low 1980). Tandem duplications also arise during certain cloning operations described in the next section.

Three properties that characterize a tandem duplication can be used to establish its structure: segregation, restriction map, and transduction of the novel joint. Tandem duplications are unstable: Homologous recombination

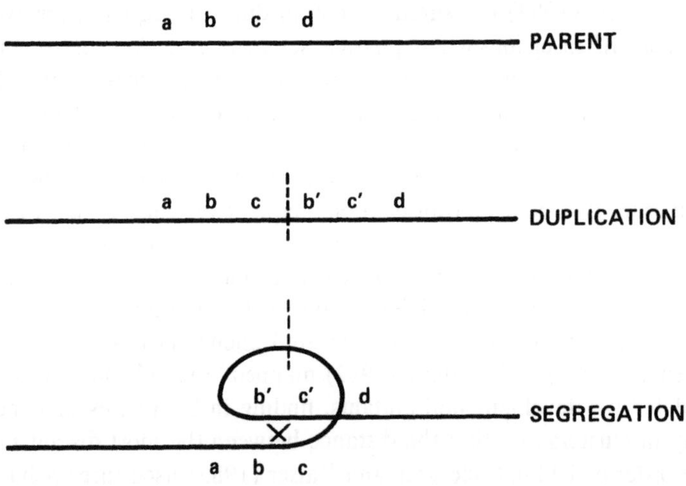


**Figure 8.5.** Formation and decay of a tandem genetic duplication. The first line shows the parental sequence and the second line a duplication of the b–c segment. The duplicate copy is indicated by primed letters (i.e, b′, c′). The dashed verticle line represents the novel joint, the only unique sequence in the tandem duplication. A tandem duplication decays (segregates) by homologous recombination indicated by a cross anywhere in the duplicated segment, that is, between primed and unprimed letters. No matter where the recombination occurs in the duplicated segment, the novel joint is lost and the segregants are stable.

between the duplicated parts excises a circular DNA molecule and restores the unduplicated sequence (Fig. 8.5). Segregation always eliminates the novel joint, and the resulting segregant is stable. It is, in fact, a completely normal chromosome. When the duplicated copies carry different alleles of a locus, one allele at each duplicate locus is lost during segregation. Segregants may carry parental combinations of alleles or new combinations. The frequency of loss of b or b′ and of c or c′ in Fig. 8.5 is directly proportional to their distance from the novel joint. The map position of the novel joint can thus be determined by measuring the frequency of each segregant class.

Restriction mapping helps to establish the structure of duplication strains. Comparison of the restriction map of the parental strains in the vicinity of the ends of Tn*5* with a duplication strain can be made with radiolabeled Tn*5* as the DNA-hybridization probe (Kuner and Kaiser 1981). The Tn*5* "end" restriction fragments contain both Tn*5* DNA and adjacent chromosomal DNA, thus identifying the locus of Tn*5* insertion. The Tn*5*-wt and Tn*5*-132 at allelic sites, at nonallelic sites, and one or the other at the novel joint of a tandem duplication give predictable sets of Tn*5* end fragments (Avery and Kaiser 1983a,b). The particular set of end fragments observed from the duplication strain and from its segregants gives detailed information about the structure of the duplication.

The novel joint of a tandem duplication uniquely determines its structure (Fig. 8.5). If a novel joint sequence is transduced into a new strain, a duplication of the same chromosome segment as the donor is recreated in the transductants (Anderson and Roth 1977). Even duplications that are longer than the amount of DNA packageable in a single phage particle can be transduced into new strains because the duplicated DNA arises from the recipient. Transduction of a novel joint not only helps to characterize a tandem duplication, it also is important in constructing partial diploid strains having the genetic composition necessary for dominance and complementation tests.

Genetic duplications in *M. xanthus* have been isolated in two different ways, using insertions of Tn*5*. Both ways are thus applicable to any region of the genome in which Tn*5* insertions are located, as well as to any organism with transposons and a method for gene transfer. One way, called tandem duplication trapping, is the simultaneous selection in a transduction experiment of two different alleles that occupy the same chromosomal site. Only a duplication provides two copies of the same site. Different forms of Tn*5* can be incorporated at the same site by replacement, as described above (Transposon Tn*5* in *Myxococcus xanthus*). A transduction is then performed from a donor containing Tn*5*-wt ($Km^r$), for example, into a recipient that contains Tn*5*-132 ($Tc^r$) at the same site, the latter obtained by replacement of the former and transductants having both $Km^r$ and $Tc^r$ are selected (Avery and Kaiser 1983a). The duplications are distinguished from phenotypically similar transpositions by segregation and by their restriction maps. One particular duplication was trapped among approximately 10,000 $Km^r$ transductants (Avery and Kaiser 1983a). The frequency of $Km^s$ and $Tc^s$ segregants measured the distance between the site of Tn*5* insertion and the novel joint. The duplication spanned at least 45 kb, a length too great to be packaged in a single transducing particle. Nevertheless, a new duplication with the same properties as the old was generated by transduction of the novel joint to a new strain (Avery and Kaiser 1983a).

A second way, called construction, permits the formation of tandem duplications with predetermined end points. By "trapping" any duplication that covers a predetermined site can be isolated; these duplications may have random endpoints, provided only that they flank the predetermined site. To construct a tandem duplication, Tn*5*-wt is placed at one end of the segment to be duplicated, and Tn*5*-132 is placed at the other end in the same strain. The desired duplication forms during growth of this double insertion strain by unequal recombination between the two transposons (Avery and Kaiser 1983b).

To visualize the appropriate recombination, imagine a cell that has recently replicated the chromosomal segment containing both transposons. Homologous recombination between the distal Tn*5* in one replica and the proximal Tn*5* in the other generates two recombinants. One recombinant is the desired tandem duplication for the segment between the transposons, and the other is a deletion of the same segment. The duplication is then isolated from the mixture by transduction into a new strain, selecting for the copy of Tn*5* at the novel joint. Duplications of the segment between Tn*5* inserted at Ω925 and Ω2224 were found in a double insertion strain at a frequency of about 1%. These duplications, which span the *rif, tgl, cglB* region, show the expected segregation and restriction maps around their constituent Tn*5* insertions. Using the duplication, it was shown that $rif^s$ was dominant to $rif^r$ as in *E. coli*, that $cglB^+$ was dominant to $cglB^-$, and that $tgl^+$ was dominant to $tgl^-$. The latter two dominance relationships are consistent with the ability of *cglB* strains to be stimulated by $cglB^+$ strains and of *tgl* strains to be stimulated by $tgl^+$ (Hodgkin and Kaiser 1979b).

## Cloning Myxobacterial Genes

An enormous variety of cloning strategies has been used for genes and many of them could be used for myxobacteria. Since several reviews of these procedures are available (Davis *et al.* 1980; Maniatis *et al.* 1982), only the two strategies that have already proven successful in the cloning of *M. xanthus* genes in *E. coli* will be discussed here. One uses synthetic oligonucleotides to identify a fragment of *M. xanthus* DNA carrying the coding sequence for protein S (Inouye *et al.* 1983). Another uses a linked insertion of Tn*5* and selection for kanamycin resistance in *E. coli* (Shimkets *et al.* 1983; O'Connor and Zusman 1983).

Inouye *et al.* (1983), knowing the amino acid sequence of the carboxyterminal region of protein S, obtained a mixture of eight tetradecameric oligonucleotides representing all possible DNA sequences that would code for a segment of the known sequence, Met–Asn–Asn–Asn–Thr. They screened several restriction digests of DNA from *M. xanthus* with radioactive mixed tetradecamers for hybridization in Southern gel blots (Southern 1975). They selected a 9.7-kb fragment produced by digestion with restriction endonuclease *Hind*III that hybridized with the tetradecamers. This restriction fragment, cloned in *E. coli* plasmid pBR322, proved to have two copies of a DNA sequence that would specify the amino acid sequence –Ala–Met–Asn–Asn–Asn–Thr–Ser–Ser–Ile– (which is the sequence near the C terminal of protein S) (Inouye *et al.* 1981). The two copies of the protein S gene lie in the same orientation, are separated by 1.2 kb of another sequence, and have several differences in DNA sequence.

The Tn*5*, which came to *Myxococcus* from *E. coli*, expresses kanamycin resistance in *E. coli* (Berg *et al.* 1975) as well as in *M. xanthus*. Consequently, if we clone random DNA fragments in *E. coli* from a *M. xanthus* strain containing Tn*5* inserted near a gene of interest, the resulting kanamycin-resistant *E. coli* strains may also contain the desired *Myxococcus* gene. With this procedure, a 21-kb segment of *M. xanthus* DNA in the *spoC* region (Shimkets *et al.* 1983) and a 37-kb segment in the *aggR* region (O'Connor and Zusman 1983) have been cloned, and several similar selections are in progress (cited in O'Connor and Zusman 1983). The *spoC* and *aggR* clones have been shown to contain the genes of interest by transferring them from *E. coli* back into *M. xanthus* where they complement the corresponding defective *spoC* or *aggR* mutants and produce phenotypically wild-type cells (Shimkets *et al.* 1983; O'Connor and Zusman 1983). In addition, Gill (unpublished) has developed an in situ cloning vector that includes a fragment of Tn*5*-wt. When it is transduced by P1 from *E. coli* into *M. xanthus,* it integrates by sequence homology into the left or right IS50 of a resident Tn*5*-132. The vector design ensures that cleavage of DNA from the kanamycin-resistant transductant releases a single DNA fragment containing both the entire vector and *M. xanthus* DNA from one side of the Tn*5* in xanthus. This single DNA fragment can form a circle by a monomolecular reaction. The *spoB* (Gill unpublished), *spoD* (Gill and Cheng, unpublished), and *mgl* (Stephens, unpublished) have been cloned in this way.

Two different methods, both of them efficient, return the cloned genes from *E. coli* to *M. xanthus;* they are specialized transduction by P1 and generalized transduction by P1. Transfer by specialized transduction was developed by Gill (Shimkets *et al.* 1983); he derived a cloning vector by adding a 6.7 kg *Eco*RI fragment encoding P1-specific incompatibility to pBR322 (Sternberg *et al.* 1981). When an infecting P1 establishes lysogeny in an *E. coli* strain carrying Gill's plasmid, P1 cannot lysogenize in the usual way due to the preestablished incompatibility (Hedges *et al.* 1975) and instead forms a cointegrate with the plasmid by recombination in the homologous P1-incompatibility sequences. Upon induction to lytic growth, P1 begins packaging DNA within the P1 segment of the cointegrate, and many of the resulting phage particles contain the entire plasmid (Shimkets *et al.* 1983; Gill, unpublished). This P1 stock transfers the plasmid to *M. xanthus* by specialized transduction, and when the chimeric P1-plasmid DNA is injected into *M. xanthus,* homologous recombination between P1 incompatibility sequences regenerates the original circular plasmid. The plasmid apparently cannot replicate in *M. xanthus,* but a single homologous recombination anywhere within its inserted *M. xanthus* DNA integrates the plasmid into the host chromosome as shown in Fig. 8.6. This is the only event that can give a $Km^r$ transductant because the *M. xanthus* insert terminates within Tn*5*. The entire plasmid integrates into the chromosome, duplicating in tandem those sequences common to the cloned DNA and to the chromosome. The structure of the resulting tandem duplications has been analyzed by restriction enzymes and by segregation and has been found to conform to the expectations shown in Fig. 8.6 (Shimkets *et al.* 1983). *M. xanthus* DNA segments as large as 40 kb and as small as 1.5 kb have been cloned in *E. coli* and returned to *M. xanthus* in this way.

The route from *E. coli* back to *M. xanthus* via generalized P1 transduction was developed by O'Connor and Zusman (1983), and was also observed by

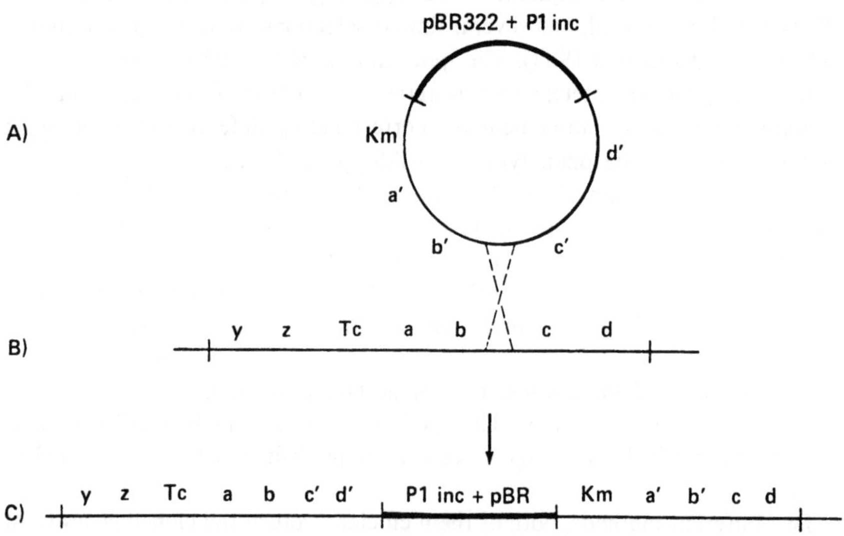


**Figure 8.6.** Integration by homologous recombination of a circular DNA element transduced from *E. coli* into *M. xanthus* by bacteriophage P1. Line A shows a plasmid with vector sequences, indicated by a heavy line, and cloned *M. xanthus* sequences (light line) a′b′c′d′ with adjacent Tn*5* (Km). Line B shows an *M. xanthus* recipient with Tn*5*-132 (Tc) at the same site occupied by Tn*5* in the donor, generated as described in Fig. 8.4, and *M. xanthus* sequences abcd homologous to the cloned segment. Recombination (dashed lines) between the circular DNA element from the donor and the recipient chromosome anywhere in the homologous segment generates a tandem duplication (c) with vector sequences at the novel joint. Modified from Shimkets *et al.* (1983).

Gill (unpublished). O'Connor and Zusman (1983) cloned *M. xanthus* genes in *E. coli* on a pBR322 plasmid by selecting kanamycin resistance of a Tn*5* insert linked to those genes, a selection similar to the first method. They discovered that P1 grown on *E. coli* that carried the plasmid could tranduce the plasmid to *M. xanthus* at a frequency of $10^{-3}$–$10^{-7}$ kanamycin-resistant transductants per plaque forming P1(O'Connor and Zusman 1983). The frequency of plasmid transduction depended on the size of the plasmid and was higher for a 43-kb plasmid than for 20–30-kb plasmids. One 10.5-kb plasmid was transducible, but another of the same size was not, suggesting that factors other than size may affect transducibility. Some of the kanamycin-resistant transductants were stable, but others were unstable. The unstable ones contained vector–plasmid sequences, as demonstrated by DNA–DNA hybridization. The instability and restriction enzyme cleavage pattern of the unstable transductants suggested that they arose from integration of the plasmid into the chromosome through homology in the cloned *M. xanthus* segment. (This integration event would be similar to that described above for the specialized P1 transduction route illustrated in Fig. 8.6.) The stable transductants contained no sequence hybridizable to the plasmid–vector and behaved like replacements of chromosomal sequences by cloned sequences (O'Connor and Zusman 1983).

When tandem duplications are generated by either specialized or generalized P1 transduction of a plasmid containing *M. xanthus* sequences, gene conversion is frequently observed (Shimkets *et al.* 1983; O'Connor and Zusman 1983). For example, about 20% of the kanamycin-resistant transductants from a *spoC*$^+$ donor into a *spoC*$^-$ recipient have a *Spo*$^-$ phenotype. These transductants are *spo*$^-$/*spo*$^-$ partial diploids, having restriction fragments characteristic of both the donor and the recipient. They segregate like the 80% phenotypically *Spo*$^+$ majority except that all of their segregants are *spo*$^-$. Shimkets *et al.* (1983) suggested that in these cases the *spo*$^+$ allele of the donor was gene-converted to *spo*$^-$ by a mismatch repair that used the *spo*$^-$ strand in a *spo*$^+$/*spo*$^-$ heteroduplex as repair template. In these experiments, a + allele can be gene-converted to a −, or a − allele can be gene-converted to a +. Similarly, conversion can go to the donor or to the recipient genotype. The substantial frequency of gene conversion in *M. xanthus* is consistent with the high frequency of repair of ultraviolet-induced lesions observed by Grimm and Herdrich (1980).

## Extracellular Complementation

Cellular interactions can be explored by a type of complementation test that uses mixtures of whole cells. We call such a test extracellular complementation to distinguish it from a genotypic complementation test that is carried out inside a cell. A genotypic complementation test performed in a partially diploid bacterium measures the complementary quality of immediate gene products, ones that have not had to pass through any membrane. Extracellular complementation, since it is observed in mixtures of whole cells, by contrast, concerns substances on the cell surface or substances secreted by cells.

To see the close connection between cell interaction and extracellular complementation, consider an (idealized) interaction and distinguish a source cell from a target cell. Consider further a mutant defective as a source but still able to function as a target; it would be defective in any process that depended on the interaction. Such a mutant mixed with wild-type cells or with a complementary mutant would still be able to interact, and the mixture might have a phenotype approaching that of pure wild type. Extracellular complementation satisfying these expectations has been observed in *M. xanthus* among mutants defective in motility and among mutants defective in fruiting body formation.

Certain nonmotile mutants of *M. xanthus* are conditional mutants of a novel kind: The mutant cells become transiently motile when mixed with nonmutant cells or with cells of a different mutant type (Hodgkin and Kaiser 1977). Direct contact between cells appears to be required, because these cells move when contact is permitted but not when contact is prevented by spatial separation on an agar surface or by the interposition of a membrane filter (Hodgkin and Kaiser 1977). This phenomenon, called motility stimulation, has generated six different types of stimulatable mutants. A brief description of the genetics of gliding in *M. xanthus* may clarify the significance of different stimulation types (Hodgkin and Kaiser 1979a,b; Sodergren and Kaiser 1983). The interested reader will find a description of gliding motility in Chapter 7.

Investigation of more than 100 motility mutants has revealed two components in the movement of wild-type cells, an A component, for adventurous, and an S component, for social. Mutants derived in one mutational step from wild type defective in the A-motility component express pure S motility. These $A^- S^+$ cells move when they are very near to other cells, but they are unable to move when they are more than about one cell's length from any other cell (Kaiser and Crosby 1983). Pure S motility results in a spatial distribution of cells with many cells in clusters and very few isolated single cells. The normal function of several loci is required for S motility, including the loci *tgl,* several *sgl* loci, and the locus *mgl.*

The alternate type of single mutant, $A^+ S^-$, is defective in S motility and expresses pure A motility. The A-motile cells can move when they are completely isolated as well as when they are close together. Pure A motility results in a spatial distribution with many isolated cells, separated by several cell lengths from their neighbors, as well as some cell clusters. The A motility requires the normal function of 22 distinct genetic loci (Hodgkin and Kaiser 1979a). There are five *cgl* loci (*cglB* through *cglF*) and 16 *agl* loci (*aglA,* etc.), and one *mgl* locus that is involved in both A and S motility. Mutant cells that are defective in any of these 22 loci lack A motility, whether or not they have S motility.

Double mutants that are defective in any one of the 22 loci required for A motility and in any one of the loci known to be required for S motility are nonmotile under all conditions examined. Mutation and transduction tests have revealed that only two transitions between A-motile, S-motile, and nonmotile types are possible: A motile to nonmotile and S motile to nonmotile. No single-step transitions between A motile and S motile have been observed, nor (*mgl* apart) have single steps between $A^+S^+$ and nonmotile. This implies that motility is controlled by two multigene systems, system A (the 22 loci specifically required for the expression of A motility) and system S (the loci required for the expression of S motility). Each system of genes can function without the other to render cells motile. Wild-type cells have a combination of A and S motility. With the sole exception of *mgl* mutants, single mutations affect either A motility or S motility but not both. The genetic control of motility in *M. xanthus* is summarized in Table 8.2.

Extracellularly complementing (stimulatable) mutants are found in both motility systems. Stimulatable mutants, which are initially nonmotile, regain motility when they are mixed with wild-type cells or with other mutants. Stimulation is phenotypic and not genetic in that the stimulated cell moves transiently, perhaps 100 μm, and no stably motile strains are thereby generated.

In gene system S, *tgl* is the only stimulatable type and any *tgl*$^+$ strain can stimulate a *tgl* mutant (Hodgkin and Kaiser 1977, 1979a,b). The motility of a stimulated cell depends on its entire motility genotype. In the simplest case, a nonmotile *agl*$^-$ *tgl*$^-$ recipient is stimulated to exhibit pure S motility (Hodgkin and Kaiser 1979b). When a *tgl*$^-$ strain is stimulated, its phenotype becomes temporarily $Tgl^+$. All $S^+$ strains of *M. xanthus* have pili and *tgl* mutants lack pili (Kaiser 1979). When an *agl*$^-$ *tgl*$^-$ is stimulated it not only becomes transiently S motile but also transiently piliated (Kaiser 1979). With further cell growth, the fraction of piliated cells decays exponentially, as if no new pili were being produced, with only the old ones remaining after stimulation has ceased.

In motility system A there are five distinct types of stimulatable or *cgl* (for conditional gliding) mutants: *cglB, cglC, cglD, cglE,* and *cglF* (Hodgkin and Kaiser 1977). Any mixture of two mutants of different types causes movement, but in mixtures of two different mutants of the same type the cells remain nonmotile. Either stimulatable or nonstimulatable mutants can be stimulation donors, but nonstimulatable mutants cannot be recipients. Thus, for example, any *cglB*$^+$ strain, whether motile or nonmotile by virtue of mutations at other

**Table 8.2.** Genetic Control of Motility in *Myxococcus xanthus*

| | Movement pattern | Genes |
|---|---|---|
| A motility | Isolated cells move | System A 23 loci, *cgl, agl, mgl* |
| S motility | Cells move only when close together | System S *tgl, sgl, mgl* |
| Wild type | $A^+$ $S^+$ Has both sets of genes and exhibits a combination of A and S motility | |

motility loci, can stimulate a *cglB*⁻mutant. When a *cglB*⁻ mutant is stimulated, it temporarily becomes phenotypically *CglB*⁺ (Hodgkin and Kaiser 1979a). Its motility depends on the rest of its genotype: In the simplest case, a nonmotile *cgl*⁻ *sgl*⁻ recipient will be stimulated to pure A motility (Hodgkin and Kaiser 1979a).

The genetic significance of stimulation is revealed by the fact that all mutants of a given stimulation type map to the same genetic locus (Hodgkin and Kaiser 1977, 1979a,b; Sodergren and Kaiser 1983). The mutations found in all 14 mutants, classified by stimulation tests as type *cglB,* map in a cluster that is 90% cotransducible with Tn*5* insertion Ω1932 and define a unique *cglB* locus. Similarly, all *cglC, cglD, cglE,* and *tgl* mutants map to four discrete loci. Only one *cglF* mutant exists, and it maps to a distinct site near *cglE.* Only stimulatable mutants and no other mutant types have been found at these loci; therefore, stimulation is locus and not allele specific (Hodgkin and Kaiser 1977, 1979a,b; Sodergren and Kaiser 1983).

In summary, there is a one-to-one correspondence between stimulation type and genetic locus. The temporary restoration of normal motility depends on specific genetic complementarity between two motility mutants. It is not yet known whether stimulation plays a natural role in the swarming or fruiting of wild-type *M. xanthus.* Experimentally, stimulation is observed to be very efficient, and wild-type cells can be stimulation donors (Hodgkin and Kaiser 1977).

Extracellular complementation is also found among mutants defective in the formation of spore-filled fruiting bodies. Since intercellular signaling is important in the multicellular development of fruiting bodies, it is discussed in the chapters on fruiting and its initiation, but some aspects may be discussed here.

A set of mutants were analyzed by Hagen *et al.* (1978) and by LaRossa *et al.* (1983) that failed to form heat-resistant spores under conditions that would induce the wild type to form spore-filled fruiting bodies. These mutants could sporulate, however, if they were mixed with wild-type cells or with other mutants. Pairwise testing of 57 mutants divided them into four groups, and sporulation would occur in mixtures of two mutants belonging to different groups but not in a mixture of two mutants belonging to the same group (Hagen *et al.* 1978). Each group represented a different extracellular complementation group. Complementation did not seem to be the result of crossfeeding of essential intermediary metabolites since the mutants could grow on a minimal defined medium (LaRossa *et al.* 1983). Rather the groupings defined different units of developmental physiologic function because members of the four groups were also distinguished from each other by the stage in the fruiting process beyond which they failed to develop or by the genetic locus that had mutated (LaRossa *et al.* 1983). Some *rif*ʳ mutants of *M. xanthus* that had greatly reduced levels of developmental sporulation (Rudd and Zusman 1982) were found to belong to group B (LaRossa *et al.* 1983). Complementation for sporulation in fruiting body development can occur when the wild-type and mutant cell populations are on opposite sides of a membrane filter, implying that extracellular complementation can occur through materials that diffuse between cells (LaRossa *et al.* 1983). Thus, either cell contact, as in motility stimulation, or diffusible substançes can be the basis for extracellular complementation. Biochemical evidence that extracellular complementation does reflect intercellular signaling awaits chemical identification of the signals and proof that they are normally part of motility or fruiting body formation.

## Summary

The genetic tools now available in myxobacteria, particularly in *M. xanthus,* appear sufficient to investigate the genetics of multicellular development and other complex processes.

1. Because the genome complexity of *M. xanthus* is comparable to that of other bacteria (*M. xanthus* has 5700 kb of unique DNA), it should be possible to isolate mutations in most of the genes involved in a particular function of interest.
2. Efficient generalized transducing myxophages are available for transferring genes between strains of *M. xanthus*. Multiple mutant stocks can be constructed and mutant loci can be mapped.
3. Genes can be transduced by coliphage P1 from *E. coli* to *M. xanthus* either by generalized or specialized transduction.
4. The transposon Tn*5*, carrying a gene that confers kanamycin resistance, can transpose from bacteriophage P1 to many different sites within the *M. xanthus* genome. Once inserted, Tn*5* remains at the same chromosomal position during growth and subsequent transduction.
5. It is possible to identify a Tn*5* insertion near a gene or near a set of interesting mutations (from a library of many independent Tn*5* insertions). This Tn*5* provides a selectable marker for strain construction, localized mutagenesis, and chromosome mapping. Genetic work can be focused on a chromosomal region of particular interest.
6. Tandem duplications of predetermined regions of the *M. xanthus* chromosome can be constructed and can serve as partial diploids for tests of dominance and genetic complementation.
7. Using kanamycin resistance of an inserted Tn*5* for selection, nearby *M. xanthus* genes can readily be cloned in an *E. coli* plasmid. The plasmid and its insert can be transduced back to *M. xanthus* by bacteriophage P1.
8. Cellular interactions can be investigated directly in mixtures of whole cells by extracellular complementation.

These techniques, with the possible exception of the last, are general and do not depend on particular properties of myxobacteria. Since transduction is the most common mode of gene exchange in bacteria and since transposons occur widely, these techniques may be applicable to other interesting but genetically underdeveloped bacteria.

## Acknowledgment

Unpublished work described here from the author's laboratory was supported by grants from the National Institute of Health (GM 23441 and AGO 2908) and the National Science Foundation (PCM 80-10635).

# 9

# Developmental Program of *Myxococcus xanthus*

David R. Zusman*

The myxobacteria are social organisms. Throughout their life cycles, individual bacteria maintain close contacts with each other. Although bacteria are capable of independent growth on an appropriate medium, they tend to aggregate into swarms and move as a group. Presumably, the swarm provides the critical cell mass needed by the myxobacteria to attack and digest other microorganisms or to degrade complex organic matter. One consequence of this sociability is that the myxobacteria show characteristic behaviors in response to certain stimuli. The most striking behaviors are associated with starvation. Outward movement of the swarms stops and cells begin a complicated series of interactions that result in cellular aggregation and sporulation. Profound changes in the metabolism, biochemistry, and ultimately the morphology of the cells occur. How are these changes triggered? How is the pattern of gene expression modified? What specific biochemical events occur? The developmental program of the myxobacteria is a fascinating subject for investigation. But research in this area has only recently begun. Since most of the investigations have been performed with *Myxococcus xanthus*, development in this organism will be the focus of this chapter.

## Development in *Myxococcus xanthus*

### Initiation of Fruiting Body Formation

Since fruiting body formation is never observed to occur in a rich growth medium, nutrition has long been thought to be of critical importance in sig-

*Department of Microbiology and Immunology, University of California, Berkeley, California 94720 U.S.A.

naling the aggregation and fruiting response (Dworkin 1963; Hemphill and Zahler 1968b). Bretscher and Kaiser (1978) carefully examined the specific nutritional requirements of *M. xanthus* and devised a defined minimal medium that supports vegetative growth of this organism. Manoil and Kaiser (1980a) found that fruiting of *M. xanthus* could be induced on this medium by limitation (but not total elimination) of any one of the essential or strongly growth-limiting amino acids. Starvation for nonessential amino acids, for example, tryptophan (in a tryptophan auxotroph) or serine (by adding serine hydroxamate) could induce fruiting as well. However, starvation for carbon, energy, or inorganic phosphate also induced fruiting, suggesting that the link between nutrition and fruiting was not limited to amino acid metabolism. Since the starvation conditions that induce fruiting were also found to lead to increased levels of guanosine tetraphosphate and even larger increases in guanosine pentaphosphate, Manoil and Kaiser (1980a) suggested that guanosine polyphosphate accumulation may trigger the initiation of fruiting. However, subsequent experiments by Gill and Kaiser (personal communication) showed that mutants that are defective in the synthesis of guanosine polyphosphate are capable of normal fruiting body formation. Since these mutants do not accumulate guanosine polyphosphates, a role for these compounds in the initiation of fruiting is still unclear.

### Developmental Aggregation in *Myxococcus xanthus*

The first morphologic indication that the developmental program has been triggered in *M. xanthus* is cell–cell aggregation. Aggregation in *M. xanthus* is relatively simple. Cells glide toward aggregation centers, often in a series of waves or ripples (Shimkets and Kaiser 1982a). The aggregates grow into raised mounds that are simple in structure and lack a specific shape (Zusman 1980). The cells within the mounds are later converted into spores. In *M. xanthus,* fruiting bodies consist of mounds of spores (Fig. 9.1). On many fruiting media, *M. xanthus* shows vegetative swarming at the colony edge and fruiting at the colony center where nutrients are presumably depleted (see Fig. 9.2). Intermediate stages showing aggregation but not sporulation can be observed between these areas.

Aggregation occurs in *M. xanthus* when cells are plated on a nutrient-limiting medium provided that a solid surface is present (the myxobacteria are nonmotile in liquid broth) and that cells are present in sufficient density (Wireman and Dworkin 1975). Usually, fruiting bodies are obtained on an agar–air surface or on the surface of moist rabbit dung pellets (Peterson 1969). However, Fluegel (1964) and more recently Kuner and Kaiser (1982) found that apparently normal aggregation and fruiting body formation can be obtained on glass or plastic surfaces under a layer of liquid (submerged culture) provided that the strain of *M. xanthus* has functional motility systems and that calcium ions and neutral pH are maintained. This observation is very impor-

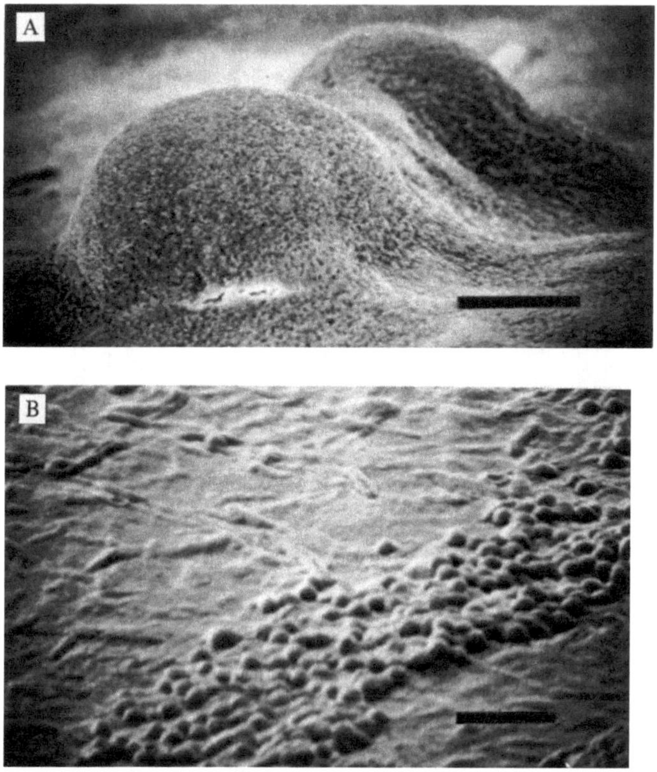

Figure 9.1. (A) Scanning electron micrograph of fruiting bodies of *M. xanthus* strain DZ2. The bar represent 50 $\mu$m. (B) Detail of agar surface showing spore formation within a slime trail. The bar represents 5 $\mu$m. From Campos and Zusman (1975).

tant because it indicates that whatever factors or attractants may function in *M. xanthus* aggregation, they are unlikely to be freely soluble. The submerged culture technique causes convectional and vibrational mixing during incubation and handling that would disturb soluble gradients over distances comparable to the diameter of a fruiting body (0.1 mm) or to the distance between fruiting bodies (0.5 mm).

The importance of high cell density in aggregation and fruiting body formation was described by Wireman and Dworkin (1975). Vegetative cells washed with distilled water and plated on Bacto agar without any added nutrients constructed fruiting bodies rapidly and synchronously. The most rapid fruiting (2 days) was observed when cells were spotted at high cell density (10 $\mu$l/spot; $2 \times 10^9$ cells/ml); cells spotted at $1 \times 10^9$ cells/ml required three days to fruit. At lower cell densities ($2.5 \times 10^8$ cells/ml), no fruiting bodies were observed after 7 days, although there were sufficient total cells to make perhaps 50 fruiting bodies per spot. Thus a critical cell–cell distance may

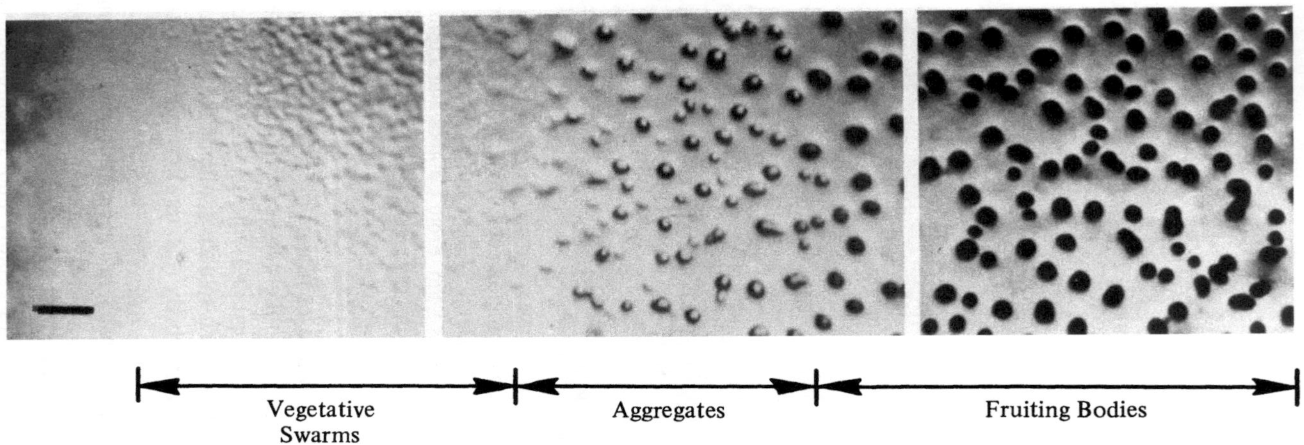


**Figure 9.2.** A section through a fruiting colony of *M. xanthus* strain DZ2. All stages of fruiting body formation can be seen under the dissecting microscope. Cells grow vegetatively at the colony edge. As nutrients become limiting, aggregation occurs. In the colony center, mature fruiting bodies can be seen. Aggregates appear dark after sporulation. The bar represents 1 mm. Courtesy of Ken Rudd. From Zusman (1980).

be necessary to promote intercellular signaling of aggregation. Shimkets and Dworkin (1981) found that adding adenine or various adenine-containing nucleotides stimulated aggregation and fruiting at low cell densities. They suggest that adenosine may act as a cell density signal.

## Sporulation in *Myxococcus xanthus*

After aggregation of cells into raised moundlike structures, individual cells sporulate. In *M. xanthus,* the process of sporulation involves cellular morphogenesis. In contrast to endospore formation in *Bacillus* sp., the whole cell is converted to a myxospore. The spores are resting cells resistant to various environmental stresses such as drying, heating (60°C for 1 hr), detergents, and sonication (Sudo and Dworkin 1969). However, they are much less resistant to heat than are *Bacillus* spores, which can survive boiling.

Wireman (1979) observed that cells plated on a fruiting medium undergo a change to a "committed" state before the appearance of changes in cell shape. Cells harvested at 32 hr from development plates have a rod shape but are committed to sporulate, even if they are subsequently incubated in a rich medium. Eighteen-hour cells become committed if they are prestarved in liquid before incubation in a rich medium. Cells harvested at earlier times cannot be induced to form spores in rich medium. These results indicate that there is a sequence of physiologic changes in developing cells that eventually leads to a commitment to sporulate.

Wireman and Dworkin (1975,1977) noted that only about 20% of the cells present at 24 hr formed spores by 72 hr; the remaining cells lysed. The massive cell lysis appears to be linked to the developmental program since some mutants blocked in sporulation (e.g., SpoC mutants) do not show this lysis (Shimkets and Kaiser 1982b). Stimulation of these mutants to sporulate results in the normal level of lysis. Wireman and Dworkin (1975) suggested that cell death by one portion of the population may occur to provide sufficient nutrition to the surviving cells. An alternate hypothesis is that sporulating cells undergoing morphogenesis are osmotically delicate and subject to lysis. Although it seems likely that lysed cells provide some nutrition, it is not clear if lysis is a necessary condition for sporulation. There is no evidence for multiple cell types in *M. xanthus* such as prelysis cells or prespore cells.

When spores are placed on a rich growth medium, germination occurs. The large concentration of germinating cells derived from a single fruiting body may ensure the critical cell mass needed by the myxobacteria to function as an effective predator or food gatherer. The specific nutrients or chemicals that trigger germination are unknown. Almost nothing is known about the germination program or its regulation. This area remains an interesting one for future investigations.

## Sporulation Induced by Organic Solvents

Dworkin and Gibson (1964) observed that the addition of high levels of glycerol (0.5 *M*), several other alcohols, or dimethylsulfoxide (Zusman, reported in Bacon and Rosenberg 1967) to exponential phase cultures of *M. xanthus* caused the synchronous conversion of rod-shaped cells to refractile, sonication-resistant, heat-resistant, spherical "spores." The conversion occurs in rich media and is very rapid: Phase-bright spheres are observed in about 2 hr at 30°C (about half the generation time). Removal of the glycerol results in synchronous germination (Ramsey and Dworkin 1968) and cell division (Zusman and Rosenberg 1968). The glycerol induction technique had a very significant impact on the course of experimentation in the myxobacteria. Research on fruiting body formation virtually ceased for almost a decade following its discovery. The convenience of the rapid and synchronous induction of sporulation in liquid culture invited numerous investigations of changes in DNA synthesis (Rosenberg *et al.* 1967; Zusman and Rosenberg 1968), RNA (Bacon and Rosenberg 1967; Okano *et al.* 1970; Foster and Parish 1973; Juengst and Dworkin 1973), and enzyme systems (Bland *et al.* 1971; Hanson and Andreoli 1973; Kottel *et al.* 1974; Westby and Tsai 1974; Orlowski and White 1974; Filer *et al.* 1977). For example, during glycerol induction, DNA synthesis continues until chromosomes are completed (Rosenberg *et al.* 1967). Net RNA synthesis immediately stops, but extensive RNA turnover occurs that includes all classes of RNA (Bacon and Rosenberg 1967).

Unfortunately, the mode of action of glycerol induction is still unknown. But, more importantly, the sporelike cells that result from solvent induction differ in many respects from fruiting body spores. Morphologically, glycerol spores lack the three heavy spore coats described by Inouye *et al.* (1979b); in its place they have a very thin one-layer coat that is not resistant to boiling in detergents (see Fig. 9.3). The interior of the glycerol spores show massive convoluted vesicular and lamellar structures, many of which are continuous with each other (Bacon and Eiserling 1968). Bacon *et al.* (1975) showed that glycerol-induced spores have rather high rates of respiration, even after 24 hr in glycerol: 20% of the respiratory rate of vegetative cells. The "spores" accumulate neutral carbohydrate, which, after 24 hr in glycerol, amounted to 20% of the dry weight of the cells. In contrast, Dworkin and Niederpruem (1964) were unable to detect respiration in a suspension of fruiting body spores. Thus, changes observed during glycerol induction are not necessarily relevant to sporulation in fruiting bodies.

Despite the numerous differences between fruiting body spores and glycerol spores, it is clear that at least part of the sporulation program is triggered in these cells. For example, although glycerol-induced cells do not produce protein S or myxobacterial hemagglutinin (see "Biochemical Markers of Development"), they do produce protein U (Komano *et al.* 1980). Further, Morrison and Zusman (1979) found that four out of five mutants that were found to be temperature sensitive for sporulation in fruiting bodies were also temperature

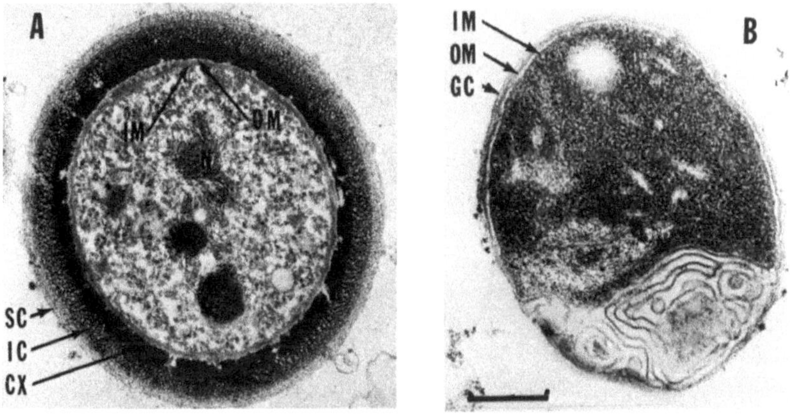


**Figure 9.3.** Electron micrographs of thin sections from *M. xanthus* strain FB fruiting body spores (A) and glycerol spores (B). The bar represents 0.5 μm. Abbreviations: N, nucleoid; IM, inner membrane; OM, outer membrane; CX, cortex; IC, intermediate coat; SC, surface coat; GC, glycerol spore coat. Micrographs prepared by Robert Silverstein. From Inouye *et al.* (1979b).

sensitive for glycerol induction. Thus, functions that are operative both in normal sporulation during fruiting body formation and in glycerol induction are affected in these mutants. Whatever the mechanism of glycerol induction [unregulated transcription (Zusman 1980), membrane perturbation (Sadler and Dworkin 1966, etc.], the technique does permit the synchronous expression of many sporulation genes. In this respect it can be useful.

## Biochemical Markers of Development

### Pattern of Gene Expression during Development

Since fruiting body formation involves complex changes in cellular shape, metabolism, and intercellular interactions, we might anticipate regulated changes in the pattern of gene expression during development. Inouye *et al.* (1979a) investigated the pattern of protein synthesis during fruiting body formation using polyacrylamide gel electrophoresis. Conditions were established for synchronous development in *M. xanthus* (Fig. 9.4). At various stages of development, cells were pulse-labeled with a $^{14}$C-labeled amino acid mixture. Although the "pulses" were relatively long (4 hr) in order to ensure sufficient incorporation of label, the labeling period was still a small fraction of the entire developmental cycle of this organism (about 48–72 hr). The soluble and membrane fractions from the labeled cells were analyzed by sodium dodecyl sulfate (SDS)-polyacrylamide gel electrophoresis. Of the 30 major soluble polypep-

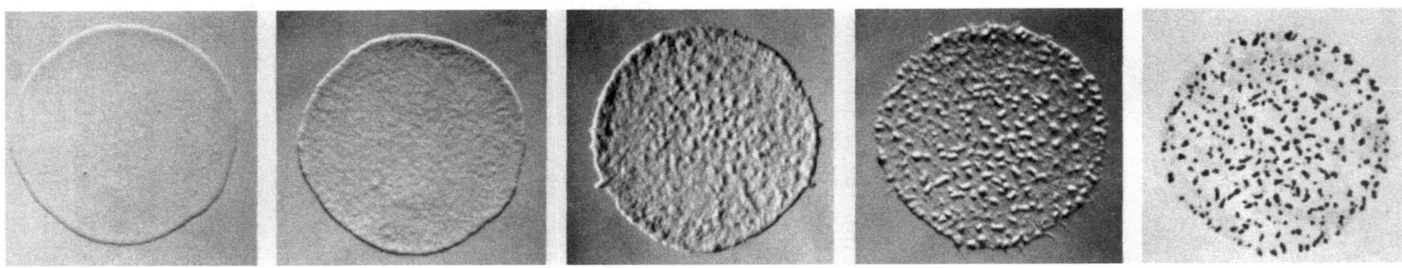


**Figure 9.4.** Morphology of *M. xanthus* strain FB at different times after spotting on fruiting agar. The spots are about 7 mm in diameter. Courtesy of Ken Rudd.

tides that could be resolved in a one-dimensional system, at least 25% showed significant changes in their rates of production during development. Several significant changes were also found in the membrane proteins as analyzed by two-dimensional gel electrophoresis. The most striking finding in this study was that one protein, called protein S, was tremendously overproduced during early development, reaching a maximum of 15% of total soluble protein synthesis (Fig. 9.5). The shear abundance of this protein and the fact that it was clearly subject to developmental regulation suggested that it would be an interesting subject for further study.

Another protein of interest that was identified in this study was called protein U. It is an abundant protein that appears only at late times of fruiting body formation (about 45 hr), a time when sporulation is first observed. This protein has been purified and found to be associated with mature spores. [It was released from spores, along with protein S, after they were boiled in 1% SDS (Inouye *et al.* 1979b), Fig. 9.6.]

It is interesting to compare the pattern of gene expression of these two spore-associated proteins, S and U. Protein S can first be detected in the cytoplasmic fraction at about 3–6 hr of development. The rate of synthesis appears to accelerate until 24 hr when production reaches a peak. At later times, synthesis of protein S declines so that after 48 hr essentially all production has ceased (Nelson and Zusman 1983b). In contrast, protein U cannot be detected until very late in development, about the time when protein S production ceases. Protein U (molecular weight about 16–18,000) is not a proteolytic cleavage product of protein S (molecular weight 19,000), since antisera raised against protein S

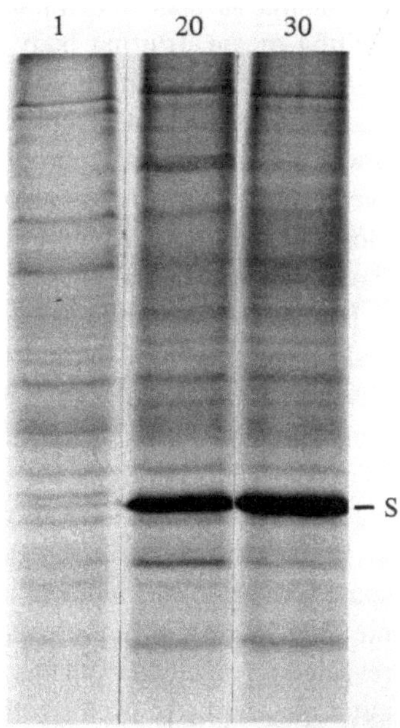


**Figure 9.5.** Pattern of protein synthesis during development of *M. xanthus* strain FB. Cells spotted on fruiting agar were labeled for 4 hr with a $^{14}$C-labeled amino acid mixture at 1, 20, and 30 hr. The soluble proteins were analyzed by SDS-polyacrylamide gel electrophoresis followed by autoradiography. From Inouye *et al.* (1979a).

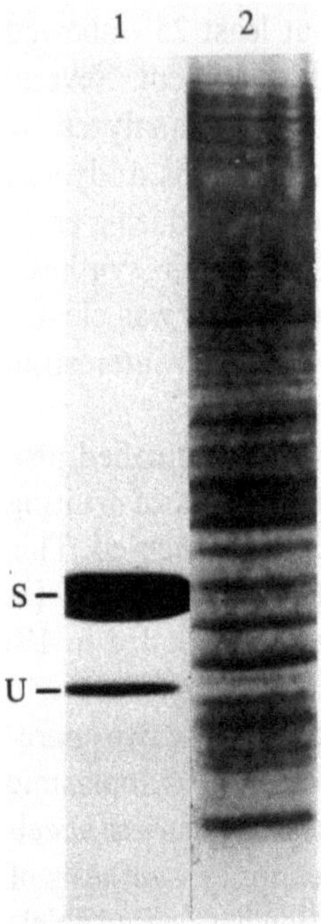


**Figure 9.6.** Analysis of proteins released by boiling fruiting body spores (lane 1) or glycerol spores (lane 2) in 1% SDS. The proteins were analyzed by SDS-polyacrylamide gel electrophoresis. S, protein S; U, protein U. Courtesy of Sumiko Inouye.

did not cross-react with protein U and antisera against protein U did not cross-react with protein S (Komano *et al.* 1980).

The expressions of protein S and protein U also differ under conditions of methionine surplus. When high levels of methionine (200 μg/ml) are added to fruiting media, fruiting body formation is inhibited (Rosenberg *et al.* 1973; Campos and Zusman 1975). Although initial aggregates are observed, formation of mounds and spores are not observed. Under these conditions, protein S was produced at the appropriate times of plating and at normal levels. However, protein U was not observed (Inouye *et al.* 1979a). It should be noted that most of the other early developmental proteins are also produced normally in the presence of methionine.

It is interesting to compare the pattern of gene expression during fruiting body formation with the pattern observed after glycerol induction of spores. Komano *et al.* (1980) examined the pattern of protein production after glycerol, dimethyl sulfoxide (DMSO), or phenethyl alcohol induction of spores. They pulse-labeled cells for 30 min with [$^{35}$S]-methionine in A1 minimal medium (Bretscher and Kaiser 1978), a defined medium that permits growth of *M. xanthus* with a 22-hr generation time. Synchronous and rapid (4–6 hr) spore formation was induced under these conditions. Examination of the soluble and membrane proteins by SDS-polyacrylamide gel electrophoresis revealed no significant changes in the staining patterns of the gels. However, autoradiography showed significant changes in the levels of synthesis of many polypeptides. Several new protein bands were observed at early, middle, and late times of glycerol spore induction. The banding pattern observed in these autoradiograms differed significantly from the pattern observed with cells labeled at different times of fruiting body formation. It is of particular interest that protein S was undetected under these conditions even with antiprotein S serum; however, protein U could be detected at late stages of glycerol induction with antiprotein U serum. Thus, the general pattern of protein synthesis during glycerol (dimethyl sulfoxide or phenethyl alcohol) induction of spores is clearly different from that observed during fruiting body formation. However, some steps in these processes (e.g., the induction of protein U) are apparently similar. Okano *et al.* (1970) studied transcription during glycerol induction. Hybridization competition experiments showed that certain messenger RNA (mRNA) species were synthesized only during vegetative growth, whereas others were synthesized only during glycerol-induced sporulation. However, these authors did not compare the RNA synthesized during glycerol induction with that made during fruiting body formation.

## Biosynthesis and Self-Assembly of Protein S

During early fruiting body formation, the most abundant soluble protein labeled in pulse experiments was protein S (Inouye *et al.* 1979a). Protein S was purified from 40-hr developmental cells and found to be a soluble protein with an apparent molecular weight on SDS gels of 23,000 (Inouye *et al.* 1981; the actual molecular weight deduced from the DNA sequence is 18,792 (Inouye *et al.* 1983a). The protein was first detected in the cytoplasmic fraction of developing cells at very early times (3–6 hr). The protein accumulated in this fraction until about 15–18 hr when it was transported through the cytoplasmic membrane. Thus protein S is unusual among translocated proteins in that it accumulates as a soluble cytoplasmic protein before its translocation. In addition, translocation appears to occur in the absence of a cleaved signal peptide (Nelson and Zusman 1983b; Inouye *et al.* 1983a). Biosynthesis of protein S ceased after 48 hr, at which time most protein S has been translocated through the outer membrane and polymerized onto spore material. Protein S can be extracted from mature spores with 1 *M* sodium chloride (NaCl) or ethylenediaminetetraacetate (EDTA); extraction was inhibited by $Ca^{2+}$ and to a lesser extent by $Mg^{2+}$ (Inouye *et al.* 1979b). The viability of salt-extracted (protein S deficient) spores, however, was unchanged. Examination of thin sections of mature spores and extracted spores by electron microscopy suggested that the protein S deficient spores lacked a spore surface coat about 300 Å thick.

Purified protein S was found to self-assemble spontaneously onto protein S deficient spores after removal of the NaCl by dialysis or by simply adding 10 m*M* $Ca^{2+}$ to undialyzed samples (Inouye *et al.* 1979b). Quantitation of the self-assembly process showed almost stoichiometric binding of protein S to the protein S deficient spores until saturation at $3.3 \times 10^6$ molecules per spore, a value 1.35 times higher than the normal level of protein S found in mature spores (Fig. 9.7). Electron microscopy of the reconstituted spores revealed the assembly of new material on the spore surface. The reconstituted coat had a different appearance than the original outer surface coat. Adjacent spores were often observed to be fused to each other through a common protein S layer. These results suggest that protein S may serve a function in holding spores together within the fruiting body. Any spore–spore associations would be destroyed by the sonication step normally used to purify spores. It is interesting that glycerol spores, which do not contain protein S, cannot serve as primers for assembly of protein S. This suggests that protein S recognizes a specific receptor (which must be missing from glycerol spores) for the initiation of the polymerization process.

Since protein S appears to require $Ca^{2+}$ for assembly onto protein S deficient spores, one might predict conformational changes in the protein following $Ca^{2+}$ addition. Inouye *et al.* (1981) examined circular dichroism spectra of purified protein S in the absence and presence of 10 m*M* calcium chloride ($CaCl_2$), but no significant effect of $CaCl_2$ was observed. The spectra did show that the protein had a very high content of $\beta$- structure and almost no $\alpha$ helix. However, the crystallization of protein S was very much influenced by $Ca^{2+}$. Without added $CaCl_2$, rhombic polyhedron crystals were found, but in the presence of

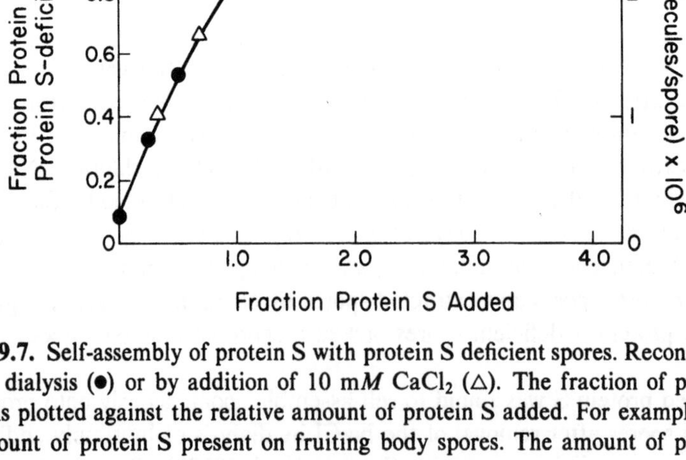


**Figure 9.7.** Self-assembly of protein S with protein S deficient spores. Reconstitution was by dialysis (●) or by addition of 10 m*M* $CaCl_2$ (△). The fraction of protein S bound is plotted against the relative amount of protein S added. For example, 1.0 is the amount of protein S present on fruiting body spores. The amount of protein S bound to spores is also expressed as the number of molecules per spore. From Inouye *et al.* (1979b).

$CaCl_2$, prismatic needles were formed (Fig. 9.8). The x-ray structure of the crystals is currently under investigation (Inouye *et al.* 1980b).

When protein S was treated with trypsin, a core peptide that was resistant to further trypsin digestion was obtained. The purified core peptide had a molecular weight of about 10,000. It was able to self-assemble on the surface of protein S deficient myxospores in the presence of 10 m*M* $CaCl_2$. Competition-binding experiments showed that the core peptide and protein S bound to the same site on the spore surface.

Recently, Inouye *et al.* (1983b) prepared a synthetic DNA probe corresponding to the C-terminus of protein S and used this probe to locate and clone the gene for protein S. They found that the probe hybridized to two tandemly repeated homologous genes separated by a DNA region of approximately 1.3 kilobases. They determined the DNA sequence of 3692 base pairs encompassing both genes and deduced the amino acid sequence of the two gene products (Inouye *et al.* 1983a). From their structures (which showed extensive amino acid homology, 88%), the gene 2 (downstream) copy was identified as the gene for protein S. Interestingly, the primary structure of protein S shows striking similarities with calmodulin, suggesting that protein S may have some calmodulin-like properties.

Downard *et al.* (1984) independently cloned the protein S genes and isolated mutations throughout the region using the transposon Tn*5*. The Tn*5* insertion mutations were introduced into *M. xanthus* by selecting for gene replacements following P1 mediated generalized transduction (O'Connor and Zusman, 1983). *M. xanthus* strains with mutations in gene 2 showed no detectable protein S on spores and lacked the mRNA bands which hybridize with the cloned genes. In contrast, *M. xanthus* strains with mutations in gene 1 exhibited normal levels of protein S. Thus gene 2 is responsible for most if not all of the production of protein S during *M. xanthus* development. Surprisingly, *M. xanthus* strains containing Tn*5* insertions in gene 1, gene 2, or both genes were found to aggregate and sporulate normally even though they made no protein S. The protein S deficient spores were indistinguishable from the wild type spores in their resistance to heat, ultraviolet radiation, cycles of freezing and thawing, desication and in germination under low nutrient conditions. Therefore, the function of protein S, at least in the laboratory, remains an enigma. It may have an important role during fruiting body formation under more natural conditions, however.

Another interesting problem relating to protein S is the mechanism for its huge overproduction during development. Fruiting body formation is a time of great deprivation for *M. xanthus,* yet 15% of all protein synthetic capabilities are mobilized toward the biosynthesis of a single protein. How is this accomplished? One mechanism for overproducing protein S might relate to the unusual stability of its message. Nelson and Zusman (1983a) measured the half-life of the mRNA for protein S by inhibiting RNA polymerase with rifampin and then measuring the rate of $^{35}S$-methionine incorporation into protein during a series of 2-min pulses. The amount of $^{35}S$ incorporated into

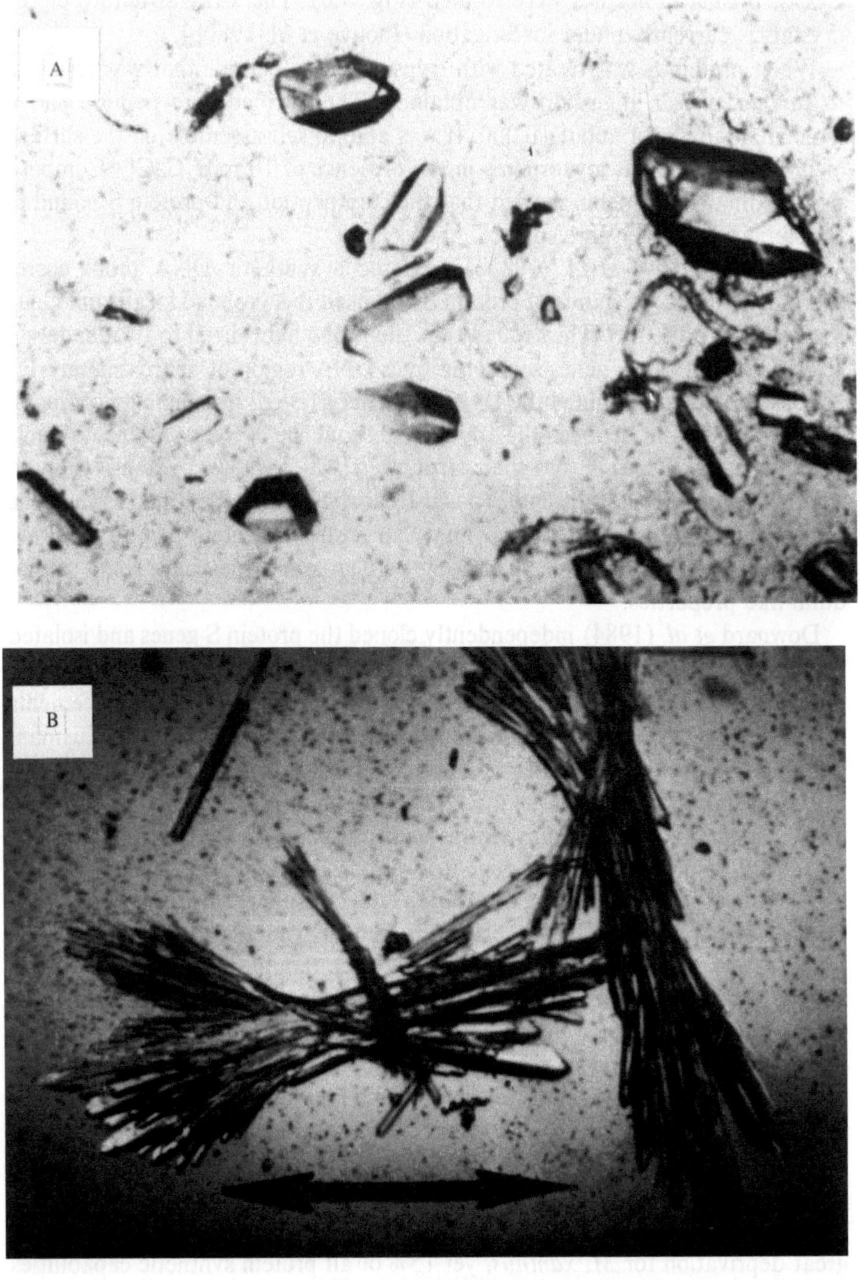


**Figure 9.8.** Crystals of protein S that formed in the absence of $CaCl_2$ (A) or presence of 10 m*M* $CaCl_2$ (B). The bar represents 1 mm. From Inouye *et al.* (1981).

protein S was determined by immunoprecipitation of labeled proteins with antisera specific for protein S and then analyzing the precipitates by SDS-polyacrylamide gel electrophoresis. The results showed that the message for protein S was unusually stable, exhibiting a half-life as long as 30 min. Hybridization studies of protein S mRNA with the cloned protein S gene showed that the message has a physical half-life of 38 min (Downard *et al.* 1984). In contrast, most mRNA species from *M. xanthus* show average half-lives of about 3 min. Thus, the unusual stability of the protein S message may facilitate its overproduction while at the same time conserving energy by not rapidly turning over the mRNA. Nelson and Zusman (1983a) found that about one-third of all developmental protein synthesis is coded for by relatively stable mRNA. The presence of relatively long-lived mRNA in developing cells of *M. xanthus* was first reported by Smith and Dworkin (1981).

## Myxobacterial Hemagglutinin: A Development Specific Lectin

Cell-surface carbohydrate binding or "lectin like" proteins have been implicated in cell–cell interactions and intercellular cohesion in the cellular slime molds *Dictyostelium discoideum* (Rosen *et al.* 1973; Siu *et al.* 1976) and *Polysphondylium pallidum* (Rosen *et al.* 1974), as well as in other eukaryotic systems (Burger and Jumblatt 1977). Extracts of *M. xanthus* were therefore examined for a hemagglutinating activity. Although no activity was found in extracts of vegetative cells, cells starved in liquid buffer, or in glycerol-induced cells, a large amount of activity was found in cells starved on a solid medium in which aggregation occurs (Cumsky and Zusman 1979). Extracts were able to agglutinate sheep or guinea pig erythrocytes but not horse, ox, chicken, or human erythrocytes. Thus, the *M. xanthus* hemagglutinin appears to recognize a specific receptor that is not present on all erythrocytes in their native form.

The kinetics of appearance of the hemagglutinating activity in soluble extracts of *M. xanthus* is illustrated in Fig. 9.9 for strain DZ2. No detectable hemagglutinating activity was observed in very early developmental cells. As development proceeded, the specific activities increased reaching a peak at the time of tight cell cell aggregation (translucent mound stage). As the cells continued their differentiation into mature spores, the specific activities declined. At peak times, strain DZ2 contained eightfold more hemagglutinating activity than did strain FB. This difference is attributable to the *sgl*Al mutation in FB, since restoring this locus to the wild-type allele results in a corresponding increase in the level of hemagglutinin production.

Myxobacterial hemagglutinin (MBHA) was purified from developmental extracts of *M. xanthus* (Cumsky and Zusman 1981a). The purified protein exists as a monomer in solution with an apparent molecular weight of 28,000 and an isoelectric point of 8.3. The amino acid composition shows a very high content of glycine (19%) as well as aromatic amino acids (9%); it has a low percentage of charged amino acids. No detectable carbohydrate was found in

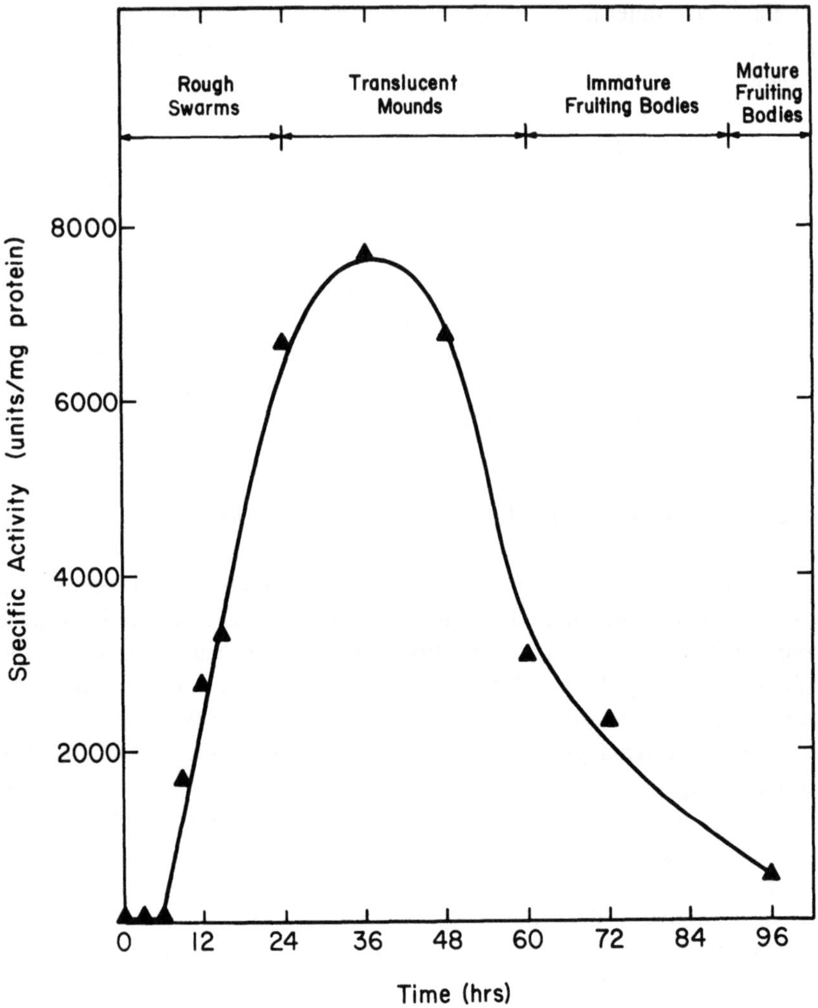


**Figure 9.9.** Kinetics of appearance of hemagglutinating activity in soluble extracts of *M. xanthus* strain DZ2. Cells were spotted on fruiting agar and harvested at various times during development. Extracts were prepared and the soluble fractions assayed for hemagglutinating activity with formalinized sheep erythrocytes. The labels at the top depict the stages of development for this strain. From Cumsky and Zusman (1981a).

a large sample (50 $\mu$g) of MBHA. The far-ultraviolet (UV) circular dichroism (CD) spectrum of MBHA indicates a secondary structure that contains very little $\alpha$ helix, 50 $\pm$10% $\beta$ sheet and 50 $\pm$ 10% random coil. The MBHA comprises 1–2% of soluble protein of *M. xanthus* at the time of cellular aggregation.

Because MBHA binds to erythrocytes and agglutinates them, a large number of simple sugars and amino sugars was tested at concentrations up to 0.1 *M* in an effort to inhibit the hemagglutinating activity, but no inhibition was detected (Cumsky and Zusman 1979). However, when the glycoprotein fetuin was tested, a large amount of inhibition was observed. The *O*-glycosidically linked trisaccharide glycopeptide of fetuin was shown to be inhibitory by itself. The penultimate galactose of this glycopeptide was directly implicated in the inhibitory activity, because the inhibition by asialofetuin was reduced to one-sixtieth by periodate oxidation and to one-fifteen after $\beta$-galactosidase treatment. Thus MBHA binds to complex receptors that probably contain a $\beta$-linked galactose residue. It should be noted that glycophorin and rabbit IgG both contain the same *O*-glycosidically linked tetrasaccharide sequence found in fetuin. They also inhibit MBHA.

The nature of the receptor for MBHA on the outer surface of *M. xanthus* was investigated by studying the binding of $^{125}$I-MBHA to vegetative and developmental cells (Cumsky and Zusman 1981b). The amount of binding and hence the number of binding sites per cell appeared to increase fourfold during development, to $2.1 \times 10^4$ sites per cell. Further, the apparent association constant ($K_a$) for MBHA increased threefold to $3 \times 10^7$ $M^{-1}$. Fetuin blocked the binding of $^{125}$I-MBHA to vegetative cells but not developmental cells. Thus, the MBHA binding sites from developmental cells clearly differ from the vegetative binding sites. The $K_a$ for MBHA binding to sheep erythrocytes ($3.5 \times 10^6$ $M^{-1}$) was one order of magnitude lower than that for developmental *M. xanthus* cells. The erythrocyte binding sites are also much more sensitive to concanavalin A inhibition than are the *M. xanthus* sites.

Although hemagglutinating activity was first detected in the soluble fraction of sonicated developmental cells, some activity was also found in the membrane fraction (Cumsky and Zusman 1979). Nelson *et al.* (1981a,b) therefore examined the localization of MBHA. Cells were fractionated into shock, cytoplasmic, and membrane fractions. As much as 90% of the total hemagglutinating activity was recovered in the wash and shock fractions. Analysis of these fractions by SDS-polyacrylamide gel electrophoresis showed that these fractions are enriched in MBHA. Thus MBHA is a periplasmic protein or a protein loosely bound to the surface of the cell.

The MBHA was detected on the surface of developmental cells but not vegetative cells by immunofluorescent staining procedures (Nelson *et al.* 1981b). The fluorescence was localized in distinct patches that were usually located at one or both of the cell poles, though patches of fluorescence could also be seen at additional sites (Fig. 9.10). The presence of MBHA on the cell surface was also detected by electron microscopy of developmental cells stained with ferritin conjugated antibody. Most of the cells showed distinct patches of ferritin staining at one or both of the cell poles; nonpolar staining, which was also observed, was always accompanied by membrane protuberances. The amino acid sequence of the *N* terminus of MBHA was found to be extremely hydro-

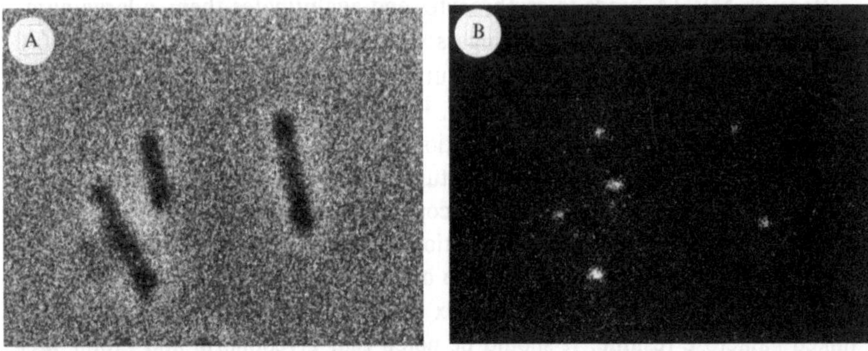


**Figure 9.10.** Immunofluorescent staining of *M. xanthus*. Developmental cells (24 hr) were treated with rabbit anti-MBHA IgG followed by fluorescein-conjugated goat antirabbit IgG. Cells were photographed under phase contrast (A) and then under fluorescence optics (B). Control experiments performed with vegetative cells showed no fluorescence. The cells are 4–6 μm in length. From Cumsky and Zusman (1981a).

phobic, suggesting that it may be important in transmembrane transport or localization. The site-specific localization of MBHA at the cell poles suggests that MBHA may function in end-to-end cellular interactions during aggregation. The actual role of MBHA in development is not known at this time. It probably can best be determined with the isolation of structural gene mutants.

## Analysis of the Developmental Program with Mutants

### Phenotypes of Nonfruiting Mutants of *Myxococcus xanthus*

Mutants of *M. xanthus* defective in fruiting body formation are relatively easy to obtain. When cells are plated on clone fruiting (CF) media consisting of low levels of nutrient broth (McVittie *et al.* 1962), Casitone (Burchard and Parish 1975), or Casitone plus pyruvate (Hagen *et al.* 1978), small vegetative colonies arise, which, after 5–8 days, aggregate and form fruiting bodies in situ. Thus it is possible to screen colonies visually for nonfruiting phenotypes without using direct selection procedures (Fig. 9.11). However, since wild-type colonies sometimes do not fruit under these circumstances, it is usually necessary to retest nonfruiting clones to confirm their developmental phenotype. The frequency of appearance of spontaneous nonfruiting mutants of *M. xanthus* strain FB is about $10^{-4}$ (Morrison and Zusman, unpublished data). This relatively high frequency of nonfruiters probably reflects the large number of genes that is necessary for fruiting. This frequency can be increased at least 10-fold by chemical mutagenesis. Using transposon Tn*5* induced mutagenesis (Kuner and Kaiser 1981), about 4% of 3000 independent insertion mutants showed some defects in fruiting (Zusman, unpublished data).

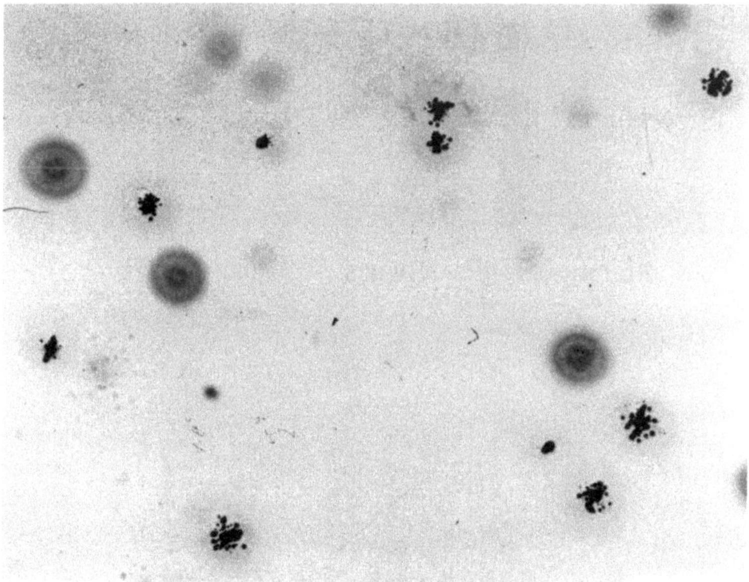

**Figure 9.11.** Colony morphology of fruiting and nonfruiting mutants of *M. xanthus* strain FB plated in a top agar overlay on clone fruiting agar. Fruiting clones appear as clusters of darkened fruiting bodies. Nonfruiting clones appear as diffuse flattened colonies.

Nonfruiting mutants of *M. xanthus* fall into the following three general categories:

1. *$Agg^-$ $Spo^+$ mutants.* These mutants are defective in developmental aggregation but not sporulation. Some of these mutants show defects in motility (Hodgkin and Kaiser 1979a,b) that can be readily observed by the formation of colonies with *smooth* surfaces and edges. Others are defective in cell–cell recognition or chemotaxis toward aggregation centers. The morphology of these mutants have been described as follows (Morrison and Zusman 1979; Fig. 9.12):

    a. *Rough.* Mutants that show no evidence of aggregation. Since they are motile, they move about and redistribute themselves so that cells spotted on CF media have a rough textured appearance.
    b. *Swirls.* Mutants that form ridges of cells in a ring or doughnut pattern. Wild-type cells can sometimes be seen to form swirls at the edges of fruiting clones, particularly at low density. Swirl mutants, however, never progress beyond this stage.
    c. *Flat mounds.* These mutants aggregate into a relatively small number of broad flattened mounds.
    d. *Frizzy.* Zusman (1982) observed that some mutants initially form lumps or flat mounds that later degenerate into ribbonlike filaments with a tan-

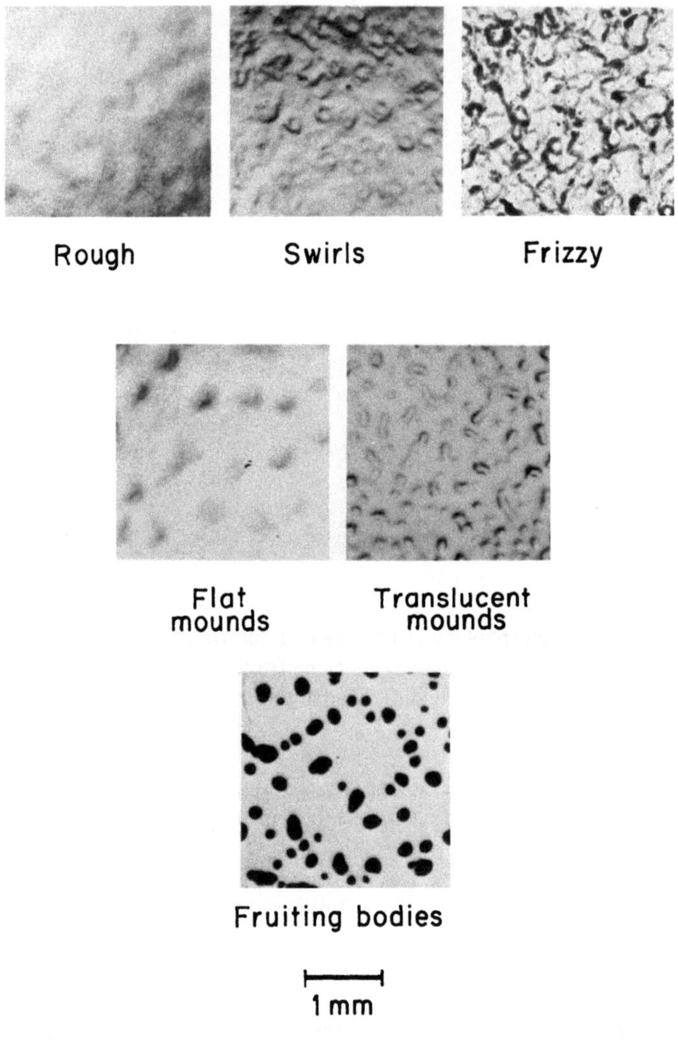


**Figure 9.12.** Morphologies frequently observed in nonfruiting mutants of *M. xanthus*.

gled or frizzy appearance. These mutants show an unusually high number of swirls. Observation of single cells of these mutants by time-lapse video microscopy reveals that, unlike wild-type cells, they move almost exclusively in one direction, rarely, if ever, reversing their direction of gliding (Blackhart and Zusman, unpublished data).

2. *Agg*$^+$ *Spo*$^-$ *mutants.* These mutants aggregate normally but are defective in sporulation. Since the cells never sporulate, the raised mounds (which are about 0.2 mm in height and have the same shape and dimensions as mature fruiting bodies) usually have a *translucent* appearance. On prolonged incubation, the mounds often deliquesce.

3. *Agg⁻ Spo⁻ mutants.* These mutants neither aggregate nor sporulate. These mutants must be blocked in the initiation of fruiting or in some developmental steps that are common to both aggregation and sporulation.

## Cell–Cell Signaling During Fruiting Body Formation

McVittie *et al.* (1962) isolated a small number of nonfruiting mutants of *M. xanthus* and noted that certain pairs of these mutants could successfully complete the developmental cycle when mixed. They spotted aliquots of two nonfruiting mutants a few millimeters apart on fruiting agar and allowed the cells to grow and spread out from each spot until the colonies merged. In several of the mixtures, bands of fruiting bodies formed at the interface of the two colonies. Other pairwise mixtures of mutants did not form fruiting bodies. The mutants fell into two synergism (extracellular complementation) groups: Members of one group could not synergize with other members of that group but could synergize with mutants from the other group. Analysis of the spores formed in the fruiting bodies showed only the parental phenotypes, indicating that cell–cell stimulation during synergism did not involve recombination. Unfortunately, these mutants were never characterized further nor were they saved.

Hagen *et al.* (1978) used a similar approach in the characterization of developmental mutants. In a preliminary study, they selected 14 mutants that were motile but appeared to be arrested at stages observed in normal development. Cultures were grown vegetatively and then mixed and spotted on a fruiting agar. Most pairwise mixtures of mutants looked like a superimposition of the phenotypes of the two constituent mutants. However, the morphology of some combinations suggested synergestic or antagonistic interactions (Fig. 9.13). For example, mutant strain DK510 exhibited synergism with mutants DK603 and DK642 in that fruiting bodies were formed in the mixtures but not by each mutant alone. In contrast, mutant strain DK631 not only was unable to fruit on its own but prevented the fruiting of wild type, DK101. These experiments clearly demonstrate that intercellular signaling is an important component in fruiting body formation.

Since the morphology of synergizing mutants is not always consistent nor identical with normal fruiting, Hagen *et al.* (1978) decided to focus on sporulation because spore formation can be assayed by a functional test, heat resistance. They selected about 100 mutants that were unable to sporulate by themselves and tested them in mixtures with wild-type cells. The genotypes of the spores were determined by using appropriately marked "wild-type" (red pigment mutant) and "nonsporulating" (yellow pigment) strains. They found that about half of the "nonsporulating" strains could actually sporulate when tested in mixtures with wild-type cells. Thiry-eight of these mutants were further tested in pairwise mixtures with each other. The mutants fell into four synergism groups (called SpoA to SpoD) such that, with a few exceptions, mutants

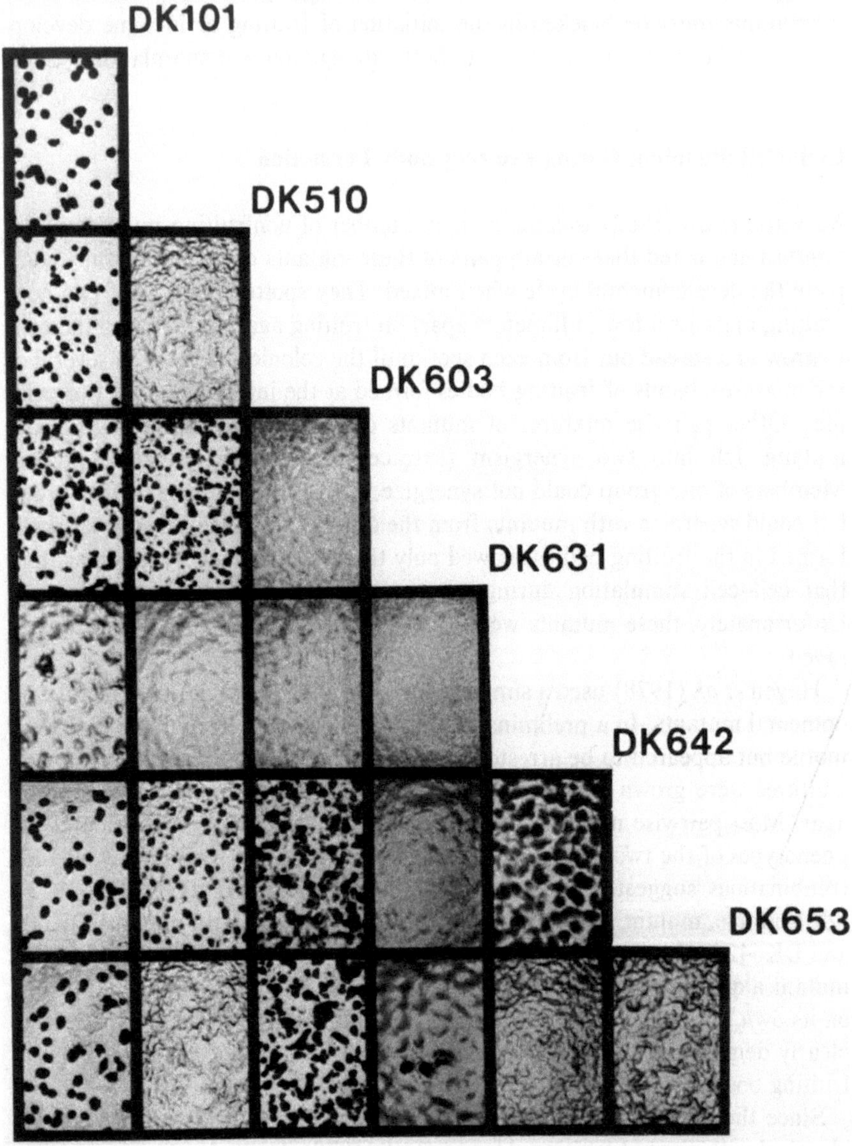


**Figure 9.13.** Synergistic interactions in mixtures of nonfruiting mutants. DK101, wild-type, and the nonfruiting mutants DK510, DK603, DK642, and DK653 were grown in Casitone (CT) broth at 33°C to cell densities of about 5 × 10$^8$ cells/ml. Equal volumes from the cultures were mixed to form all pairwise combinations. Each mixture was spotted (10 $\mu$l volumes) on coli agar and incubated at 25°C for 5 days. The photographic frames are arranged so that DK101 (wild type) is present in all of the mixtures of the first column, DK510 is present in all of the mixtures of the second row and the second column, and so forth. Each strain mixed with itself is thus presented along the diagonal. The dimensions of each frame are 3.2 × 2.1 mm. From Hagen *et al.* (1978).

sporulated in the presence of a member of any other group but not with members of the same group. Hagen *et al.* (1978) postulated that wild-type cells may require at least four different stimuli to initiate sporulation and that these may serve as cell–cell communication signals for synchronous sporulation.

Physiologic investigations performed by LaRossa *et al.* (1983) indicated that the mutants within each class arrested at different distinct points in the developmental program: SpoA and SpoD mutants were blocked early; SpoB and SpoC mutants were blocked later (LaRossa *et al.* 1983; Shimkets and Kaiser 1982b). The four classes of mutants accumulated guanosine tetraphosphate when starved for carbon and amino acids and therefore were all able to respond to certain starvation-related signals. However, mutants of group A continued to grow vegetatively on plates under phenylalanine limitation, a physiologic condition that causes wild-type and the SpoB–SpoD mutants to arrest. Further, mutants in group A delayed the production of protein S (compared to wild type) and did not produce MBHA. Thus, LaRossa *et al.* (1983) concluded that the SpoA mutants have a reduced sensitivity to starvation and are blocked early. The SpoD mutants respond normally to starvation and synthesize protein S at the normal time but fail to synthesize MBHA. These mutants were therefore thought to be blocked at a somewhat later time than the SpoA mutants. The SpoB and SpoC mutants make normal levels of protein S and MBHA but do not show developmental lysis. These mutants must therefore be blocked at a later stage of development. Mapping studies facilitated by transposon Tn*5* insertions indicated that the signal mutations are clustered at a few sites on the chromosome but that mutations affecting the different signals map at different sites. The localization of Tn*5* insertions near these genes should facilitate the further mapping and cloning of these genes.

The analysis of the synergizing mutants and their cell–cell signals should provide an important handle for the investigation of the triggering of the sporulation program in *M. xanthus*. The purification and characterization of some of the signaling substances have already begun. Shimkets and Kaiser (1982b) reported that the sporulation defects in SpoC mutants and about one-third of the SpoB mutants can be overcome by a mixture of peptidoglycan (murein) components: *N*-acetylglucosamine, *N*-acteylmuramic acid, diaminopimelic acid, and D-alanine. When all four components were added to fruiting agar, the number of spores produced by the mutants was raised to the wild-type level. Thus the SpoC mutations, which result in translucent mound formation in strain DK1622, can be induced to to form mature fruiting bodies in that strain by addition of peptidoglycan components. The spores produced were identical to wild-type spores in their ultrastructure, protein composition, and resistance to heat and sonication. Peptidoglycan components would normally be released during fruiting body formation as a large proportion of the cells lyse at the time of sporulation (Wireman and Dworkin 1977). SpoC mutants are blocked before developmental lysis occurs. Thus, ironically, addition of peptidoglycan components, by bypassing this block, causes substantial developmental lysis and still further release of peptidoglycan components. Since *M. xanthus* produces at least three extracellular murein hydrolases: an endo-*N*-acetylglucosaminidase, an *N*-acetylmuramyl-L-alanine amidase, and a peptidase active on D-alanyl-diaminopimelate (Sudo and Dworkin 1972), the bacteria could reduce murein at least to the level of peptides and disaccharides that might later be incorporated into spores. Shimkets and Kaiser (1982b) suggest that these substances may be an important signal for synchronous spore formation. The SpoC loci have recently been cloned (Shimkets *et al.* 1983). All four independent SpoC mutations were found to be in the same gene or unit of transcription. Analysis of the clone and the SpoC gene product should yield more specific data on the signaling process.

## Analysis of Aggregation and Sporulation Functions

Fruiting body formation in *M. xanthus* consists of a temporal sequence of aggregation followed by sporulation. Although aggregation normally precedes sporulation, it is not in itself a necessary precondition for sporulation. Most scanning electron micrographs of fruiting *M. xanthus* show maverick individual cells or small groups of cells that have sporulated directly on the surface of the agar outside the raised mound structure (e.g., see Fig. 9.1b). Furthermore, nonmotile mutants or aggregation-defective mutants retain their ability to sporulate. Aggregation and sporulation probably represent separate parellel pathways in development, since most mutants blocked in aggregation (rough, swirl, and frizzy mutants) show normal levels of sporulation, and most mutants blocked in sporulation (translucent-mound mutants) show normal aggregation (Morrison and Zusman 1979; Zusman 1982). Some linkage between the pathways must exist, however, since some mutants are blocked in both aggregation and sporulation. Since many developmental mutants are double mutants, this category of mutants may actually be less frequent than originally observed.

Further evidence for the existence of separate aggregation and sporulation pathways during fruiting body formation in *M. xanthus* comes from the study of temperature-sensitive periods in mutants. Morrison and Zusman (1979) isolated a collection of temperature-sensitive nonfruiting mutants that formed normal fruiting bodies at 28 °C, the permissive temperature, but were arrested at some intermediate stage at 34 °C, the nonpermissive temperature. The temperature-sensitive periods for the mutants were determined by shifting developing cells from 28 to 34 °C at regular intervals during the fruiting process. Mutants that arrested at the formation of translucent mounds (Fig. 9.14) showed temperature-sensitive periods that ended at 9, 15, 18, and 33 hr (depending on the mutation), although the phenotype of translucent mounds is not usually observed until 24 hr at 28 °C. Since translucent mounds are essentially fruiting bodies that lack spores, these mutants must define early steps in a sporulation pathway. Mutants that were temperature sensitive for aggregation also showed temperature-sensitive periods that ended at various times (6, 9, 12, 15, and 21 hr). Thus, both aggregation and sporulation func-

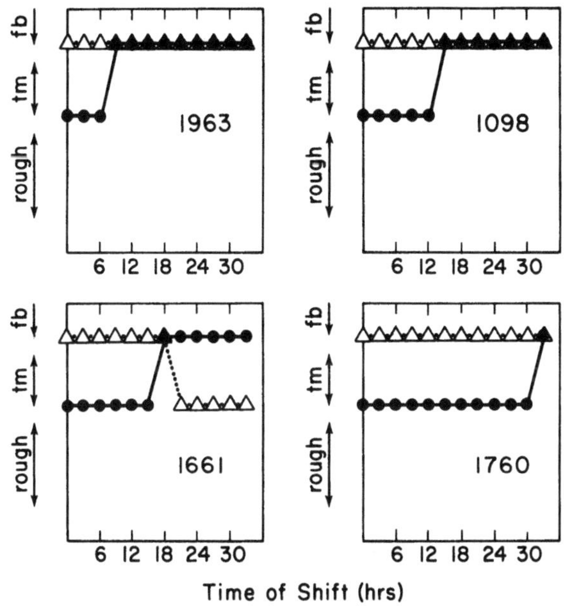


**Figure 9.14.** Temperature-shift analysis of translucent-mound mutants. At 3-hr intervals, plates were shifted from 28 to 34°C or from 34 to 28°C. Individual plates were subsequently maintained at the second temperature for the remainder of the experiment. Each point represents the morphology of the strain at 72 hr. The numbers refer to the strains used. Abbreviations: tm, translucent mounds; fb, fruiting bodies. Symbols: ●, plates shifted from 28 to 34°C; △, plates shifted from 34 to 28°C.

tions are expressed (and needed) at early times in development. These results clearly show that many sporulation functions occur concurrently with aggregation functions and are not triggered only at the completion of mound formation. In agreement with these results is the finding of LaRossa *et al.* (1983) that some of the synergizable sporulation mutants (SpoA and SpoD) actually exchange their signals at early times in development.

Another approach to the isolation of developmental mutants is selected mutagenesis. Rudd and Zusman (1979) used the antibiotics rifampin and streptolydigin to obtain mutants of *M. xanthus* with altered RNA polymerase molecules, the target molecules for these antibiotics. The rationale for these experiments was that if mutants can be isolated that show defects in fruiting body formation in addition to drug resistance, then the target molecule becomes implicated in the developmental program. Rudd and Zusman (1979) found that as many as 20% of the spontaneous rifampin-resistant mutants exhibited defects in fruiting, but none of the streptolydigin-resistant mutants were defective. Since rifampin-resistance and nonfruiting phenotypes showed cotransduction frequencies close to 100% ($\geq$99%), the Rif$^r$ and Fru$^-$ phenotypes were probably the result of a single mutation. The nonfruiting mutants were defective in both aggregation and sporulation. These results suggest that

RNA polymerase plays a significant role in the regulation of development in *M. xanthus*, since mutations that cause no apparent changes in vegetative growth result in striking defects in fruiting body formation.

## Use of Recombinant DNA Technology to Analyze Development in *Myxococcus xanthus*

One area of special promise that has recently emerged in the study of development in *M. xanthus* is nontraditional genetic techniques using transposable genetic elements and recombinant DNA technology (see Chapter 8). Although the amount of accumulated genetic information about *M. xanthus* remains fragmentary, even miniscule compared to *Escherichia coli* or *Bacillus subtilis* (e.g., no genetic map has been made for *M. xanthus* except for several small regions that are unlinked to each other), the technology for sophisticated analyses of specific genetic regions is excellent. For example, the transposon Tn*5* can be introduced into *M. xanthus* using coliphage P1 as a vector (Kaiser and Dworkin 1975; Kuner and Kaiser 1981). Insertions near or in particular genes allow for the isolation of developmental mutants, mapping them using generalized transduction, or for more complex manipulations of the genes. Some of the more powerful approaches involve cloning developmental genes in *E. coli* and then reintroducing the DNA back into *M. xanthus* by P1 mediated transduction. The general strategies used for these experiments are as follows:

1. *Clone genes that code for known developmental proteins.* For example, the gene(s) for protein S were cloned by Inouye *et al.* (1983b) using synthetic hybridization probes predicted to be complementary to part of the sequence. An alternate approach is to screen a bank of recombinant lambda clones for expression of a developmental protein using a solid-phase radioimmunoassay specific for that protein (Broome and Gilbert 1978). Once the genes have been cloned, specific mutations in the genes can be introduced using Tn*5* mutagenesis. The mutated DNA can then be transduced back into *M. xanthus* by selecting for kanamycin resistance. In this manner it is possible to use structural and regulatory mutations to explore the role of the developmental proteins in fruiting body formation. The latter approach was used by Downard *et al.* (1984).
2. *Clone genes linked to a known and characterized Tn5 insert.* If a strain of *M. xanthus* can be isolated that contains a Tn*5* insert closely linked to one or several developmental genes, the genes can be cloned in a plasmid vector in *E. coli* by direct selection for the drug-resistance element (kanamycin resistance) present in Tn*5*. This method was used to clone the *spoC* loci (Shimkets *et al.* 1981), several *aggR* loci (O'Connor and Zusman 1983), and the "frizzy" mutant loci (Blackhart and Zusman, unpublished data). Confirmation that the genes of interest are in fact present on the cloned DNA can be obtained by transducing the DNA back into *M. xanthus*

(using bacteriophage P1) and selecting for kanamycin resistance (O'Connor and Zusman 1983; Gill and Kaiser, personal communication). If appropriate *M. xanthus* mutants are used as recipients for these transductions, many will coinherit the $Fru^+$ allele either by direct recombination or by the formation of tandem duplications. Since tandem duplications are partial diploids, analysis of these transductants can be used for complementation studies of the original mutations.

The recent availability of cloned developmental genes presents tremendous opportunities for future research. For example, in vitro transcription of the cloned DNA with RNA polymerase from vegetative cells (Rudd and Zusman 1982) or from various developmental stages should reveal what role polymerase plays in gene regulation. The analysis of promoter sequences in the DNA as well as promoter mutations should help us understand the regulation of these genes. The subcloning of certain developmental genes in expression vectors should allow the identification of the gene products in minicell (Reeve 1979) or maxicell (Sancar *et al.* 1979) systems. In short, the advent of recombinant DNA technology permits molecular studies of gene regulation and the developmental program that would be intractable by other methods.

## Outlines of a Developmental Program

At this time, it is easier to list what we do not understand about development in *M. xanthus* that what we do understand. We do not know what triggers development (although it must somehow be linked to nutrition) or how changes in gene expression are regulated. We do not know what all, or even most, of the developmental changes are. We do not know how hundreds of thousands of bacteria are signaled to aggregate to specific foci, nor do we understand how sporulation occurs. What we do know is only a bare outline of certain selected topics that have been approached by relatively few laboratories. In the interest of correlating some of these findings and stimulating further research, I offer the following hypothetical model of development in *M. xanthus* (Fig. 9.15). In this model, the developmental program is viewed as having two major parts: an aggregation pathway and a sporulation pathway. The basis for this division rests primarily on the study of mutants. Most mutants that are defective in sporulation, show normal aggregation; in contrast, most aggregation-defective mutants show normal levels of sporulation. Further, the timing of spore formation in these latter mutants is approximately the same as that observed within fruiting bodies. Although sporulation is the last step in fruiting body formation, many sporulation functions are needed at early times, that is, the temperature-sensitive periods for some of these mutants end well before aggregation occurs (Morrison and Zusman 1979). Thus many aggregation and sporulation functions must occur at the same times of development, in parallel. However, since some mutants (e.g., RNA polymerase mutants) appear to be

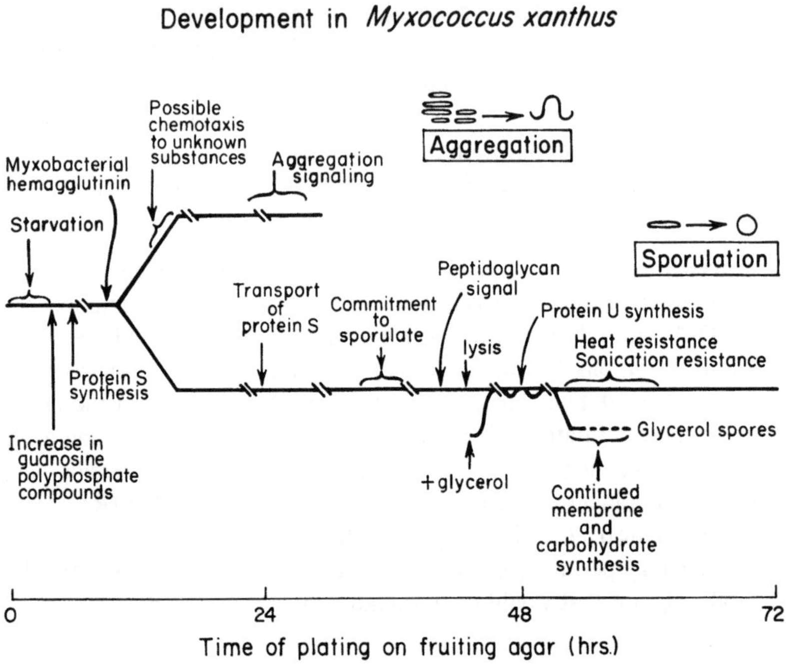


**Figure 9.15.** Model of development in *M. xanthus*. The model depicts two developmental pathways: aggregation and sporulation. Both pathways are triggered by starvation. Glycerol induction bypasses the starvation signal resulting in rapid formation of glycerol spores. The major events in the developmental program are outlined in the text. Broken lines \\ represent sites in the pathway represented by mutants.

blocked in both functions, it is likely that some steps must be common to both aggregation and sporulation. This is probably particularly relevant at very early times of development.

In Fig. 9.15, the two developmental proteins, protein S and MBHA are represented as being in a common pathway. We do not actually know if this is true. We know that both proteins can be detected at early times in development (protein S, 3–6 hr; MBHA, 6–9 hr). A relatively large fraction (about 10%) of nonfruiting mutants of all phenotypes (Agg$^-$Spo$^+$, Agg$^+$Spo$^-$, and Agg$^-$Spo$^-$) are defective in the synthesis of protein S (Blackhart and Zusman unpublished data). This suggests that protein S may be involved in a common pathway. The case for MBHA is even weaker. The MBHA presumably is involved in aggregation because it is a lectin and is localized on the cell surface at the cell poles. However, two of the sporulation signal defective mutant groups (SpoA and SpoD) apparently do not make MBHA (LaRossa *et al.* 1983). Thus, MBHA may be on a common aggregation and sporulation pathway. It is hoped that the specific roles for these developmental proteins will soon be clarified with the isolation and further analysis of mutants.

The aggregation functions are almost completely uncharacterized. The periodical movements of *M. xanthus* and the directed movements toward aggregation centers suggest chemotaxis, yet chemotaxis has still not been demonstrated in this organism. Aggregation occurs in submerged culture, further indicating that whatever signaling molecules may be functioning, they are probably not freely soluble (Kuner and Kaiser 1982). The only protein so far suggested to be involved in cell–cell recognition is myxobacterial hemagglutinin. However, no specific data have directly implicated this protein in aggregation.

Sporulation not only requires the synthesis of new proteins (e.g., protein S and protein U), but also changes in cell shape and metabolism. At a particular time in development, cells become committed to sporulate. Shimkets and Kaiser (1982b) have shown that peptidoglycan components are an important signal for sporulation. During the morphogenesis, lysis of 80–90% of the cells occurs (Wireman and Dworkin 1977). In the absence of starvation or early sporulation functions, cells of *M. xanthus* can be induced to form "glycerol spores" by various organic solvents. The glycerol spores clearly differ from fruiting body spores in their morphology and metabolism. However, they also share many properties of fruiting body spores (e.g., the synthesis of protein U). The time frame for glycerol spore induction is very rapid compared to fruiting. No lysis or cell–cell signaling has been demonstrated in glycerol induction. Removal of glycerol at any time after addition results in germination and the resumption of vegetative functions. The glycerol induction technique must bypass the normal regulatory controls of sporulation since most mutants that have lost the ability to form glycerol spores can still sporulate normally in fruiting bodies (Burchard and Parish 1975), and many mutants that can no longer form spores in fruiting bodies can be induced to sporulate with glycerol (Hagen *et al.* 1978; Rudd and Zusman 1979).

In conclusion, fruiting body formation in *M. xanthus* appears to show an ordered program of developmental changes that result in aggregation and sporulation. Although many of the specific steps in the developmental process are yet to be described, the availability of sophisticated genetic techniques and several biochemical markers should facilitate rapid advances in this field.

## Acknowledgments

I would like to thank Dale Kaiser and Masayori Inouye and their laboratories for sharing their unpublished findings with me. My laboatory is supported by grants from the National Institutes of Health (GM 20509) and the National Science Foundation (PCM-7922249).

# 10

# Social Adaptations in Myxobacteria

Amotz Zahavi and Dina Ralt*

Myxobacteria are unique amongst the procaryotes in displaying a wide range of social adaptations (Kaiser et al. 1979). They live in compact masses, grow better on polymeric substrates when in dense populations (Rosenberg *et al.* 1977), and collaborate in producing fruiting bodies. Moreover, they generally sporulate only when in dense populations (Chapter 4); while sporulating, most of the cells lyse and presumably nourish those that sporulate (Wireman and Dworkin 1977).

Sporulation in myxobacteria as well as their movements in packs are most probably synchronized by a communication system. Although accumulation of various compounds has been correlated with development (Chapter 4), it is difficult to determine, based on our present knowledge, which of those compounds serve as the signals in the communication system.

The above characteristics make the myxobacteria a suitable model system for the study of social behavior. Due to the immense difficulties in conducting experimental evolutionary studies, much of the theory of evolution in macroorganisms is based on comparative studies. On the other hand, bacterial evolution is relatively easy to follow experimentally owing to their short life cycle and large population densities that increase the chance of finding appropriate mutations and the possibilities of controlling their environment. Using microbial model systems have proven to be fruitful in many genetic and physiologic fields and, we hope, will contribute to the understanding of social behavior as well. Moreover, looking at the social interactions of myxobacteria from an evolutionary point of view might suggest new approaches to experimentation with myxobacteria. We hope that the approach presented in this chapter will con-

*Institute for Nature Conservation Research, The George S. Wise Faculty of Life Sciences, Tel Aviv University, Ramat Aviv, Israel.

tribute to the understanding of both the evolution of social adaptations and the complex mechanisms in the life cycle of the myxobacteria.

## Group Selection and Individual Selection

It is generally accepted that an individual bacterium that grows and survives better than another bacterium would ultimately prevail in a culture if the two types are mixed and grown together. The success of one type of bacterium over the other types is a consequence of the advantage of the individual over the other cells. The selection mechanism involved is refered to as *individual selection,* because the adaptation spreads in the population by the selection of individuals.

Individual selection, however, is not the only type of selection that is theoretically possible. Wynne-Edwards (1962) and Wilson (1975) have suggested that adaptations may be selected for by the selective elimination of groups or populations, that is, *group selection.* Hamilton (1964) has suggested that social adaptations may be selected for through investments in individuals that are genetically related to one another, that is, *kin selection.* Among organisms such as myxobacteria, which reproduce asexually and do not move over large distances, there is not much difference between models based on group selection as suggested by Wynne-Edwards (1962) or those based on kin selection as suggested by Hamilton (1964).

There is no question that a group composed of individuals, all of which are superior to others, would outcompete other groups composed of inferior individuals. But is it possible that an adaptation that enables inferior individuals to cooperate better would enable them to outcompete, as a group, individuals that are otherwise superior to them? Group selection theory suggests that under certain conditions that could be the case.

Maynard-Smith (1976) discussed the conditions necessary for the operation of group selection. The main weakness in any group selection theory is the possibility that a group composed of cooperating individuals that are otherwise inferior would be infiltrated by superior types that are not cooperative. The latter individuals would prevail and outcompete the cooperating types through their superior adaptations as individuals. Group selection would succeed only if new cooperating groups are generated quicker than those destroyed by superior noncooperative individuals. Bacteria and other microorganisms may fulfill the conditions necessary for the operation of group selection more than macroorganisms. Although Maynard-Smith (1976) did not believe that group selection plays a primary role in evolution, he nevertheless suggested it as the mechanism by which less virulent myxomatosis has been selected from the virulent type.

Group selection is especially attractive in interpreting the evolution of social adaptations in myxobacteria. Group selection theory furnishes an apparently easy answer. Easy explanation for why individual myxobacteria (1) invest in secreting enzymes whose products are used by other bacteria and (2) participate in a sporulating mass in which they are most likely to perish. Populations of myxobacteria that possess these adaptations thrive and outcompete groups that do not possess them. Zahavi (1981) suggested that what appears to be group selection in animals can be interpreted as a more complex mode of individual selection. Could this interpretation be applied to myxobacteria as well? We will suggest some mechanism to explain the social adaptations of myxobacteria as a consequence of individual selection and present some preliminary experimental data that are in accordance with this notion.

## Social Behavior During Growth

It has been suggested that the cooperation among myxobacteria facilitates their growth because together they can use their substrates better than if each of them operated alone (Chapter 11). This suggestion was supported by experiments that showed that dense cultures of *Myxococcus xanthus* grew on casein better than the sparse population did (Rosenberg *et al.* 1977). This phenomenon of cell density dependent growth did not occur when the bacteria were grown on casitone (casein hydrolysate). The bacteria produce much less protease activity under this latter condition (Guespin-Michel personal communication). Although the myxobacteria have the capability of regulating their protease activity, it has repeatedly been observed that when myxobacteria are growing on solid casein medium they hydrolyze casein over a large area (several centimeters) around the colony. Thus, the myxobacteria produce and secrete the proteolytic enzymes in sufficiently large amounts so that the population as a whole may benefit from the breakdown of the protein. Although this phenomenon can be explained by either group or individual selection theory, each theory makes a different prediction regarding the properties of individual cells within the population. If casein use is a group characteristic selected for on the basis of the group, a mutant deficient in protease activity may exploit the niche within cultures of wild-type cells growing on casein. If, on the other hand, protease excretion is individually advantageous, mutants deficient in protease activity would be less competitive than the wild type would be. Recently, Guespin-Michel (unpublished) and Zahavi and Ralt (unpublished) have independently isolated protease deficient mutants of *M. xanthus*. It is interesting to note that neither laboratory was able to isolate a mutant with no activity; the mutants still retain 5% of the wild-type protease activity on casein. A protease-deficient mutant enabled Zahavi and Ralt to conduct competition experiments with the wild type and the mutant on liquid and solid casein media. In preliminary experiments Zahavi and Ralt have shown that the wild type outcompeted the mutant on casein medium and had no advantage on casitone medium. These results are compatible with the assumption that the adaptation to produce extracellular protease is selected for on the basis of individual selection.

This conclusion raises an interesting question: What is the advantage to the individual in the production of excessive amounts of extracellular protease? We may argue that there is no other way for myxobacteria to secrete a protease beside that which is involved with the waste, later used by other members of the population. Although such an explanation is possible, it is known that other bacteria can hydrolize casein efficiently, and their cooperation is not required for the use of the protein. For example, individual *Bacillus subtilis* cells plate on casein medium and show no cell density dependent growth (Fattum and Rosenberg 1977). An alternate explanation to cooperation in the use of casein is that there is an additional benefit to the individual from the secretion of the protease besides its function in breaking down the casein. Zahavi (1977) developed a model to explain allo feeding amongst cooperating babblers. This model suggests that the feeding is used by the babblers as a status signal (i.e., only a strong bird can afford to feed another bird). Following this model, we suggest that the protease or the breakdown products produced by its action function as an attractive signal attracting the myxobacteria into the pack. A mutant that does not excrete the protease, cannot "signal," is thus less attractive, and may be left outside the mass of cells with no collaborators.

Why is it necessary for the myxobacteria to live in packs in order to exploit their niche? One answer could be that feeding on certain substrates, or predation on other bacteria, requires the cooperative effort of more than a single bacterium. Alternatively, it could be that living in compact masses enables the myxobacteria to withstand desiccation better and thus enable them to exploit comparatively dry environments. It is interesting to note that extracellular polysaccharides, similar to those found in the slime (Kaiser *et al.* 1979), have been suggested to function in protection against desiccation (Dudman 1977).

Hence, assuming that it is advantagous for the myxobacteria to live in a pack, it is reasonable to suggest that individual myxobacteria would rather move toward a collaborator that produces excessive extracellular enzyme (or break down products) than toward a cell that produces less of them. It follows that it is advantagous for myxobacteria to produce and excrete enzyme since they thus ensure collaborators.

## Social Behavior During Development

### Synchronization of Sporulation

The selective advantage of collaboration for survival requires that the bacteria should synchronize their sporulation. Otherwise those that sporulate out of phase will not participate in fruiting body formation and will find themselves solitary or in smaller groups. It is thus reasonable that myxobacteria would signal their "intention" to sporulate and respond to the signal by initiating developmental sporulation. Bacteria that respond to the signal would benefit from that signal in proportion to the reliability of the signal. The reliability of the signal would be maintained if bacteria that produce the signal *must* sporulate. Such a link between the signal and its message, as suggested theoretically by Zahavi (1977), would reduce mistakes brought about by growing cells that would try to exploit the tendency of other cells to respond to the signal without themselves participating in the collaborative sporulation. For example, a reliable signal may be a chemical that harms growing cells; it would be reliable because it cannot be produced by growing cells. A toxic chemical signal has another advantage because it forces other cells to stop their growth. It is interesting to note that recently it has been shown that myxobacteria produce antibiotics against myxobacteria just before the lysis of the culture (Chapter 5). Such antimyxoantibiotics may serve as sporulating signals and fulfill the demand of a reliable signal.

**Developmental Lysis**

During the sporulation process on solid media many of the myxobacteria cells lyse; the survivors have all of the enzymatic machinery necessary to feed on the dying cells and hence can exploit them as substrates (Chapter 5).

Myxobacteria aggregate as a response to signals generated by starvation even though most of them will lyse as a consequence of this response. The response is still reasonable if the likelihood of surviving as a resistant spore is even smaller when the cell remains outside the mass of cells. It is important to emphasize that during development, cell death is followed by lysis that can provide nutrients for the survivors.

Using group selection theory, it is easy to suggest an hypothesis to explain the lysis during sporulation. It is better for the group as a whole to have some cells lyse and feed others under starvation conditions then to have the entire population suffer. If a mass of cells can either make many deficient or incomplete spores or fewer mature and resistant spores, a smaller mass of mature and resistent spores has a clear advantage. Thus, a group in which some members lyse to nourish the survivors has an advantage over a group in which all cells sporulate independently when nutrients are limiting.

As pointed out earlier, group cooperation is always endangered by individual exploitation, that is, a lysis-resistant mutant will have a clear advantage during the sporulation stage. Hence it would be more logical to assume that individuals actually benefit from their adaptation to lyse upon death. Lysis is tightly coupled both in time and position with sporulation (Chapter 9). Complementation of SpoC mutants deficient in lysis and sporulation with other Spo mutants or wild-type cells restores both activities (Janssen and Dworkin 1983; Shimkets and Kaiser 1982a); so far no one has reported successful complementation of sporulation mutants that does not also result in their lysis. These observations are consistent with individual adaptation theory that suggests that the ability to lyse is correlated with the ability to sporulate.

One explanation may be that the lysis process is simply a by-product of the attempt to sporulate. Those cells that fail to sporulate, die, and their lysis is just a consequence of the specific biochemical reactions of the myxobacteria.

The assumption that individuals benefit in the cooperation may suggest an alternate explanation for the specific lytic process. The following is such a speculative hypothesis.

The lytic process is an individual adaptation that serves to attract other cells to the sporulating bacterium. Under severe starvation conditions some cells die and their lysis products can serve to nourish the others; their ability to lyse might serve as a signal to other bacteria to collaborate with it. Other things being equal, a lytic genotype should be more attractive as a collaborator for sporulation than another genotype that does not lyse when dead; a cell that would survive may benefit more from such a collaborator. We suggest that myxobacteria signal their commitment to lyse if they die and thus attract other cells to sporulate with them. A nonlytic mutant that may arise occasionally cannot signal its commitment to lyse and thus would be left out of the sporulating mass. The signal could be an altered cell wall that breaks more easily. It is interesting to note that while sporulating, rearrangements in cell wall are necessary, and the cell goes through a stage in which the cell wall sensitivity to antibiotics is altered (White *et al.* 1968). Moreover, cell wall debris have been reported to restore both lysis and sporulation in the deficient mutant C (Janssen and Dworkin 1983). Alternately, the occurrence of adenosine among sporulating cells (Shimkets and Dworkin 1981) might indicate, for example, a hyperpermeability of the cell envelope, serving as a signal to indicate that such a cell would lyse easily upon death. Other signals are obviously possible as long as they are reliable, that is, the signal has to be linked to the "message" in such a way that false signals could not benefit the signaler.

## Conclusion

We hope that the highly abridged and speculative ideas presented in this chapter will be helpful in stimulating research on the social behavior of myxobacteria. We have attempted to understand some of the mechanisms responsible for the social interactions of myxobacteria in terms of evolution theories of social adaptation by individual selection. They were not intended to be post hoc explanations to facts that have already been established by experiments but rather to serve as a logic construction based on general assumptions concerned with the evolution of social interaction (Zahavi 1981). Such a theoretical construction may help to suggest new approaches to experimentation with myxobacteria.

Most evolutionary theories have arisen as a result of logic or mathematical models based on ideas generated from comperative observation; thus we believe that the use of microorganisms to test experimentally those theories would be most fruitful. It will be hard to consider any better candidate among bacteria than myxobacteria for testing theories concerning the evolution of social adaptation.

# 11

# Research on the Myxobacteria: Past, Present, Future

Martin Dworkin*

Examination of each of these periods of time—the past, the present, and the future—has its own peculiar quality and its own problems. We examine the past for a variety of reasons. One of the most important is that it is satisfying in some strangely undefined way to understand about the intellectual and scientific ground into which our roots are sunk. ("Life must be lived forwards but can only be understood backwards", Kierkegaard.) Furthermore, as scientists, our stock in trade is ideas. If we are fortunate, we leave a bit of understanding behind us; and it is important that the contributions of those who preceded us be recognized, recorded, and acknowledged as part of the intellectual history of our profession. In a more practical sense, the past often contains buried in it nuggets that when rediscovered prove to be of value. And finally, an examination of the past has heuristic value—in the words of George Santayana, "Those who cannot remember the past are condemned to repeat it." My brief discussion of the historic background of studies on the myxobacteria will not satisfy all these needs. It will not be a review of the literature. For such, the reader is directed to a number of recent reviews (Kaiser *et al.* 1979; Zusman 1980; Reichenbach and Dworkin 1981b; White 1981). I hope instead to try to point out highlights and turning points.

The present is the most troublesome. We do not have the test of time to confirm and validate what seems exciting and important today. However, I shall let this book itself deal with that problem and shall instead outline those general areas that seem to be the major concerns of researchers in the field.

In general terms, the future is, of course, the easiest. It requires little stretching of the imagination to foresee some of the immensely exciting areas of development that are opened through the myxobacteria. They are, among the pro-

*Department of Microbiology, University of Minnesota, Minneapolis, Minnesota 55455 U.S.A.

karyotes, the most complex, behaviorally and structurally. They manifest a rudimentary multicellularity, with the attendant need for cell-to-cell communication and interactions. Some of them construct elaborate fruiting bodies. And all of this takes place on a stage that is free of much of the experimental complexity and inconvenience that characterizes most eukaryotic multicellular systems. In the myxobacteria we can separate growth and development; thus it is possible to grow the cells as we do most other bacteria—in liquid or solid culture, exponentially with relatively short generation times and subject to the same sort of analysis that has been so successful in other, more conventional prokaryotes. It has, therefore, been possible to amass a considerable amount of information describing the nutrition, metabolism, fine structure, and genetics of the cells that are in the growth mode. Thus an examination of the nature of the transition from that mode to the developmental one becomes practicable. Their domestication has made it possible to subject the fascinating and often mysterious subject of cell interactions to the kind of rigorous genetic and physiologic analysis that has been so powerful in describing other prokaryotic phenomena.

## The Distant Past

"A few years since, while collecting fungi at Kittery and in several other localities in New England and the Southern states, the writer's attention was attracted by a bright orange colored growth occurring upon decayed wood, fungi and similar substances . . ." (Thaxter 1892). Thus began the studies on the group of bacteria that were eventually to be referred to colloquially as "the myxobacteria." This is not the place for a detailed and scholarly exegesis, but three names, in addition to Roland Thaxter stand out as pioneers in the field.

In 1924 the German botanist, E. Jahn, published his own classic monograph on the myxobacteria (Jahn 1924). This was the first real review of the myxobacteria and dealt with their morphology, growth, nutrition, spore formation and germination, fruiting body formation, and gliding motility. In addition, Jahn presented a taxonomic scheme that firmly placed the myxobacteria among the bacteria.

In the late 1920s and early 1930s the Polish microbiologists, Helena Krzemieniewska and Seweryn Krzemieniewski, (1926, 1927a, b, 1930) published their monumental studies on the myxobacteria of Poland. This work was essentially concerned with the ecology of the myxobacteria, and its focus was a systematic description of the various species found in the soil. However, what emerged that has proven to be more important, was their attempts to improve and refine techniques for the isolation of myxobacteria from nature. Their method using rabbit dung pellets embedded in soil is still the method of choice for isolating many myxobacteria. In addition, they recognized that there were two different vegetative cell morphologies among the myxobacteria—the long, thin rods that now characterizes the suborder *Cystobacterineae* and the shorter, blunt-ended rods characteristic of the *Sorangineae* (Krzemieniewska and Krzemieniewski 1928; Reichenbach and Dworkin 1981b).

Hans Kühlwein was perhaps the first person to develop a sustained physiologic analysis of the myxobacteria. His research on myxobacteria began in 1950 (Kühlwein 1950) and continued until his retirement in the late 1970s. During that time he not only trained a number of young investigators who themselves have become major figures on myxobacterial research, but he also examined a wide variety of subjects such as nutrition (Kühlwein and Finck 1950), motility (Schmidt-Lorenz and Kühlwein 1968), slime production (Grimm and Kühlwein 1973b), lytic activities (Harcke *et al.* 1971), induction of fruiting bodies (Kühlwein 1953), cell structure (Kühlwein 1963), taxonomy and ecology (Kühlwein 1960). Kühlwein represented the transition from the distant past into what may be referred to as the recent past.

## The Recent Past

### The Organism: *Myxococcus xanthus*

Nellie Angelina Woods, in her Master's Thesis in 1948 (Studies on the Myxobacteria, Masters Thesis, University of Washington) described a strain of *Myxococcus xanthus* that grew in a dispersed state. This observation went unrecognized for a decade, but given the preference by most bacteriologists for working with cultures growing in a dispersed fashion in liquid medium, it is not surprising that the approach surfaced once again with the work of Dorothy Powelson. She and one of her students, using a dispersed-growing strain of *M. xanthus*, published a paper on the lysis of *M. xanthus* including the first published electron micrographs of thin sections of the cells (Mason and Powelson 1958a).

I had already decided to begin to study the myxobacteria, and two things were immediately obvious. First, I settled on the strategy of focusing intensely on one species of the group, hoping to collect sufficient detailed information on its fundamental biology to be able then to ask questions about its development. This represented for the myxobacteria a break with the past. Most of the work up until then had been done by microbiologists who had been trained as botanists and whose approach was monographic—to attempt to study the myxobacteria as a group. Second, it was obvious that the possibility of using a dispersed-growing strain offered the opportunity to subject the organism to the powerful technology of bacterial physiology; we could study growth, nutrition. fine structure, and intermediary metabolism in a convenient and familiar fashion. At the time, none of this information was available for *M. xanthus*. I obtained a culture of *M. xanthus* from Powelson's lab, and this culture was designated vc to indicate that it had lost the ability to undergo development and could only exist in the vegetative growth phase. With this organism I was able to work out the nutritional requirements for vegetative growth (Dworkin 1962). I subsequently (1961) obtained another culture of *M. xanthus* from the culture collection of the Department of Bacteriology and Immunology at the University of California in Berkeley. This strain, designated FB to indicate that it was able to form fruiting bodies in a normal fashion, was originally isolated by Erling Ordal at the University of Washington. This strain, which is also capable of dispersed growth, is the progenitor of many of the laboratory strains of *M. xanthus* currently in use. Using this strain, my laboratory was able over the next 10 years to study the nutritional induction of morphogenesis (Dworkin 1963), the fine structure of *M. xanthus* during myxospore morphogenesis (Voelz and Dworkin 1962), the physiology of myxospore development and germination (Dworkin and Voelz 1962; Ramsey and Dworkin 1968), terminal and intermediary metabolism (Dworkin and Niederpreum 1964; Watson and Dworkin 1968), photobiology (Burchard and Dworkin 1966a), cell wall composition (White *et al.* 1968), myxospore resistance (Sudo and Dworkin 1969), and peptidoglycan–lytic enzymes (Sudo and Dworkin 1972) and to isolate the first bacteriophage reported for the myxobacteria (Burchard and Dworkin 1966b).

## Glycerol Induction

The fact that myxobacterial development took place on only a solid surface was a deterrent to attempts to study the physiology and biochemistry of the process. Thus it came as a most welcome finding that it was possible to induce vegetative cells growing dispersed in a liquid culture to convert to myxospores rapidly, quantitatively, and relatively synchronously (Dworkin and Gibson 1964). This induction could be carried out in a reliable fashion by the addition to the culture of 0.5 *M* glycerol or a number of other related compounds. With this discovery, it became possible to isolate a portion of myxobacterial development and to subject it to analysis. For the 10 years or so following the discovery of glycerol induction, essentially all of the work on myxobacterial development was done with this system. It has subsequently become obvious that there are a number of aspects of glycerol-induced myxospores that differ from those of bona fide fruiting body myxospores (Chapter 9). However, they are sufficiently similar so that it has been possible to investigate the sequence of morphologic events involved in their development and germination, resistance properties of myxospores, chemical composition of the spore coat, relationship between shape change and peptidoglycan, and certain aspects of their intermediary metabolism. Perhaps more importantly, the technique led to the recognition that the myxobacteria were no longer exotic, arcane, and experimentally impractical organisms, but rather that they were now clearly amenable to biochemical and genetic analyses.

## Chemotaxis

It has been tacitly assumed for many years that the myxobacteria possess a chemotactic mechanism and that this mechanism is involved in the formation of the aggregates that ultimately become fruiting bodies. There have been a number of demonstrations of diffusible factors that induce fruiting (Jennings 1961; Lev 1954; Fluegel 1963b; McVittie and Zahler 1962), but in no case has it been shown that these factors actually function as chemoattractants rather than as inducers of fruiting body formation. One group (Dworkin and Eide 1983; Dworkin 1983) has recently presented experimental and theoretical arguments that suggest that *M. xanthus* may perceive objects rather than chemical gradients.

## Molecular Analyses

The earliest attempts to generate a molecular analysis of development in myxobacteria seem rather crude in the light of today's sophisticated molecular techniques and concepts. The first paper by Gogol and Rosenberg (1964) on DNA synthesis was followed by measurements of macromolecular synthesis during glycerol induction of myxospores (Sadler and Dworkin 1966), the DNA cycle during growth and development (Rosenberg *et al.* 1967; Zusman and Rosenberg 1968, 1970), and RNA synthesis during growth and development (Bacon and Rosenberg 1967; Ramsey and Dworkin 1970, Okano *et al.* 1970) and studies of ribosome synthesis during development. Most of this work was essentially descriptive and had relatively little explanatory content, however, it laid the groundwork for what was to come years later.

The first attempt to determine the genome size of *M. xanthus* was based on a simple correlation of the DNA content and the chromosome number in myxospores. The value obtained was $4.9 \times 10^9$ daltons (Zusman and Rosenberg 1968). Subsequently, Zusman *et al.* (1978) reported that the genome of *M. xanthus* was $8.4 \pm 1.2 \times 10^9$ daltons. This was an indirect measurement based on the amount of DNA in exponential phase cells and in newborn cells. It was also based on a number of assumptions about the nature of the DNA synthesis cycle in *M. xanthus*. The value was supported by a viscoelastic measurement that showed the molecular weight to be $8.2 \pm 1.5 \times 10^9$ daltons. These latter data were unpublished, but the finding that the genome size was from 3 to 4 times the size of *Escherichia coli* and *Bacillus subtilis* was the basis of much discussion over the years. In 1981, Yee and Inouye reexamined the genome size using an ingenious new method based on quantitation of individual restriction fragments. They also measured the size by renaturation analysis. Their results indicated quite clearly that the size of the genome of *M. xanthus* was $3.1–3.8 \times 10^9$ daltons. The genome size of *Stigmatella aurantiaca* was also determined and found to be essentially the same as that of *M. xanthus*.

Thus, the genomes are 25–50% larger than *E. coli*—far smaller than had once been believed but still large enough to suggest either that a great deal of extra information is necessary for the complicated social behavior of the myxobacteria or that there is a great deal of nontranscribed DNA.

There have been many attempts among workers in prokaryotic development to implicate stable mRNA analogous to that commonly encountered in developing eukaryotic systems. Some of these early attempts have been far less convincing than others. However, Smith and Dworkin (1981) reported that during fruiting body formation, *M. xanthus* synthesized a population of messenger RNA (mRNA) whose half-life was substantially longer than that characteristic of vegetative cells, that is 13 versus 4 min. This observation was repeated and confirmed by Nelson and Zusman (1983a) who also showed that the stable mRNA was probably coding for protein S.

## Taxonomy

An area of major activity of the 1960s and early 1970s was that of myxobacterial taxonomy. It all began in 1965 with a seminal paper by Mandel and Leadbetter (1965) in which it was reported that the percent guanine plus cytosine (G + C) ratios of a number of diverse myxobacteria all fell within a narrow range of 68–71%. Even more interesting, the ratios for *Cytophaga* and *Sporocytophaga,* long considered to be nonfruiting myxobacteria, were shown to fall in a range of 34–43%. McCurdy's laboratory systematically examined the genera of the myxobacteria in a series of papers leading to his revised taxonomic scheme that appeared in the eighth edition of *Bergey's Manual* (1974). (The new respectability of the myxobacteria was demonstrated by the appearance of a picture of a fruiting body of *S. aurantiaca* on the cover of *Bergey's Manual.*) An alternative taxonomic scheme has been published by Reichenbach and Dworkin (1981), and time will determine which of the two is the more useful or phylogenetically accurate.

## Cinematography

An important contribution to the biology of the myxobacteria has been a series of time-lapse motion picture films of the myxobacteria made by Reichenbach in collaboration with Encyclopedia Cinematographica, a commercial organization in Göttingen, Germany (Reichenbach 1974b, Reichenbach *et al.* 1965a, b,c,d, 1975–1976). These films, done with dazzling technical expertise, illustrate the gliding motility, fruiting body formation, spore and cyst germination, rhythmic oscillations, and a variety of cell interactions in a way that words or still photos could not achieve. The films emphasize the behavioral complexity of the myxobacteria and help to define the biologic problems that we can now begin to dissect.

### *Stigmatella*

As indicated earlier, a conscious decision was made in my lab to focus essentially all our attention on *M. xanthus*. However, we could not resist the allure of *S. aurantiaca* and, with Reichenbach, published three papers on *S. aurantiaca* (Reichenbach and Dworkin 1969, 1970; Reichenbach *et al.* 1969). These papers and another on the fine structure of *Stigmatella* fruiting bodies (Voelz and Reichenbach 1969) made it clear not only that *S. aurantiaca* was experimentally tractable, but also that it was too interesting to be ignored. During the next 10 years White's laboratory worked out most of the practical details of growing and handling a dispersed-growing strain of *S. aurantiaca* and for inducing the relatively rapid and synchronous formation of fruiting bodies (Qualls *et al.* 1978a). In addition, he has demonstrated a morphogenetic effect of visible light (Qualls *et al.* 1978b) (a unique phenomenon for prokaryotes) and a diffusible pheromone required for development (Stephens *et al.* 1982).

### Bacteriophage

The first bacteriophage isolated for *M. xanthus* was phage Mx-1 (Burchard and Dworkin 1966b). Mx-1 is a virulent phage, and although attempts to use it for genetic transduction were unsuccessful, it showed an interesting property with regard to the developmental cycle of the host cell. If cells were infected a short time after myxospore formation had been induced, the phage was able to enter the cell, but its subsequent development was preempted by the myxospore formation. Mature spores contained a trapped immature phage that, upon subsequent germination of the spore, released mature virulent phage. This phenomenon was rediscovered in *Bacillus* some years later and led to the exciting work on the role of RNA polymerase accessory factors as developmental regulators.

Campos and Zusman (1978) reported the isolation of a phage for *M. xanthus* that was capable of generalized transduction (Mx-4). This extremely important accomplishment was followed by the isolation of a series of other generalized transducing phages, Mx-8 and Mx-9 (Martin *et al.* 1978). Rodrigues *et al.* (1980) isolated additional phages similar in morphology to Mx-1, Mx-4, and Mx-8 and obtained preliminary indications that one of the isolates established lysogeny in *M. xanthus*. Recently, Orndorff *et al.* (1983) have provided convincing evidence that Mx-8 is capable of lysogenizing *M. xanthus*.

No bacteriophage have yet been isolated for *Stigmatella aurantiaca* or for that matter for any other myxobacteria.

## Submerged Fruiting

In 1964, Fluegel described a method for inducing *Myxococcus fulvus* to form fruiting bodies on the glass surface of a Petri dish covered by liquid medium. His method entailed growing the cells on a skim milk medium during which they attached to the bottom of the Petri dish, and then replacing the growth medium with a $Ca^{2+}$ solution. Under these conditions they formed fruiting bodies analogous to those formed on the surface of an agar medium. It was an interesting observation but 10 years too early. The field was not yet at the point at which the method could be used to ask the kinds of questions about cell interactions that have slowly evolved. Furthermore, the method does not work with the typical dispersed growing laboratory strains that do not adhere to surfaces. In 1982, Kuner and Kaiser resurrected Fluegel's technique using a strain of *M. xanthus* (DK-1622) that had been genetically constructed so as to contain $A^+$ and $S^+$ motility (see Cell Interactions—Motility) and that adhered more strongly than the fully dispersed-growing strains to the Petri dish. The method will be useful for detecting and collecting any diffusible signals involved in fruiting body formation and will also facilitate microscopic examination of fruiting body formation.

## Antibiotics and Lytic Factors

The ability of myxobacteria to lyse or inhibit other microorganisms has long been recognized (Chapter 5). This fundamental property of the myxobacteria has been of interest for the following three reasons:

1. It seems obvious that the ability to inhibit the growth of other organisms by the excretion of an antibiotic is of ecologic interest. Whether or not this property functions as a natural weapon is no more clear in the case of the myxobacteria than with other antibiotic producers. However, the ability of the myxobacteria to excrete cell-lytic enzymes and to feed on the solubilized products of the lysed cells certainly seems to go hand in hand with antibiotic production.
2. It has long been considered a possibility with other bacteria that antibiotics function in a developmental rather than an ecologic sense. That this is so for the myxobacterial autolytic enzymes seems clear (Wireman and Dworkin 1977); that it may also hold for myxobacterial antibiotics has been suggested by the work of Rosenberg's group (see Chapter 5).
3. The practical implications of antibiotic production are enormous. The myxobacteria represent an untapped source of new antibiotics (Chapter 5).

## The Present

There are about 20 laboratories actively working on the myxobacteria. One-half to two-thirds of them are in the United States; the rest are scattered among England, Canada, France, West Germany, and Israel. I shall cover the topics that are presently the focus of the most general interest. These fall into four general areas:

1. Genetic analysis of development
2. Biochemical correlates of development
3. Cell interactions
4. Motility and tactic behavior

### Genetic Analysis of Development

#### Generation of Mutants

One of the first things a geneticist does in attempting to apply a genetic analysis to a problem is to obtain mutants. This allows us to do a number of things: (1) We can determine if a physiologic or biochemical property is causally related to the process we wish to analyze. (2) We can determine the complexity of a process by determining the number of genes involved in it. (3) We can order events. (4) We can reveal interactions between genes that regulate the process.

There has been no shortage of mutants in *M. xanthus*, and obtaining them has proven to be a relatively routine matter. Thus there are available a fairly wide variety of conventional physiologic mutants as well as a large number of different types of developmental mutants. Of especial interest in this latter category is two groups of mutants that are blocked in their ability to manifest the normal cell interactions. The members of one of these groups (Hagen *et al.* 1978) were unable to sporulate and seemed to be blocked at a variety of stages during fruiting body formation. The important feature of these mutants was that some of the mutants when mixed with other mutants could, in partnership, undergo complete development. The progeny of this collaborative development retained their original developmental mutation. Thus, the complementation was phenotypic rather than genetic. The mutants fell into four complementation groups referred to as A, B, C, and D, defined by the fact that members of any group could complement members of another group but not of its own. These mutants thus implied the existence of intercellular signals involved in fruiting body development.

The second group of cell interaction mutants revealed an interesting and unexpected feature of the motility of *M. xanthus*. There were two classes of mutants, referred to as A (for adventurous) and S (for social). Mutants in an A locus were unable to move as single cells but could only glide as a group;

mutants in an S locus moved as single cells or as small, thin flares. Both of these groups of mutants are discussed in more detail in Chapter 8. The important point to emphasize here is that they offer an invaluable approach for analyzing the mechanisms and functions of cell interactions.

Kaiser and Dworkin (1975) showed that it was possible to infect *M. xanthus* with the *E. coli* phage, P1. The infection was nonproductive in terms of progeny phage, but it was possible to transduce chloramphenicol resistance. Subsequently Kuner and Kaiser (1981) showed that they could infect *M. xanthus* with P1 carrying the Tn*5* transposon, which, in turn, bears the gene for kanamycin resistance. This discovery has had a profound effect on myxobacterial genetics, because it can not only generate a single-site, null mutant at any position on the chromosome, but also because it can place a selectable marker (kanamycin resistance) adjacent to a nonselectable gene (e.g., a developmental gene). Thus the latter becomes, in effect, selectable by virtue of cotransduction or cotransformation and subject to mapping, complementation, or whatever procedure that requires that we be able to pick out a rare recombinant. Sodergren and Kaiser (1983) have pointed out that the usefulness of Tn*5* as a linked genetic marker is based on the following five features:

1. The level of resistance to kanamycin generated by the presence of Tn*5* is higher than that found in spontaneous kanamycin-resistant mutants. Thus it can be efficiently selected.
2. Tn*5* can be located sufficiently close to any *M. xanthus* gene for it to be cotransducible with it.
3. The inserted Tn*5* is stable, further transposition occurring with a frequency less than $10^{-3}$ per cell.
4. It is possible to replace the Tn*5* (kan) with a Tn*5* (tet) at the same site.
5. The frequency of cotransduction between two Tn*5* insertions is a valid parameter of the physical distance between the two sites.

## Mapping and Complementation of Genes

There are, generally speaking, three sets of genes in *M. xanthus* that are being subjected to genetic analysis in terms of mapping. These are the genes for motility, the genes involved in the phenotypically complementable fruiting mutations, and those in the aggregation loci.

Sodergren and Kaiser (1983) have recently mapped seven types of the motility mutants; these included one mgl mutant, which is believed to be a mutant with a defect in its motility machinery, and six conditional mutants from six different complementation groups that can be stimulated to move by physical contact with a wild-type cell or by a mutant from a different complementation group. The important result from these experiments was that the six stimulatable mutant groups mapped in six distinct loci. This suggests that each locus is coding for a different protein and that six such proteins are needed for motility. Since stimulation requires cell-to-cell contact, it suggests the exciting possibility that these six proteins are cell surface components that can be transferred from cell to cell.

Kaiser's lab has also been mapping the various loci associated with the sporulation mutations that can be phenotypically complemented. Again using the Tn*5* transposon and three-factor crosses, three group C mutations have been shown to map in a single cluster distant from the group B mutations.

One of the problems that has limited the utility of genetic approaches to studying prokaryotic development has been the difficulty in obtaining stable partial diploids with which to perform complementation and dominance tests. Recently, however, Shimkets *et al.* (1983) in Kaiser's lab has constructed a tandem duplication of the spoC locus of *M. xanthus*. The authors used phage P1 to transduce a Tn*5* carrying the spoC gene into a recipient that already contained a spoC gene adjacent to another Tn*5* transposon. This extremely valuable approach now makes it possible to do complementation and dominance tests—an extremely useful genetic tool. (See Chapter 8 for additional details.)

### Recombinant DNA Techniques

Cloning of developmental genes of *M. xanthus* into *E. coli* is now routinely feasible. See Chapter 8 for a more detailed discussion of the techniques and approaches now available.

## Biochemical Correlates of Development

The most sophisticated genetic analysis is not completely useful in describing the control of a process unless biochemical approaches can be developed to complement the genetics. Thus the thrust of this section will be to point out those areas in which developmentally relevant gene products have been defined.

### Protein S

During fruiting body formation, *M. xanthus* directs a considerable proportion (15%) of its biosynthetic capacity toward the synthesis of protein S. This 23,000 mol wt protein is synthesized and accumulates in the cytoplasm during early stages of development and is later transported to the surface of the myxospore (Inouye *et al.* 1979a,b). Zusman (Chapter 9) has suggested that protein S functions as a surface adhesive to hold the spores together in the fruiting body. Protein S has been purified, crystallized, and characterized physically, and part of its structure has been determined (Inouye *et al.* 1980b,

1981). In addition, its gene has been cloned and has been shown to exist as a natural tandem duplication (Inouye *et al.* 1983). It is thus a prime candidate to serve as a model system for investigating developmental regulation.

### Myxobacterial Hemagglutinin

Cumsky and Zusman (1979) have shown that developing cells of *M. xanthus* synthesize a protein with hemagglutinating activity. The protein has been isolated, purified, and characterized (Cumsky and Zusman 1981a), and its biosynthesis shows developmental kinetics—appearing in the periplasm during cell aggregation and disappearing during spore formation (Cumsky and Zusman 1981a). Although the myxobacterial hemagglutinin (MBHA) binds to the galactose residue of the glycoprotein fetuin and less avidly to cells of *M. xanthus* (Cumsky and Zusman 1981b), essentially nothing is known of its function. An interesting clue has been provided by Nelson *et al.* (1981b) who showed, by means of immunofluorescent labeling, that the protein was concentrated at the cell ends suggesting that it may play a role in end-to-end interactions.

### Membrane Proteins

Orndorff and Dworkin (1982) took a structural rather than a functional approach to the problem of defining relevant gene products and examined the changes in membrane proteins during fruiting body formation. They had originally hoped to be able to do so in outer and cytoplasmic membranes and devised a method for separating the two fractions in vegetative cells (Orndorff and Dworkin 1980). Unfortunately, some change in the structural organization of the outer and cytoplasmic membranes occurs during development, so that the separation techniques effective for vegetative cells proved ineffective for developing cells. In any case, they were able to show, not surprisingly, that during development the protein composition of the membranes changed, with new proteins appearing and some vegetative proteins disappearing from developing cells. This was consistent with an early paper by Inouye *et al.* (1979a) that showed that, in general, the biosynthetic pattern of the proteins of *M. xanthus* changed during development.

## Cell Interactions

Cell interactions represent the social behavior of the myxobacteria. It is this aspect of their biology that distinguishes them from all other prokaryotic systems that are studied from a developmental perspective. And it is this aspect of their biology that captures the attention of many of those who now work with the myxobacteria. Cell interactions pervade the behavior of these organisms whether they are engaged in the growth mode or the developmental mode of their life cycle. It has been suggested that the function of social behavior in the myxobacteria is to optimize feeding (Dworkin 1973). In nature, the myxobacteria presumably feed on insoluble macromolecules such as protein, polysaccharide, and peptidoglycan. They do so by excreting a variety of hydrolytic enzymes that solublize the target, whereupon the low molecular weight products then diffuse back to the cell. It seems intuitively correct that such a process would benefit by the cells being present at a high cell density, thus optimizing the efficiency with which they can solublize the particulate substrate. Rosenberg *et al.* (1977) showed that this was in fact the case by demonstrating that the rate of growth of *M. xanthus* on casein (but not on a low molecular weight substrate) was a function of the cell density. In fact, single cells were unable to grow on casein yet were perfectly able to form a colony on an enzymatic hydrolysate of casein.

Thus the behavior of the myxobacteria, whether reflected by their group movement, their group germination or aggregation and the formation of multicellular fruiting bodies, seems always directed toward maintaining or reestablishing a high cell density population. In a number of cases that will be discussed (e.g., phenotypic complementation, social motility, cell density effects), the evidence points to the existence of cell–cell signals. However, in no case has this been rigorously demonstrated. Such rigorous demonstration requires that, in addition to the type of evidence already available, the existence of signal receptor sites on a recipient cell be demonstrated.

## Motility

It has always been clear to anyone who has observed the gliding movement of the myxobacteria that there is a social dimension to their motility. This was first obvious upon examination of Reichenbach's time-lapse films; the first experimental handle came with Burchard's isolation of a motility mutant he referred to as SM for semimotile. This mutant was able to move as a group of cells but not as single cells. Hodgkin and Kaiser (1979b) then showed that motility in *M. xanthus* was controlled by two systems that they designated as A for adventurous and S for social. The A system controlled movement of individual cells and the S system movement of groups of cells. This unexpected result was the outcome of a series of elegant genetic experiments described in more detail by Kaiser in Chapter 8. The subject will be considered again in the discussion of motility and tactic behavior. At this point I want to emphasize the interactive aspects of the phenomenon. I have already referred to the conditional mutants whose motility can be temporarily restored by physical contact with wild-type cells. Kaiser (1979) has shown that there is a close correlation between S motility and the presence of polar pili on the cells. (A motility seems to be independent of the presence of pili; thus motility per se is not involved.) Furthermore, when the conditional S motility mutants are stimulated to move, they synthesize pili for a brief period of time. The physiologic basis of such an interaction is a mystery, and until we know more about the mechanism of gliding, it is likely to remain a mystery. However, it seems reasonable to suggest that piliation plays some sort of a role in the social interactions involved in group movement.

## Signaling

The phenotypically complementable mutants described by Hagen *et al.* (1978) and referred to earlier have proven to be an extremely useful experimental probe into the problem of cell interactions. It has been implied, but as pointed out earlier not rigorously demonstrated, that the complementation groups reflect an equivalent number of sets of signals that must normally be exchanged for development to take place. La Rossa *et al.* (1983) have attempted to demonstrate the presence of extracellular signals that can affect the complementation. They have done so by separating the complementing pairs by a filter that will allow the passage of subcellular material but not of cells themselves. Under these conditions, the wild-type cells can complement each complementation group, but unfortunately the groups will not complement each other across the filter. One interpretation of these results is that the complementing signal is, in fact, not diffusible but rather is cell bound. The ability of the wild-type cells to complement across the membrane may be a reflection of the fact that the wild-type cells (but none of the mutant groups) is able to undergo developmental autolysis, thus releasing the signal to diffuse across the membrane. The developmental autolysis referred to was described by Wireman and Dworkin (1977) and involves the autolysis of 80–90% of cells of *M. xanthus* during the process of fruiting body formation. One never sees development without autolysis, either in *M. xanthus* or in *S. aurantiaca,* even under a variety of different conditions leading to fruiting body formation. It has been suggested that the autolysis is an integral and necessary part of the developmental process providing nutrient and/or some specific signals that coordinate subsequent events. However, in the absence of an autolysis negative mutant, it is difficult to subject the notion to a definitive analysis.

Janssen in my laboratory has been able to isolate a fraction from wild-type cells of *M. xanthus* that, when added to group C mutants, allows them to complete development, including autolysis and sporulation. The signal is tightly cell bound and appears to have the properties of a polysaccharide or a glycopeptide. The activity of the factor can be replaced by mannosamine or glucosamine but not by any combination of fragments of peptidoglycan (Janssen 1983). Shimkets and Kaiser (1982a), on the other hand, have reported that a mixture of *N*-acetylglucosamine, *N*-acetylmuramic acid, diaminopimelic acid, and D-alanine is effective in rescuing sporulation in group C mutants. This apparent discrepancy is due to the fact that the group C mutants (and probably the other groups as well) represent at least a double mutant. Work in my laboratory has clearly indicated that Janssen's cell-bound signal is complementing a different developmental mutation than that being rescued by the murein components.

It is a common observation that a number of processes during development of myxobacteria is dependent on the cell density. For example, the dependence of fruiting body formation on cell density has been clearly demonstrated (Wireman and Dworkin 1975; Shimkets and Dworkin 1981). Thus, the obvious question is, how does a cell know what its cell density is. Ramsey and Dworkin (1968) observed that the germination of glycerol-induced myxospores in distilled water was cell density dependent and showed that the spores excreted orthophosphate whose external concentration they then used as a parameter of cell density. This simple but cunning strategy seems also to be used by the cells in determining their cell density prior to fruiting body formation; the cells excrete adenosine whose extracellular concentration serves as their cell density parameter (Shimkets and Dworkin 1981). As is the case for all the other signals used by the cells, the obvious questions have to do with the timing and regulation of the signal, its reception and perception by the cells, and how that perception is transduced into developmental behavior.

In 1969, Reichenbach and Dworkin reisolated *S. aurantiaca* that had, since Thaxter's time, been included among the *Chondromyces* as *Chondromyces aurantiacus*. Reichenbach and Dworkin resurrected *S. aurantiaca,* proposed it as an independent species, described a number of aspects of its growth and physiology, and, most importantly, showed that it could be grown dispersed in liquid culture.

The fruiting bodies of *S. aurantiaca* are among the most complicated that are formed by the myxobacteria (Voelz and Reichenbach 1969). Thus at first glance, even the most courageous investigators might hesitate before choosing *Stigmatella* as a subject for investigation. However, White's laboratory has since then almost single-handedly developed *S. aurantiaca* as a model system for examining certain kinds of cell interactions among the myxobacteria (White 1981). There are four aspects of the behavior of *S. aurantiaca* that, at the present time, are more appropriately studied in this organism than in *M. xanthus;* three of these are currently being investigated, a fourth lies in the future. They are as follows:

1. *The developmental effects of light* (Qualls *et al.* 1978b). When *S. aurantiaca* is incubated at a low cell density but under conditions that would otherwise lead to fruiting body formation, no fruits are formed unless the cells are exposed to a brief period of irradiation with light in the blue region of the visible spectrum (410–470 nm) (White *et al.* 1980b). Inouye *et al.* (1980a) have shown that there is a different pattern of protein synthesis in the light and in the dark suggesting that the effect extends to the level of control of protein synthesis.
2. *Cell clumping and adhesiveness.* When cells of *S. aurantiaca* are removed from an agar surface on which they are undergoing aggregation, but before aggregation has occurred, and placed in liquid suspension, they will form clumps that eventually become shear resistant. These clumps are specific in

that they essentially exclude cells of *E. coil, Pseudomonas,* or *M. xanthus* included in the suspension (White 1981). The clumps presumably reflect the normal acquisition of adhesiveness during aggregation and represent a nice model system for examining the nature of the cell-to-cell interactions that are involved.

3. *Stigmatella pheromone.* If low-density populations that ordinarily will not fruit are placed adjacent to a fruiting, high-density population, the low-density cells are induced to fruit. A low molecular weight lipidlike material has been isolated that will duplicate the effect and has been referred to as a myxobacterial pheromone. The pheromone will also induce dark grown cultures to form fruiting bodies (Stephens *et al.* 1982).
4. *Stigmatella fruiting bodies.* No one who has seen a fruiting body of *S. aurantiaca* or, even more dramatically, watched the time-lapse films of its formation can resist wondering how a prokaryotic cell can form so complex a multicellular structure. There are no experimental approaches currently available to examine this fascinating question.

It has, by now, been clearly demonstrated that there is a set of questions pertaining to myxobacterial interactions for which *S. aurantiaca* is useful, if not uniquely suited. Furthermore, the routine laboratory technology for doing so is now in place, and the essential limitations to working with this fascinating myxobacterium would be limitations with regards to creative ideas rather than techniques.

In summary, there is little doubt that cell interactions occur in the myxobacteria. Furthermore, it is highly likely that these include both cell surface contact interactions and those mediated by diffusible, extracellular signals. However, the functions of diffusible signals are not obvious for the *M. xanthus*. First, the fact that it is possible to obtain fruiting body formation submerged in liquid culture (Kuner and Kaiser 1982) suggests that, although the process may involve diffusible signals, it is unlikely that these are in the form of concentration gradients that control directional processes. Second, the very nature of myxobacterial development is such that the cells, in order to develop properly, are literally piled atop each other on the agar surface. Unlike the case with *Dictyostelium,* when the myxamebae occupy about 1% of the surface area prior to aggregation, the myxobacteria have no need to call each other in over vast distances—they are usually literally in touch with each other. Thus the roles of the adenosine cell density signal or the *Stigmatella* pheromone are somewhat enigmatic.

## Motility and Tactic Behavior

### Mechanism of Gliding Motility

The mechanism of gliding motility in the myxobacteria, or for that matter in the gliding bacteria in general, remains an enigma. There are simply no compelling explanations at hand. Pate and Chang (1979) and more recently Lapidus and Berg (1982) have described a tantalizing observation using *Flexibacter* and *Cytophaga.* Their observations concern the ability of these gliding bacteria to move small polystyrene latex beads over their surface in a way that suggests that it may be related to the gliding mechanism. The one myxobacterium that has been examined in my laboratory, *M. xanthus,* does not move polystyrene latex beads at all. It is quite possible that there are classes of different mechanisms of gliding among the group as a whole.

With regard to myxobacterial motility, there are two approaches that have been seriously pursued. One focuses around the presence in the cytoplasm of myxobacteria of structures referred to as cytoplasmic fibrils or filaments. These were first described in a number of myxobacteria by Kühlwein's laboratory (Schmidt-Lorenz and Kühlwein 1968) and subsequently shown to be present in *Chondromyces crocatus* (MacRae and McCurdy 1975) and in *M. xanthus* (Burchard *et al.* 1977a). These seem not to be the same structures; they differ in morphology, cellular organization, and in the conditions under which they are found in cells. There is no direct evidence that any of these are related to motility. Keller *et al.* (1983) and Dworkin *et al.* (1983) have proposed that the driving force for motility in *M. xanthus* is the polar and reversible excretion of surfactant. They have presented a theoretical analysis to show that it is feasible for a cell to generate a sufficient gradient of surfactant, by polar excretion, to propel an object the size of a cell at the speed and under the conditions characteristic of myxobacteria (Keller *et al.* 1983). Furthermore, their analysis is consistent with the observed motility behavior of pairs of cells in close juxtaposition to each other. Finally, the analysis predicts that groups of cells should be able to move more rapidly than individual cells. In an accompanying paper (Dworkin *et al.* 1983), they have provided experimental evidence for the model. The idea that myxobacterial motility is driven by a cell-generated gradient of surface tension makes certain predictions. For example, since agar is 98.5% water, the interface between agar and an aqueous layer should have a negligible interfacial tension. Since the cell is essentially therefore not at an interface (with regard to surface tension), a surfactant gradient should have no effect. Thus, movement of cells on agar overlaid with aqueous medium or embedded in 0.5% agar should be inhibited; this has, in fact, been demonstrated (Dworkin *et al.* 1983). Such behavior is difficult to understand if motility is a result of some mechanical process. In addition, the model predicts that if the cell-generated surfactant gradient is swamped out by a high background of added surfactant, motility would be inhibited. And, in fact, the addition of two nontoxic surfactants, albumin and lignosulfonate, have that effect. Finally, it has been shown that the cells, while gliding, do produce and excrete surfactant (Dworkin *et al.* 1983).

Movement driven by a surface tension gradient, in contrast to mechanical models, is difficult to intuit; yet the model has the virtue of having been shown to be theoretically feasible; the experimental data, although not compelling, are permissive and, in some cases, difficult to rationalize by a mechanical model.

### Tactic Behavior

The myxobacterial literature abounds with casual references to the role of chemotaxis in myxobacterial feeding or development. In addition, there are a number of papers that either explicitly claim to have demonstrated chemotaxis (McVittie and Zahler 1962; Fluegel 1963b) or that have been cited as having demonstrated chemotaxis and induction of fruiting body formation. In one case in which a direct claim was made for a chemotactic effect per se (Ho and McCurdy 1979), it was claimed that cells of *M. xanthus* showed biased movement toward higher concentration of cyclic guanosine 5′-monophosphate (cGMP) and adenosine 5′-monophosphate (5AMP). No evidence was presented for the latter; for the former, the data are simply not convincing. Shimkets *et al.* (1979) also claimed to have demonstrated a chemotactic response to AMP; however, Dworkin and Eide (1983) have recently reported an extensive and systematic attempt to demonstrate a chemotactic response by *M. xanthus*. Using three different approaches, a variety of different compounds and undefined mixtures over a wide range of concentrations, and responder cells prepared under different physiologic conditions, the authors have been unable to demonstrate a chemotactic response. They have, in addition, presented ex post facto arguments that suggest that chemotaxis is not appropriate for myxobacteria from a mechanistic, a functional, or an ecological point of view. That is not to say that myxobacteria do not manifest directed movement. The demonstration of elasticotaxis by Stanier (1942), the time-lapse films of Reichenbach *et al.* as well as the informal observations of many workers suggest that such directed movement does indeed take place. Using time-lapse videomicroscopy, I have been able to show that directed movement does in fact occur; however, it is toward inert objects rather than toward concentration gradients. Thus cells of *M. xanthus* gliding on agar surfaces will perceive polystyrene latex and glass beads of 5 and 8 μm diameter, respectively, at a distance and move directly toward them. Furthermore, one mutant lacking A-type motility seems to be unable to carry out such directed movement (Dworkin 1983). The mechanism whereby the cells can perceive objects at a distance remains unknown, but the phenomenon may be similar to contact guidance reported for mammalian cells (Weiss 1961) in which cells can perceive and follow stress lines on the substrate.

In view of the fact that myxobacterial cells in nature probably feed on insoluble particulate material, the ability to perceive objects rather than concentration gradients makes a certain kind of sense.

## The Future of Myxobacterial Research

When I told one of my colleagues that part of a chapter I was writing had to do with the future of work in my field, he wondered, perhaps facetiously, why we should devote time and energy worrying about the future when the present is so fully demanding. It reminded me of the Irish legislator who, when asked to vote for a bill that promised some benefit or other for future generations, wondered, "What has posterity ever done for us"?

Nevertheless, the remaining part of the section will cover those areas of future research that reflect the interests and activities of existing major laboratories, as well as those areas that may or may not be addressed, but that represent fundamental aspects of myxobacterial biology.

## Molecular Details of the Initiation and Regulation of Developmental Events

It has become clear that the initiation of myxobacterial development requires that three conditions be satisfied: (1) The cells must perceive a nutritional deficiency. (2) The cells must be at a sufficiently high cell density. (3) The cells must be on a solid surface. It is not at all clear how each of these conditions is monitored, what environmental parameters are actually perceived, how they are translated via some intracellular messengers to a cellular response, and how the three intracellular responses are coordinated.

There are at least four regulatory systems that in prokaryotes have been shown to play a role in allowing the cell to adjust its physiologic and biochemical makeup to shifts up and shifts down in the nutritional environment. These are cyclic nucleotides (Rickenberg 1974), glutamine synthetase (Tyler 1978), guanosine polyphosphates (Cashel 1975), and highly phosphorylated nucleotides (HPN) (Rhaese *et al.* 1979). It is not at all evident which, if any, of these in myxobacteria signal the onset of nutritional deprivation (see Chapter 4), but it is obvious that there is a question of primary concern and of fundamental importance.

The requirement for a high cell density is a fascinating one, and the myxobacterial solution to this problem has been described earlier in this chapter. However, the evidence for the model lacks the component that would be compelling, viz., evidence that the cells have surface receptors for the signal molecules that modulate some sort of developmental response. Thus, in order to demonstrate that this strategy is indeed the one used by the myxobacteria, it will be necessary to demonstrate the receptor sites, characterize their properties, and determine how occupancy of the receptor is transduced to a developmental response. This is an exciting problem and one for which fortunately the strategic concepts and technology are already well in place, at least with regard to the receptors. The nature of the transduction may prove to be more difficult, although work on the role of protein methylation in *E. coli* chemotaxis and protein instability and phosphorylation in other systems suggests these as interesting possibilities.

The need for a solid surface could be trivial; it may only reflect the need for the cells to move into aggregation centers, a behavior that will only occur on a solid surface. If, on the other hand, there is an actual tactile perception that the cell is on a solid surface and if that perception is part of the initiation process, then that could be most interesting. In view of the fact that these cells do in fact seem to possess a tactile sensory mechanism, the possibility is not altogether unlikely.

Much of the work now underway to try to clone genes of *M. xanthus* that code for developmental gene products (e.g., S and MBHA proteins) is directed toward understanding how those genes are regulated. The dramatic advances in the understanding of the nature and role of prokaryotic promoters, RNA polymerase accessory factors, various start and stop signals, transposable elements, and so on, offer a certainty that these powerful insights will be available for understanding the regulation of myxobacterial genes. In addition, however, there is the tantalizing possibility that entire new strategies and mechanism of regulation specifically designed for controlling developmental interactions are lurking in the myxobacteria.

## Nature of Cell Interactions

There is a considerable amount of indirect evidence that cell interactions do indeed occur. Those interactions include the following:

1. Cell density effects
   a. Myxospore germination (Ramsey and Dworkin 1968)
   b. The role of adenosine in fruiting body formation (Shimkets and Dworkin 1981)
2. Cooperative motility
   a. S motility (Kaiser and Crosby 1983)
   b. Complementation of conditional nonmotile mutants (Hodgkin and Kaiser 1977)
3. Myxobacterial hemagglutinin (Cumsky and Zusman 1979)
4. Phenotypic complementation as a reflection of normal signal exchange (Hagen *et al.* 1978)
5. *Stigmatella* pheromone (Stephens *et al.* 1982) and clumping (White 1981)
6. Cell surface interaction during fruiting body formation (Janssen 1983)
7. Rippling (Shimkets and Kaiser 1982b)

All of these are likely candidates for cell–cell signaling; however, in no instance has the case for signaling been compelling. What is missing in all these is a clear and convincing demonstration that there is both a signal and a cell surface receptor. The signal may be either a diffusible one or may require cell-to-cell contact. The signal and receptor molecules could be different as in hormone–receptor interactions or lectin–ligand interactions, or they could be identical as in self-recognition systems between homotypic cells. In either case, it is necessary that a signal and a cell surface receptor be identified before we can conclude that bona fide signaling has occurred. When the signal–receptor interaction is validated and characterized, the next step is trying to answer the question of how that interaction is transduced to a developmental response. Or, put another way, what are the biochemical reactions that couple signal reception to the behavioral response? I cannot emphasize too strongly that this area of cell interactions is the one that sets the myxobacteria off uniquely from other bacteria. In addition, the opportunity to examine a set of phenomena ordinarily characteristic of higher organisms in a prokaryotic cell promises to allow a kind of molecular and genetic analysis of mechanisms of cell interactions that have, thus far, not been realized.

## Myxobacterial Genetics

Probably no single area of effort has contributed more to the recent awakening of interest in the myxobacteria than that of genetics. Myxobacterial development is amenable to genetic analysis par excellence. This is so not only because *M. xanthus* possesses all of the virtues that are, in general, a property of prokaryotic cells [e.g., relatively short generation time (about 210 min), high density populations ($>2 \times 10^9$ cells/ml), ready availability of mutants, and generalized transducing phages] but, in addition, the ability to insert randomly (and remove) a selectable marker (i.e., the Tn*5* transposon) via P1 phage transduction has been of profound importance (see Chapter 8).

One genetic phenomenon whose protean qualities has intrigued and frustrated investigators is that of phase variation (Burchard and Dworkin 1966a; Burchard *et al.* 1977b; Wireman and Dworkin 1975). Its obvious manifestations are in pigmentation, colonial morphology, and development. Although the molecular basis of phase variation of *Salmonella* flagella is understood and may be the same in myxobacteria, the function of the phenomenon is an enigma. It is certain, however, to be of considerable importance. For example, if phenotypic complementation of developmental mutants is attempted with strains that have been selected because they are phase-stable with regard to yellow pigmentation, complementation is vastly enhanced if a small percentage of the tan variant of one of the complementing pairs is included in the mixture (Janssen 1983). Any phenomenon that systematically and reversibly generates changes at a high frequency is bound to be important to a process that is essentially multicellular and depends on cell interactions. The problem of myxobacterial phase variation needs to be carefully examined.

Myxobacterial genetics needs a recombinational system that will allow large scale mapping. A transformational or mating system to obtain recombination would be most valuable, as would be a series of suppressor mutants. One or more lysogenic phages that could be used for specialized transduction would be valuable, and in this sense Orndorff *et al.* (1983) have already shown that *M. xanthus* is lysogenic for phage Mx-8; its lysogeny is of the λ-type rather than the μ-type. It will obviously be of considerable use if it were possible to use Mx-8 as a vehicle for inserting recombinant DNA into *M. xanthus.*

One last comment with regard to myxobacterial genetics is that there is, at the moment, no genetic system available for gene transfer in *Stigmatella.* The usefulness of this organism for studying certain aspects of myxobacterial development will eventually be compromised by this lack. It is thus most important that a genetic system for *Stigmatella* be developed. If we could find phages for the organism, there would at least be the possibility of transductional analysis and the use of transposons.

## Nature of Motility Mechanism

As indicated in the previous section, the models for gliding motility in myxobacteria are the filament–tubule model of Burchard and the polarized, surfactant excretion model of Dworkin and Keller. Although the evidence for neither of the models is compelling, only the latter model has been subjected to analysis. Compelling evidence for the model would include identification of the motility surfactant, demonstration that its excretion is polar and reversible, and, most important, isolation either of a nonexcreting mutant or of an excreting but nonpolarizing mutant. Both of these would be expected to be nonmotile. If the mutated property were temperature sensitive, so much the better.

## Nature of Directed Movement

If the myxobacteria do indeed perceive objects from a distance, rather than concentration gradients, the question of the nature of the mechanism whereby they are able to pull off this rather remarkable feat is a fascinating one. What are the possible mechanisms? Or, perhaps more precisely, what are the properties of an object that could be perceived at a distance? The question of how the perception is accomplished will have to await the understanding of what is perceived. Apparently, the likelihood of its ability to "smell" it has been substantially reduced by prima facie arguments as well as by the inability to demonstrate a chemotactic response in *M. xanthus.* Although it is formally possible that the cells perceive some electromagnetic, thermal, or other form of energy radiating from the object, it seems more likely that they are perceiving some deformation in the substrate generated by the presence of the object. If so, their behavior is analogous to the "contact guidance" behavior of mammalian cells, described many years ago (Weiss 1961) and still unexplained.

## Fruiting Body Morphogenesis

As indicated earlier, we cannot fail to be intrigued by the problem of fruiting body formation by *Stigmatella* (see Fig. 2.5, Chapter 2). Although it is difficult to imagine any sort of genetic blueprint for such a structure, there is a way to think about the process that suggests an experimental approach to the problem. Protein chemists and those concerned with the problems of assembly of flagella, ribosomes, and viral particles have long recognized the phenomenon of self-assembly, that is, the process whereby the primary structure of the protein, in an intrinsic sense, contains the information that leads to its own tertiary structure. Furthermore, that three-dimensional structure has a thermodynamically most-favored association with another protein. Thus a structure is assembled in a regular fashion, the only structural information that is necessary being that of the primary structure of the subunits. Some years ago, Rees (1972) in a seminal article suggested that a similar self-assembly process could take place with polysaccharides as the primary subunits instead of proteins. Agar is a familiar example of such a mixture of polysaccharides that undergoes a reversible liquid–gel transition induced by temperature changes. Even more interesting are the alginates, where the liquid–gel transition is induced instead by ligand binding and displacement; $Ca^{2+}$ causes gelling, and when the $Ca^{2+}$ is displaced by $Na^{+}$, the gel reverts to a liquid phase. It is quite feasible that what the cell needs to do in order to form a fruiting body is to make the right mixture of polysaccharides at the right time in the right place and then to cause its localized gelation by regulating the access of the appropriate ligand.

Although there has been some preliminary characterization of myxobacterial polysaccharides (Sutherland and Thomson 1975), it is critical that this approach be continued. Polysaccharides will be shown to play a role not only in fruiting body morphogenesis, but also in motility with regard to the trails that the cells lay down, in myxospore structure, and, most likely, in cell–cell signaling. It is an area that is wide open—pregnant with exciting possibilities and critical for a fuller understanding of myxobacterial biology.

## Ecology of Myxobacteria

Although there has been a great deal of work concerning whether this or that myxobacterium is present in this or that location, the work has not been quantitative. This obviously has to do with the existence of multicellular fruiting bodies, adherance of cells to each other, and the lack of strictly differential cell counting techiques. But apart from the traditional ecologic concerns of census and distribution, I suggest that there are four general ecologic problems that are of peculiar relevance to the myxobacteria:

1. The myxobacteria are strictly terrestrial organisms. They live on the surfaces of particles of soil and soil debris. They feed on living cells as well as

insolbule macromolecular debris such as dead prokaryotic and eukaryotic cells, peptidoglycan, starch, insoluble protein, and in some cases cellulose. Thus, what distinguishes the myxobacteria is that they are forced to generate their actual carbon, energy, and nitrogen sources by an action occurring outside the cell. Furthermore, because of the particulate, insoluble nature of the food, they must seek targets that are chemically invisible to them. Finally, if the nature of their habitat is the surface of a soil particle, this dictates that their movement would be a gliding rather than a swimming one. Thus, we can rationalize excreted, hydrolytic enzymes, optimization of feeding via multicellular behavior, that is, the wolf-pack effect, gliding motility, and the apparent lack of a chemotactic food-sensing mechanism. However, these are all ex post facto rationalizations based on laboratory experiments. We really do not know if those facts accurately reflect what is happening in the soil.

2. What is the role of the myxobacteria with regard to the turnover of organic material in nature? What molecules do they recycle? Is their role as scavengers of carcasses and debris unique? Is it quantitatively significant?
3. What is the nature of the interaction between environmental conditions and the myxobacterial life cycle? Living as they do on the surface of particles rather than submersed in an aqueous environment, they must be in extreme danger of desiccation. And, in fact, vegetative cells are extremely sensitive to desiccation (Sudo and Dworkin 1969). Does excreted slime play a protective role? Is it possible that as cells glide they excrete an overlying layer of slime that protects the cells from desiccation? What conditions in nature induce fruiting body formation or myxospore germination? What is the developmental response, if any, to light, temperature, moisture, nutrients?
4. What is the nature of interactions between different clonal populations or species? Do swarms merge and form mixed swarms and eventually mixed fruiting bodies? If not, how great a difference must there be? How do swarms of different species recognize each other as different? The existence of multicellular swarms and fruiting bodies allows us to ask the sorts of questions that are simply meaningless with other bacteria that do not behave in a social fashion.

## Conclusion

It is not hyperbole to predict that *M. xanthus* will be the *E. coli* of cell interactions. The past decade, dating from the first myxobacterial conference at Cold Spring Harbor, has seen tremendous progress. The genetic and molecular approaches are well under way; the biochemical definitions of the various interactive phenomena are proceeding at a rapid pace. The next decade will see them come together with an increased understanding not only of myxobacterial biology but of cell interactions in general.

## Acknowledgments

A profound debt of gratitude is owed to the National Science Foundation and the National Institutes of Health who supported work on the myxobacteria when it was still exotic and merely a promise and who continue their support now that the use of myxobacteria to study development and cell interactions is not only acceptable but even on the verge of becoming fashionable.

# Appendix A

## Selected Reviews on Myxobacteria


Dworkin, M. (1966). Biology of the myxobacteria. Annu. Rev. Microbiol. 20:75–106.

Dworkin, M. (1972). The myxobacteria: New directions in studies of procaryotic development. Crit. Rev. Microbiol. 2:435–452.

Dworkin, M. (1973). Cell–cell interactions in the myxobacteria. Symposia of the Society for General Microbiology. 23:125–142.

Kaiser, D., Manoil, C., Dworkin, M. (1980). Myxobacteria: Cell interactions, genetics and development. Annu. Rev. Microbiol. 33:595–639.

Parish, J. H. (1979). Myxobacteria, pp. 227–253. In: Developmental Biology of Prokaryotes, edited by J. H. Parish. Berkeley: University of California Press, pp. 227–253.

Peterson, J. E. (1969). Isolation, cultivation and maintenance of the myxobacteria. In: Methods in Microbiology, edited by J. R. Norris, D. W. Ribbons, London, New York: Academic Press, Vol. 3B, pp. 185–210.

Reichenbach, H., Dworkin, M. (1981). The order Myxobacterales. In: The Prokaryotes, edited by M. P. Starr, H. Stolp, H. G. Truper, A. Balows, H. G. Schlegel, New York: Springer-Verlag, pp. 328–355.

White, D. (1975). Myxospores of *Myxococcus xanthus,* In: Spores VI, edited by P. Gerhardt, H. L. Sadoff, R. N. Costilow, Washington, D.C.: American Society of Microbiology, pp. 44–51.

Zusman, D. R. (1980). Genetic approaches to the study of development in the myxobacteria. In: The Molecular Genetics of Development, edited by T. Leighton, W. F. Loomis, New York: Academic Press, pp. 41–78.

# Appendix B

## Media Useful for the Cultivation of *Myxococcus xanthus*

1. CT liquid medium (Dworkin 1962)

   2% Casitone
   0.2% $MgSO_4 \cdot 7H_2O$
   0.01 *M* Potassium phosphate buffer, pH 7.2 (autoclaved separately).
   For over 20 years this has been the standard medium for growing *M. xanthus*, providing a minimum doubling time of 3.5 hr and a maximum cell yield of $4–5 \times 10^9$ cells/ml.

2. CT agar (Dworkin 1962)

   Same as medium (1) with 2% agar.

3. 1CT liquid medium (Kimchi and Rosenberg 1976)

   1% Casitone (Difco)
   0.2% $MgSO_4 \cdot 7H_2O$
   The simplicity of this medium makes it useful for routine work.
   The two components can be autoclaved together.

4. CTT liquid medium (Bretscher and Kaiser 1978).

   Same as medium (3) with 1 m*M* $KH_2PO_4$–$K_2HPO_4$ and 10 m*M* tris·Cl buffer, final pH 7.6.

5. CF agar (Hagan *et al.* 1978).

   0.015% Casitone
   0.1% sodium pyruvate
   0.2% sodium citrate
   0.02% $(NH_4)_2SO_4$

$8$ mM $MgSO_4$
$1$ mM $KH_2PO_4$–$K_2HPO_4$
10 mM tris·HCl, final pH 7.6
1.5% agar
This medium allows single cells to form colonies and then fruiting bodies.

6. M1 medium (Zusman and Rosenberg 1971)

0.2% Leu
0.1% each of Ala, Ile, Phe, Pro, and Trp
0.5% each of Asn, Lys, and Met
0.04% Tyr
0.02% Ser, Val, $MgSO_4 \cdot 7H_2O$ and NaCl
0.01% Arg, Gly, His, Thr, Cys, and $K_2HPO_4$
2 $\mu$g/ml $FeCl_3 \cdot 6H_2O$ and $CaCl_2$
10 mM tris·Cl buffer, final pH 7.2
A chemically defined medium that allows for a generation time of 6.5 hr after the cells become adapted.

7. A1 medium (Bretscher and Kaiser 1978)

0.5% Asp·K, pyruvate·Na, and $(NH_4)_2SO_4$
0.2% $MgSO_4 \cdot 7H_2O$
0.125 mg/ml spermidine·3HCl
0.1 mg/ml Asn, Ile, Phe, and Val
0.05 mg/ml Leu
0.01 mg/ml Met
10 $\mu M$ $FeCl_3$ and $CaCl_2$
1 $\mu$g/ml vitamin $B_{12}$
1 m$M$ $K_2HPO_4 \cdot KH_2PO_4$ buffer, pH 7.6
10 m$M$ tris·Cl buffer, pH 7.6
A minimal defined medium useful for the isolation of auxotrophic mutants, generation time 22–36 hr at 29°C.

8. A1 agar (Bretscher and Kaiser 1978)

Same as medium (7) with 0.8% agarose.
(Commercial agar was avoided because it contains usable amino acids.)

# Appendix C

## Techniques for Inducing Myxospores and Fruiting Bodies in *Myxococcus xanthus*

1. *Glycerol induction of myxospores* (Dworkin and Gibson 1964). Exponentially growing *M. xanthus* cells in medium 1 are centrifuged, washed once with medium 3 and suspended to $3 \times 10^8$ cells/ml (200 Klett units with a No. 54 filter) in medium (3) containing 0.5 *M* glycerol. Refractile spheres are formed in 120 min. If the cells are grown in media 3 or 4, myxospores can be induced directly by addition of 10 *M* glycerol (final concentration of 0.5 *M*) to exponential phase cells.
2. *Fruiting body formation on agar* (Wireman and Dworkin 1977). Exponential cultures of *M. xanthus* are harvested by centrifugation, washed, and resuspended in cold medium 1 to $5 \times 10^9$ cells/ml. An 0.2-ml sample of the suspension is spread on a freshly prepared 2% agar plate with the aid of a smooth-glass rod (to avoid disruption of the agar surface). The plate should be dried at 32°C for 1 hr in a low-humidity incubator before incubation in a high-humidity incubator at 32°C. Under these conditions, cells grow for about 24 hr then aggregate into fruiting bodies and form myxospores.
3. *Fruiting body formation on clone-fruiting agar* (Hagen *et al.* 1978). Clone-fruiting (CF) agar (medium 5) allows single cells to multiply, form colonies, and then form fruiting bodies. The cells can be applied to the agar in a variety of ways such as in small drops, standard overlay techniques, and replica plating. The plates are incubated at 25–32°C for 4–10 days, depending on the initial concentration of cells.
4. *Fruiting body formation in submerged culture* (Kuner and Kaiser 1982). Exponentially growing *M. xanthus* (must be a piliated and fully motile strain) in medium 4 is diluted to $10^7$ cells/ml and dispersed into polystyrene

Petri dishes to a depth of 4–5 mm. After standing undisturbed at 32°C for 24 hr the growth medium is removed by aspiration (cells adhere to the bottom of the dish) and replaced with an equal volume of 1 m*M* $CaCl_2$ in 10 m*M* MOPS buffer, pH 6.8. Fruiting bodies containing heat-resistant myxospores are formed in 72 hr.

# Appendix D

## Techniques for Culturing *Stigmatella*

### Growth Media

Cells are grown in 1% Casitone (Difco) containing 8 m$M$ $MgSO_4$. Growth in certain batches of Bactocasitone results in cells that fruit poorly. The situation can be greatly improved if 0.35% yeast extract is added to the growth medium. Alternately, we can substitute 1% Bactotryptone for the Bactocasitone.

### Cloning Media

*Stigmatella aurantiaca* DW4 (ATCC 33878) does not clone well on ordinary Bactocasitone plates. However, we can obtain close to 100% cloning efficiency if we use spent Bactocasitone. Cells are grown in growth media and harvested in the exponential phase of growth. The media is saved and Difco agar is added to 1.5%. After autoclaving, plates are poured. Swarms appear in 8–10 days. If we do not use spent media, cloning efficiencies vary and swarms do not appear for 2–3 weeks.

### Synchronous Fruiting of Stigmatella

*Stigmatella* will fruit on 1.5% Difco agar containing 10 m$M$ $N$-2-hydroxyethylpiperazine-$N'$-2 ethanesulfonic acid (HEPES), pH 7.2 and 3.4 m$M$ $CaCl_2$ (HEPES-Ca). The agar should be poured the day before use and left at room temperature. Twenty milliliters of agar are used for 9-m$M$ Petri dishes and 300 ml of agar for 23 $\times$ 33-cm Pyrex baking dishes. The baking dishes are left either uncovered or loosely covered overnight before use. Plates or baking dishes are uncovered approximately 1 hr before use. Cells are harvested during the exponential phase of growth. The cell density should be between 1 and 2 $\times 10^8$/ml. Cells enter stationary phase at $4 \times 10^8$ cells/ml. The cells are harvested and washed once with cold HEPES-Ca. The volume of the culture medium is recorded, and the total number of cells is calculated. It is important to use plastic pipettes for resuspending the cells because they tend to stick to the glass. The cells are combined into a 50-ml centrifuge tube and cold HEPES-Ca is added so that the final cell density will be no less than $4 \times 10^{10}$/ml. A vortex mixer is used for much of the resuspension to avoid losing clumps of cells in the pipette. The total volume is then measured using a plastic pipette and the volume adjusted as necessary. The cells are either kept in the centrifuge tube in an ice bucket and mixed by vortexing whenever a sample is taken, or the cells are transferred to a small beaker and stirred on ice using a magnet. Five microliter portions are removed and spotted on standard sized Petri dishes, with no more than 20 spots per plate. Or a pasteur pipette is used to spot approximately 5 $\mu$l spots in the baking dishes. The cells in the baking dishes can be spotted closer together than the cells in the Petri dishes. After allowing the spots to dry in with the covers off, the dishes are placed at 30°C in incandescent light of approximately 20 ft-c. It is important that the temperature not rise above 32°C. In the Petri dishes, aggregates appear by 8–10 hr and fruits are fully formed by 18 hrs. We should expect approximately 300 fruits per spot.

## Development in Liquid

*Stigmatella* will also undergo development in cell suspension. Cells are harvested from the exponential phase of growth, washed once in HEPES-Ca, and resuspended to $5 \times 10^8$/ml in HEPES-Ca. They are then shaken at 30°C at approximately 140 rpm. Within 2 hr the cells have formed large orange clumps. The size of the clumps depends upon the shaking speed, the size of the flask, and the volume of cell suspension. Cells incubated this way for 6 hr will aggregate within 3–4 hr (instead of 8–10 hr) when homogenized and spotted in the usual way. They also produce pheromone in cell suspension (Stephens *et al.* 1982).

## Stimulation by Light

Aggregation and fruiting body formation in *Stigmatella* is stimulated by light. The most effective wavelengths are in the blue region, but red light and even far red light are efficient. It is important not to use a fluorescent bulb, because this causes the cells to aggregate into ridges instead of discrete aggregates. We must be sure to use fans and water traps in order to prevent the temperature from rising above 32°C. High cell densities will sometimes fruit in the dark, although the numbers of fruits in the light is usually much greater. At low cell densities ($5 \times 10^7$ cells/5/μl spot), light has a pronounced stimulation on aggregation and fruit morphogenesis (Stephens and White 1980).

# Appendix E

## Gene and Protein Structure of Protein S

Sumiko Inouye*

Protein S is the first developmental protein of *Myxococcus xanthus* purified and well characterized (see Chapter 9). The genes for protein S have been identified and cloned with the use of mixed probes consisting of eight synthetic oligodeoxyribonucleotides (Inouye *et al.* 1983). It was found that the two homologous genes for protein S (gene 1 and gene 2) are tandemly repeated in the same direction with a 1.4-kb-spacer sequence on the *M. xanthus* chromosome. The entire deoxyribonucleic acid (DNA) sequence of the region encompassing both gene 1 and gene 2 has been determined as shown in Fig. E.1 (S. Inouye *et al.* to be published).

The DNA sequences, ATGAACAACAACAC, which hybridized with the synthetic oligonucleotide probes are from residue 970 to 983 for gene 1 and from residue 2879 to 2892 for gene 2. Since these DNA sequences correspond to the those that would code for a part of the carboxyl-terminal end of protein S, Met–Asn–Asn–Asn–Thr, the amino acid sequences of the gene 1 and the gene 2 products, were deduced from the DNA sequences by extending the reading frames in both the amino and the carboxyl-terminal directions. As shown in Fig. E.1, the gene 1 and the gene 2 products were thus assigned as a protein of 175 amino acid residues (MW: 19,235) and a protein of 173 amino acid residues (MW: 18,792), respectively.

The amino terminal structure of protein S purified from differentiating cells has been determined by sequential Edman degradation to be [1]Ala–Asn–Ile–Thr–Val–Phe–Tyr–Asn–Glu–[10]Asp–Phe–Gln–Gly–Lys–Gln–Val–Asp–Leu–Pro–[20]X –Gly– X Tyr–Thr– X -Ala–Gln–Leu–Ala– (Shimonishi and Inouye, unpublished results). This result clearly indicates that the major protein S product during development is derived from gene 2 and that translation of gene

---

*Department of Biochemistry, State University of New York at Stony Brook, New York 11794 U.S.A.

CATGTGGAGC TGGCGGCGTG GATGGACTTC AGCACAGCCC GGAGCGCGCC CTGGCCCTGT CGGCGCTGGG GCAACGCTGA TGCAATCGAA CGAGGGCAGC 100

CAGTCAGCCT GGGCGCCGCG GCGACATCGC CGTCGTCTCC GACCCGCTCC TGGCGGCTGC GCCACTTCGT CATCGCACAG GGGCCCGCCG TACCGCTAGC 200

CTCCAGGCTC GCCCGTGAAC CTGTTCTGGG TCATCCCCCT GCTCGCGGCC GAGTTCGAGA TGGTGCGGCG GGCCGGCCCC CTGCACTGGG CCAACCAGCT 300

GCTGAAGGAC CCGTCCCGCA GGACGGAGCT GGTGCAGCGC TGGAGCCTGG GAGCGAACTG ATGCTCCCAG GCCGCCCGCG CGGTGCTGTA GGCGGATTGC 400

TCCTCCGGAG CGCGGTGCCC AAGCTTCCGG CGGCTTCTGG CAACGAATGG GCCGGGACCG CTCAAAGGAG GAGAACTGCA ATGGCAAACA TTACCGTTTT 500
MetAlaAsnIleThrValPh

CTACAACGAA GATTTCGGGG GTAAACAGGT CGATCTGAAG CCTGACGAAT ACAAGCGGGA CAAGCTGGAG GCGCTGGGCA TCGAGAACAA CACCATCAGC 600
eTyrAsnGluAspPheGlyGlyLysGlnValAspLeuLysProAspGluTyrLysArgAspLysLeuGluAlaLeuGlyIleGluAsnAsnThrIleSer

TCGGTGAAGG TGCCGCCTGG CGTGAAGGCT ATCCTCTACA AGAACGATGA TTTCACCGGC GACCAGATCG AAGTGGTGGC CAATGCCGAG GAGCTGGGCC 700
SerValLysValProProGlyValLysAlaIleLeuTyrLysAsnAspAspPheThrGlyAspGlnIleGluValValAlaAsnAlaGluGluLeuGlyP

CGCTGAACAA CAACGTCTCC AGCATCAAGG TCATGTCCGT GCCCGTGCAA CCCAGGGCCA GGTTCTTCTA CAAAGAGCAG TTCGATGGCA AGGAGGTGGA 800
roLeuAsnAsnAsnValSerSerIleLysValMetSerValProValGlnProArgAlaArgPhePheTyrLysGluGlnPheAspGlyLysGluValAs

CCTGCCTCCT GGTCAGTACA CCCAGGCCGA GCTGGAGCGG TACGGCATCG ACAACAACAC CATCAGCTCG GTGAAGCCGG AGGGCCTGAA GGTCGTCCTA 900
pLeuProProGlyGlnTyrThrGlnAlaGluLeuGluArgTyrGlyIleAspAsnAsnThrIleSerSerValLysProGluGlyLeuLysValValLeu

TTCAAGAACG ACAACTTCTC CGCCGGCGAC ACGCTGTCCG TGACTTCCAA CGCCCCGAGC CTGGGCGCGA TGAACAACAA CACCTCCAGC ATCAGAATCA 1000
PheLysAsnAspAsnPheSerAlaGlyAspThrLeuSerValThrSerAsnAlaProSerLeuGlyAlaMetAsnAsnAsnThrSerSerIleArgIleT

CCCCCTGACG CTGGCGCGCC GGCATTCCAG GTGACGGTCG CTGCCGCGGG GACCGCCACC CTTTCGAGCC CAGAGACTCT CGAATCACCT GCATCCACG 1100
hrPro

ACGGCAGGGG GCGGCCCCAC CCCCACTCGG TGACCCCATT CACGCTCACG GCCGAGCTAG TGCCCTGAGC GTCTGCACGC GCTCGTCCCT CCAGTCGAAG 1200

ACGGCGAGCG CACGCGCTCG CCTGCGTGAG GTGTGGAGGC AGGCTCAAGG CACTTGGCGT ACGTGATGGG GCCCCCGGGG TGCGGGCAAT TCTGAACACC 1300

TGGCCTTGCC CACGGCATGT GCGAGGAGAC CGGGCCCGGG AGTAGCCCGC ATGACCCGCC CTCAGGCGGA TCCGCTCGAG GTCGGCCACC GAACCCCGCA 1400

CCCTGCTGGA TCCAGCCCAG GTTGAGCCCG AGGGCGAGGT TCCACGGTCG GTTTCCGGTG ACTGGCATCG GCGGCCCTCC TCCCTGGGAG CAACGATGCG 1500

CACGCCGCTC GGACAGGGAC AGGCACACGG AGGGGCCCAA CTCCGGCTCA GGTGGTCAAG AGTGACGGAG GGCTCTGGCG CGTTGAGAGA TGAGCCCCCG 1600

CGCGAAGGCG CGCCGTCGCC CGGCCAGGCT CCCTGTGGCC CCCTGCACTG GGCCAACCAG CTGCTGAAGG ACCCGTCCCG CAGGACGGGC TGGTGCAGCG 1700

CTGGAGCCTG GGAGCGAACT GGCGCTCCCA GGCCCGCCCG CGCGGTACTG TGTCACCGTC CGGTGGGCGT CGTGCCGGGA GGCTTGATTG CGGACGAGGA 1800

CCCTGTTCGG CGAGCGCCGC GAAGAGCGAG CTGTTCAGGG ACTGGGCTGG AGGACTGTGT CAGGCGGCTC GGCGACGCGT CGACGTTTCC ACTCGTGAGG 1900

TAGCAGCTCA CCGATGCGTG AGTTGGGATG CGTCTGTACA CGCAGCAGCA CGTCGGCCAT GTATCCTCGG GGTTGACCTT GTTGGCCTCG CAGGTGGCCA 2000

CCAGTGCGTA GAGGCCCGCG AGGTTTTCAC CCGCGACCTC GTGGCCGACG AAGAGAAAGT TCGTACGGCC CAGGGCCGCC TTCCGCATCA CGCCGGTACA 2100

CCCACGACTG CAACGTGCTC AACCGCACGC CTCGATGCTG GGAAAACTCC TTCTGCGTCA GTCCGCTCGC CTCGAAGGCT TCAGCGACCC GGAACCACTC 2200

CTGCTTCTCC ACCGGCTTTG ACATTCCCGC CAGGGCCTCG GCGCCGACTC CTCACTCAAC GCCTCTCAGC ACGTCTTTGG CCGGACGGAT ACGGTACTGT 2300

AGACGGATTG CATTTCCGGA GCGCGGTGCC CAATGCTTCC GGCGGCTTCT GGCAACGAAT GGCCGGGGCC GCTCAACGGA GGAGCACTGC AATGGCAAA 2400
MetAlaAs

**Figure E. 1.** The DNA sequence of the fragment encompassing gene 1 and gene 2. The DNA sequence of the 3693-bp fragment is shown. The termination codons, TGA, for gene 1 and gene 2 are underlined.

```
CATTACCGTTTTCTACAACGAAGACTTCCAGGGTAAACAGGTCGATCTGCCGCCTGGCAACTACACCCGGGCCCAGTTGGCGGCGCTGGGCATCGAGAAC   2500
aIleThrValPheTyrAsnGluAspPheGlnGlyLysGlnValAspLeuProProGlyAsnTyrThrArgAlaGlnLeuAlaAlaLeuGlyIleGluAsn

AACACCATCAGCTCGGTGAAGGTGCCGCCTGGCGTGAAGGCTATCCTCTACCAGAACGATGGTTTCGCCGGCGACCAGATCGAAGTGGTGGCCAATGCCG   2600
AsnThrIleSerSerValLysValProProGlyValLysAlaIleLeuTyrGlnAsnAspGlyPheAlaGlyAspGlnIleGluValValAlaAsnAlaG

AGGAGTTGGGCCCGCTGAACAACAACGTCTCCAGCATCCGGGTCATATCCGTGCCCGTGCAGCCCAGGGCCAGGTTCTTCTACAAAGAGCAGTTCGATGG   2700
luGluLeuGlyProLeuAsnAsnAsnValSerSerIleArgValIleSerValProValGlnProArgAlaArgPhePheTyrLysGluGlnPheAspGl

CAAGGAGGTGGACCTGCCTCCTGGCCAGTACACCCAGGCCGAGCTGGAGCGGTACGGCATCGACAACAACACCATCAGCTCGGTGAAGCCGCAGGGCCTG   2800
yLysGluValAspLeuProProGlyGlnTyrThrGlnAlaGluLeuGluArgTyrGlyIleAspAsnAsnThrIleSerSerValLysProGlnGlyLeu

GCGGTCGTCCTATTCAAGAACGACAACTTCTCCGGCGACACGCTGCCCGTGAATTCCGACGCCCCGACCCTGGGCGCGATGAACAACAACACCTCCAGCA   2900
AlaValValLeuPheLysAsnAspAsnPheSerGlyAspThrLeuProValAsnSerAspAlaProThrLeuGlyAlaMetAsnAsnAsnThrSerSerI

TCAGAATCTCCTGACGCTGGCGCGCCGGCATTCCAGGTGACGGTCCCTGCCGCGGGGACCGCCCCCCTTTCGAGCCCCAGAGACTCTCGAATCACCTCGA   3000
leArgIleSer

TGCACGATGGCAGGGTGCGGCCCCACCTGAGCCTGCCCGGGTTCCGTGGACACCTGAGATGTGGGGGAAAGGGTGTCCAATGTCGGCAAGTGGCGCGAAT   3100

CAGAAGAGGGCGCGGAGGCAGCGCCGGAAGTTCACGCGCGAGTTCAAGGACGGCGCGGTGAAATTCCACGGACGCCATTCACGCAGGCAGACGTGCGGTA   3200

GCAGCTCAGAGGAGGGGAGGGATGCTCCGCTCCTGGCCGGACTCCCGGCGGGCCATGCAGGCGGCAACGCGCCGTGCCCGTGATGCTCTTTCCCGTCATG   3300

GCGACCACCTTCAGCGGAGGAGGGGCCATGTACTTCCAGGCCATCAAGACCGCGGGCCTTGCCCAAATCACCTACATCCTGGGCAGTGAAGGCGAGGCCC   3400

TCGTCGACCCCCGGCGTGATGTCGACGTCTACCTGGACGTGTTGCGGCGGCAGGAGTTGCGGCTGCGCTACGTGCTCCAGACGCACCGCCAGGAAGACTT   3500

CGTTGAGGGTACTTCCGCGCTTGCCCGCCTTGCAGGCGCGCAGGTGGTGGCTGGCCGGCACCCCATCACCGGCCATGCCGACGTGCGGATGGGCGAGCGA   3600

GAGCGGCTACACCTGGGGGCCTTGACGCTGGTGGCGCTTCATACGCCGGGCCACACGCCAGAGAGCATGAGCTGGGCCGTCTACCTGGATCC
```

**Figure E.1** (Continued.)

2 is initiated from the methionine residue immediately before the amino terminal alanine as shown in Fig. E.1. The conclusion that protein S is the gene 2 product is also supported by the carboxyl-terminal structure of protein S, which has been determined to be either isoleucine or serine (Inouye *et al.* 1981), whereas the carboxyl-terminal end of the gene 1 protein is proline (Fig. E.1).

The primary structure of the gene 2 product (protein S) appears to be composed of four unit structures or domains, which share homologous structures. In particular, as shown in Fig. E.2, there are extensive homologies between the first domain (residue 6–43; domain 1) and the third domain (residue 95–132; domain 3) and between the second domain (residue 45–84; domain 2) and the fourth domain (residue 133–172; domain 4). Domains 1 and 3 consist of 38 amino acid residues each, 26 of which are identical and 4 of which are functionally identical (79% homology; see Fig. 2). Domains 2 and 4 consist of 40 amino acid residues each, 19 of which are identical and 7 of which are functionally identical (65% homology; see Fig. 2). At the level of the DNA sequence, 88 out of 114 bases are homologous between domains 1 and 3 (77% homology). Among 38 codons in these domains, 13 codons have a single base

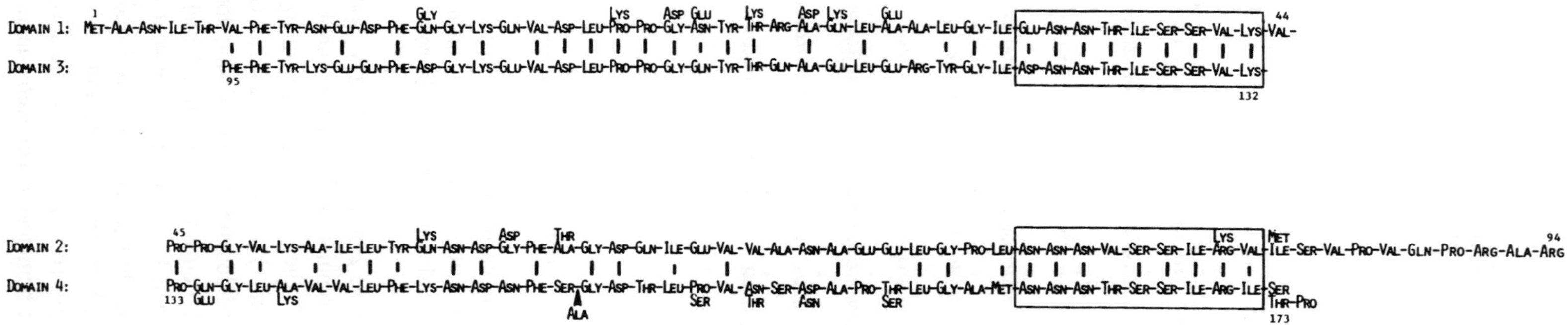


**Figure E. 2.** Four-domain structure of protein S. Putative $Ca^{2+}$ binding sites are boxed. Numbers indicate the residue numbers from the amino terminal methionine residue. Only those amino acid residues that are substituted in the gene 1 product are shown above or below the sequence of protein S. The alanine residue with an arrow in domain 4 and the carboxyl-terminal proline residue are extra amino acid residues found in the gene 1 product.

substitution (seven of them are silent mutation without changing amino acids assignments), five have two mismatches, and one has three mismatches. Between domains 2 and 4, 78 out of 120 bases are homologous (65% homology). Among 40 codons in these domains, 12 codons have a single base substitution (only two of them are silent mutations in contrast to domains 1 and 3), nine have two mismatches [one of them is a silent mutation; CGG (arg) and AGA (Arg)], and 4 have three mismatches.

Although homology between domains 1 and 2 or between domains 3 and 4 is less striking, a sequence of 10 amino acid residues at the carboxyl-terminal end of each domain shows extensive homology (boxed sequences in Fig. 1). It is of great interest to find that the sequence of nine amino acid residues in this region of domains 1 and 3, Glu (or Asp)–Asn–Asn–Thr–Ile–Ser–Ser–Val–Lys, shows a high level of homology to one of the proposed $Ca^{2+}$ binding sequences in bovine brain calmodulin, Asp*–Gly–Asn*–Gly–Thr*–Ile–Thr*–Thr–Lys (Watterson *et al.* 1980) (underlines indicate residues identical or functionally identical to the protein S sequence, and asterisks indicate putative $Ca^{2+}$ binding residues).

# References


Achenbach, H., Kohl, W., Reichenbach, H. (1978). The flexirubin-type pigments—a novel class of natural pigments from gliding bacteria. Revista Latinoamericana de Quimica 9:111–124.

Achtman, M. and Skurray, R. (1977). A redefinition of the mating phenomenon in bacteria. In: Microbial Interactions Series B, edited by J. L. Reissig, London. Chapman and Hall, Vol. 3, pp. 235–278.

Adye, J. C., Powelson, D. M. (1961). Microcyst of *Myxococcus xanthus:* Chemical composition of the wall. J. Bacteriol. 81:780–785.

Agnihothrudu, V., Barua, G. C. S., Barua, K. C. (1959). Occurrence of *Chondromyces* in the rhizosphere of plants. Indian Phytopathology 12:158–160.

Allen, R. D. (1981). Motility. J. Cell Biol. 91:148s–155s.

Anderson, R. P., Roth, J. R. (1977). Tandem genetic duplications in phage and bacteria. Annu. Rev. Microbiol. 31:473–505.

Arias, J. M., Rodriguez, C., Montoya, E., (1979). Purification and partial characterization of an antibiotic produced by *Myxococcus coralloides*. J. Antibiot. (Tokyo) 32:205–211.

Arlauskas, J., Burchard, R. P. (1982). Substratum requirements for bacterial gliding motility. Arch. Microbiol. 133:137–141.

Aschner, M. I., Chorin-Kirsh. (1970). Light-oriented locomotion in certain myxobacter species. Arch. Mikrobiol. 74:308–314.

Ashmarin, I. P., Loitsyanskaya, M. S., Kharchenko, E. P. (1969). Nucleotide composition of DNA and type of nutrition of bacteria. Microbiology (U.S.S.R.) 38:659–664.

Auerswald, E. A., Ludwig, G., Schaller, H. (1981). Structural analysis of Tn*5*. Cold Spring Harbor Symp. Quant. Biol. 45:107–113.

Avery, L. (1983). Tandem duplications as a genetic tool in *Myxococcus xanthus*. Ph.D. thesis. Stanford University.

Avery, L., Kaiser, D. (1983a). In situ transposon replacement and isolation of a spontaneous tandem duplication. Mol. Gen. Genet. 191:99–109.

Avery, L., Kaiser, D. (1983b). Construction of random genetic duplications with defined endpoints in *Myxococcus xanthus*. Mol. Gen. Genet. 191:110–117.

Bachmann, B., Low, K. B. (1980). Linkage map of Escherichia coli K-12, edition 6. Microbiol. Rev. 44:1–56.

Bacon, K., Eiserling, F. A. (1968). A unique structure in microcysts of *Myxococcus xanthus*. J. Ultrastruct. Res. 21:378–382.

Bacon, K., Rosenberg, E. (1967). Ribonucleic acid synthesis during morphogenesis in *Myxococcus xanthus*. J. Bacteriol. 94:1883–1889.

Bacon, K., Clutter, D., Kottel, R. H., Orlowski, M., White, D. (1975). Carbohydrate accumulation during myxospore formation in *Myxococcus xanthus*. J. Bacteriol. 124:1635–1636.

Bak, A. L., Atkins, J. F., Meyer, S. A. (1972). Evolution of DNA base compositions in microogranisms. Science 175:1393–1393.

Banerjee, D., Basu, M., Choudhury, I., Chatterjee, G. C. (1981). Cell surface carbohydrates of *Agrobacterium tumefaciens* involved in adherence during crown gall tumor initiation. Biochim. Biophys. Res. Commun. 100:1384–1388.

Baur, E. (1905). Myxobakterien-Studien. Archiv Protistenkunde 5:92–121.

Bayer, M. E. (1968). Areas of adhesion between wall and membrane of *Escherichia coli*. J. Gen. Microbiol. 53:395–404.

Beck, E., Ludwig, G., Auerswald, E. A., Reiss, B., Schaller, H. (1982). Nucleotide sequence and exact localization of the neomycin phosphotransferase gene from transposon Tn5. Gene 19:327–336.

Behki, R. M. Lesley, S. M. (1972). Deoxyribonucleic acid degradation and the lethal effect by myxin in *E. coli*. J. Bact. 109:250–261.

Behrens, H., Flossdorf, J., Reichenbach, H. (1976). Base composition of deoxyribonucleic acid from *Nannocystis exedens* (Myxobacterales). Int. J. Systematic Bacteriol. 26:561–562.

Bender, H. (1962). Untersuchungen an *Myxococcus xanthus*. I. Bildungsbedingungen, Isolierung und Eigenschaften eines bacteriolytischen Enzymsystems. Arch. Mikrobiol. 43:262–279.

Berg, D. E., Davis, J., Allet, B., Rochaix, J. D., (1975) Transposition of R factor genes to bacteriophage λ. Proc. Natl. Acad. Sci. USA 72:3628–3632.

Berg, D. E., Egner, C., Hirschel, B. J., Howard, J., Johnsrud, L., Jorgensen, R. A., Tlsty, T. D. (1981). Insertion, excision and inversion of Tn*5*. Cold Spring Harbor Symp. Quant. Biol. 45:115–123.

Berg, H. C., Purcell, E. M. (1977). Physics of chemoreception. Biophys. J. 20:193–219.

Bernards, A., Van der Ploeg, L. H. T. Frasch, A. C. C., Borst, P., Boothroyd, J. C. Coleman, S., Cross, G. A. M. (1981). Activation of trypanosome surface glycoprotein genes involves a duplication transposition leading to an altered 3′ end. Cell 27:497–505.

Billen, D., (1968). Methylation of the bacterial chromosome: An event at the "replication point"? J. Mol. Biol. 31:477–486.

Bird, A. P. (1980). DNA methylation and the frequency of CpG in animal DNA. Nucleic Acids Res. 8:1499–1504.

Bird, R. E., Louarn, J., Martuscelli, J., Caro, C. (1972). Origin and sequence of chromosome replication in *Escherichia coli*. J. Mol. Biol. 70:549–566.

Biro, S. L., Bekesi, I., Vitalis, S., Szabo, G. (1980). A substance effecting differentiation in *Streptomyces griseus*. Eur. J. Biochem. 103:359–363.

Bland, J., Yeh, W., White, D., Hendricks, A. (1971). Increase in glyoxylate shunt enzymes during cellular morphogenesis in *Myxococcus xanthus*. Can. J. Microbiol. 17:209–211.

Bonner, J. T. (1952). Morphogenesis. Princeton University Press, Princeton.

Bonner, J. T. (1983). Chemical signals of social amoebae. Sci. 248:114–120.

Botstein, D., Maurer, R. (1982) Genetic approaches to the analysis of microbial development. Annu. Rev. Genet. 16:61–83.

Boxari, W. L., Clarke, M. (1981). Characterization of a novel calmodulin from *Dictyostelium discoideum*. J. Biol. Chem. 256:3598–3603.

Bredt, W., Heunert, H. H., Hofling, K., Milthaler, B. (1973). Microcinematographic studies of *Mycoplasma hominis* cells. J. Bacteriol. 113:1223–1227.

Breed, R. S., Murray, E. G. P., Smith, N. R. (1957). Bergey's Manual of Determinative Bacteriology. Baltimore: The Williams and Wilkins Company, 7th ed.

Bremer, H., Churchward, G. (1977a). An examination of the Cooper-Helmstetter theory of DNA replication in bacteria and its underlying assumptions. J. Theor. Biol. 69:645–654.

Bremer, H., Churchward, G., (1977b). Deoxyribonucleic acid synthesis after inhibition of initiation of rounds of replication in *Escherichia coli* B/r. J. Bacteriol. 130:692–697.

Bremer, H., Young, R., Churchward, G. (1977). Initiation and termination of deoxyribonucleic acid replication in bacteria after a stepwise increase in the velocity of replication J. Bacteriol. 130:92–99.

Bretscher, A. P., Kaiser, D. (1978). Nutrition of *Myxococcus xanthus,* a fruiting myxobacterium. J. Bacteriol. 133:763–768.

Brockman, E. R. (1976). Myxobacters from arid Mexican soil. Appl. Environ. Microbiol. 32:642–644.

Brockman, E. R., Boyd, W. L. (1963) Myxobacteria from soils of the Alaskan and Canadian arctic. J. Bacteriol. 86:605–606.

Broome, S., Gilbert, W. (1978). Immunological screening method to detect specific translation products. Proc. Natl. Acad. Sci. USA 75:2746–2749.

Brown, D. A., Berg, H. C. (1974). Temporal stimulation of chemotaxis in *Escherichia coli*. Proc. Natl. Acad. Sci. USA 71:1388–1392.

Brown, M. R., Jr., Willison, J. H. M., Richardson, C. L. (1976). Cellulose biosynthesis in *Acetobacter xylinum:* visualization of site of synthesis and direct measurement of the in vivo process. Proc. Natl. Acad. Sci. USA 73:4565–4569.

Buckley, D. E., Anagnostopoulos, G. D. (1976). DNA synthesis by *Escherichia coli* B/r/l synchronized and grown under conditions of slow growth. Arch. Microbiol. 109:143–146.

Burchard, R. P. (1970). Gliding motility mutants of *Myxococcus xanthus*. J. Bacteriol. 104:940–947.

Burchard, R. P. (1974a). Growth of surface colonies of the gliding bacterium *Myxococcus xanthus*. Arch. Microbiol. 96:247–254.

Burchard, R. P. (1974b). Studies on gliding motility in *Myxococcus xanthus*. Arch. Microbiol. 99:271–280.

Burchard, R. P. (1975). Myxospore induction in a nondispersed growing mutant of *Myxococcus xanthus*. J. Bacteriol. 122:302–306.

Burchard, R. P. (1981). Gliding motility of prokaryotes: Ultrastructure, physiology, and genetics. Annu. Rev. Microbiol. 35:497–529

Burchard, R. P. (1982a). Evidence for contractile flexing of the gliding bacterium *Flexibacter* FS-1. Nature 298:663–665.

Burchard, R. P. (1982b). Trail following by gliding bacteria. J. Bacteriol. 152:495–501.

Burchard, R. P., Dworkin, M. (1966a). Light induced lysis and carotenogenesis in *Myxococcus xanthus*. J. Bacteriol. 91:535–545.

Burchard, R. P., Dworkin, M. (1966b). A bacteriophage for *Myxococcus xanthus:*

Isolation, characterization and relation of infectivity to host morphogenesis. J. Bacteriol. 91:1305–1313.

Burchard, R. P., Hendricks, S. B. (1969). Action spectrum for carotenogenesis in *Myxococcus xanthus*. J. Bacteriol. 97:1165–1168.

Burchard, R. P., Parish, J. H. (1975). Mutants of *Myxococcus xanthus* insensitive to glycerol-induced myxospore formation. Arch. Microbiol. 104:289–292.

Burchard, A. C., Burchard, R. P., Kloetzel, J. A. (1977a). Intracellular, periodic structures in the gliding bacterium *Myxococcus xanthus*. J. Bacteriol. 132:666–672.

Burchard, R. P., Burchard, A. C., Parish, J. H. (1977b). Pigmentation phenotype instability in *Myxococcus xanthus*. Can. J. Microbiol. 23:1657–1662.

Burchard, R. P., Gordon, S. A., Dworkin, M. (1966). Action spectrum for the photolysis of *Myxococcus xanthus*. J. Bacteriol. 91:896–897.

Burger, M. M., Jumblatt, J. (1977). Membrane involvement in cell-cell interactions: A two component model system for cellular recognition that does not require live cells. In: Cell and Tissue Interactions, edited by J. W. Lash, M. M. Burger, New York: Raven Press, pp. 155–172.

Burkholder, P. R. (1934). Movement in the cyanophyceae. Q. Rev. Biol. 9:438–459.

Burnham, C. J., Collart, S. A., Highison, B. W. (1981). Entrapment and lysis of the Cyanobacterium *Phormidium luridum* by aqueous colonies of *Myxococcus xanthus* PCO2. Arch. Microbiol. 129:285–294.

Callies, E., Mannheim, W. (1978). Classification of the *Flavobacterium–Cytophaga* Complex on the basis of respiratory quinones and fumarate respiration. Int. J. Systematic Bacteriol. 28:14–19.

Campos, J. M., Zusman, D. R. (1975). Regulation of development in *Myxococcus xanthus:* Effect of 3′:5′ cyclic AMP, ADP and nutrition. Proc. Natl. Acad. Sci. USA 72:518–522.

Campos, J. M., Geisselsoder, J., Zusman, D. R. (1978). Isolation of bacteriophage MX4, a generalized transducing phage for *Myxococcus xanthus*. J. Mol. Biol. 119:167–178.

Cashel, M. (1975). Regulation of bacterial ppGpp and pppGpp. Ann. Rev. Microbiol. 29:301–318.

Castenholz, R. W. (1973). Movements. In The Biology of the Blue-Green Algae, edited by N. G. Carr, B. H. Whitton. Berkeley/Los Angeles: University of California Press.

Chai, N. C., Lark, K. G. (1970). Cytological studies of deoxyribonucleic acid replication in *Escherichia coli*. 15T: Replication at slow growth rates and after a shift-up into rich medium. J. Bacteriol. 104:401–409.

Chandler, M., Bird, R. E., Caro, L. (1975). The replication time of the *Escherichia coli* K12 Chromosome as a function of the cell doubling time. J. Mol. Biol. 94:127–132.

Chandler, M. G., Pritchard, R. H. (1975). The effect of gene concentration and relative gene dosage on gene output in *Escherichia coli*. Mol. Gen. Genet. 138:127–141.

Chou, P. Y., Frasman, G. D. (1977). $\beta$-Turns in proteins. J. Mol. Biol. 115:135–175.

Christensen, P., Cook, F. D. (1978). *Lysobacter* a new genus of nonfruiting gliding bacteria with a high base ratio. Int. J. Syst. Bacteriol. 28:367–393.

Christison, J., Martin, S. M. (1971). Isolation and preliminary characterization of extracellular protease of *Cytophaga* sp. Can. J. Microbiol. 17:1207–1216.

Churchward, G., Bremer, H. (1977). Determination of deoxyribonucleic acid replication time in exponentially growing *Escherichia coli.* B/r. J. Bacteriol. 130:1206–1213.

Clarke, M., Spudich, J. A., (1977). Nonmuscle contractile proteins: the role of actin and myosin in cell motility and shape determination. Annu. Rev. Biochem. 46:797–822.

Clewell, D. B. (1981). Plasmids, drug resistance, and gene transfer in the genus *Streptococcus.* Microbiol. Rev. 45:409–436.

Clewell, D. B., Yagi, Y., Ike, Y., Craig, R. A., Brown, B. L. An, F. (1982). Sex pheromones in *Streptococcus faecalis:* Multiple pheromone systems in strain DS5, similarities of pAD1 and pAM$\gamma$1, and mutants of pAD1 altered in conjugative properties. In: Microbiology, edited by D. Schlessinger. Washington, D.C.: American Society for Microbiology, pp. 197–200.

Cole, R., Levitan, D., Sinden, R. (1976). Removal of psoralen interstrand cross-links from DNA of *Escherichia coli:* Mechanism and genetic control. J. Mol. Biol. 103:39–59.

Cooper, G. M., Okenquist, S., Silverman, L. (1980). Transforming activity of DNA of chemically transformed and normal cells. Nature 284:418–411.

Cooper, S., Helmstetter, C. E. (1968). Chromosome replication and the division cycle of *Escherichia coli* B/r. J. Mol. Biol. 31:519–540.

Cooper, S., Weinberger, M. (1977). Medium-dependent variation of deoxyribonucleic acid segregation in *Escherichia coli.* J. Bacteriol. 130:118–127.

Corpe, W. A. (1980). Microbial surface components involved in adsorption of microorganisms onto surfaces. In: Adsorption of microorganisms to surfaces, edited by G. Bitton, K. C. Marshall, New York: John Wiley.

Costerton, J. W. (1979). The role of electron microscopy in the elucidation of bacterial structure and function, In: Annual Review of Microbiology, edited by M. P. Starr, J. L. Ingraham, S. Raffel (eds.), Palo Alto, California, Annual Reviews Inc. pp. 459–479.

Costerton, J. W., Irvin R. T. (1981). The bacterial glycocalyx in nature and disease, In: Annual Review of Microbiology, edited by M. P. Starr, J. L. Ingraham, A. Balows, Palo Alto, California, Annual Reviews Inc. pp. 299–324.

Couche, P. (1969). Morphology and morphogenesis of *Soranguim compositum.* J. Appl. Bacteriol. 32:24–29.

Couke, P., Voets, J. P. (1967). The mineral requirement of *Polyangium cellulosum.* Zeitschrift für Allgemeine Mikrobiologie 7:175–182.

Coupin, M. H. (1923). Quelques remarques sur la locomotion des Oscillaries. Compt. Rend. Hebd. Seances L'Acad. Sci. 176:1491–1493.

Craine, B. L., Rupert, C. S. (1979). Deoxyribonucleic acid-membrane interactions near the origin of replication and initiation of deoxyribonucleic acid synthesis in *Escherichia coli.* J. Bacteriol. 137:740–745.

Cumsky, M., Zusman, D. R. (1979). Myxobacterial hemagglutinin, a development-specific lectin of *Myxococcus xanthus.* Proc Natl. Acad. Sci. USA 76:5505–5509.

Cumsky, M., Zusman, D. R. (1981a). Purification and characterization of myxobacterial hemagglutinin, a development-specific lectin of *Myxococcus xanthus.* J. Biol. Chem. 256:12581–12588.

Cumsky, M., Zusman, D. R. (1981b). Binding properties of myxobacterial hemagglutinin. J. Biol. Chem. 256:12596–12599.

Cuzin, F., Jacobs, F. (1967). Existence chez *Escherichia coli* K12 d'une unité génétique de transmission formée de différents réplicons. Ann. Inst. Pasteur Paris 112:529–545.

Daft, M. J., Stewart, W. D. P. (1973). Light and electron microscope observations on algal lysis by bacterium CP-1. New Phytol. 72:799–808.

Davis, R. W., Botstein, D., Roth, J. R. (1980). Advanced Bacterial Genetics. New York: Cold Spring Harbor Lab.

Dawid, W. (1978). Fruchkörper-bildende Myxobakterien in Böden Brasiliens. Z. Allg. Mikrobiol. 18:83–93.

Dawid, W. (1979). Vorkommen und Verbreitung Fruchtkörper-bildender Myxobakterien im Siebengebirge Z. Allg. Mikrobiol. 19:705–719.

Dawson, H., Jones, M. V. (1979). Cell wall turnover during myxospore formation in *Myxococcus xanthus*. J. Gen. Microbiol. 112:143–148.

Dazzo, F. B. (1980). Adsorption of microorganisms to roots and other plant surfaces. pp. 253–316. In: Adsorption of Microorganisms to Surfaces, edited by G. Bitton, K. K. Marshall, New York: John Wiley.

Dea, I. C. M., Morris, E. R., Rees, D. A., Welsh, E. J., Barnes, H. A. Price, J. (1977). Associations of like and unlike polysaccharides: mechanism and specificity in galactomannans, interacting bacterial polysaccharides, and related systems. Carbohydr. Res. 57:249–272.

Diener, T. O. (1979). Viroids: Structure and function. Science 205:859–866.

Diener, T., McKinley, M. P., Prusiner, S. B. (1982). Viroids and prions. Proc. Nat. Acad. Sci. USA 79:5220–5224.

Dix, D. E., Helmstetter, C. E. (1973). Coupling between chromosome completion and cell division in *Escherichia coli*. J. Bacteriol. 115:786–795.

Dobson, W. J., McCurdy, H. D. (1979). The function of fimbriae in *Myxococcus xanthus*. I. Purification and properties of *M. xanthus* fimbriae. Can. J. Microbiol. 25:1150–1158.

Dobson, W. J., McCurdy, H. D., MacRae, T. H. (1979). The function of fimbriae in *Myxococcus xanthus*. II. The role of fimbriae in cell-cell interactions. Can. J. Microbiol. 25:1359–1372.

Donachie, W. D. (1968). Relationship between cell size and time of initiation of DNA replication. Nature 219:1077–1079.

Donachie, W. D. (1969). Control of cell division in *Escherichia coli:* Experiments with thymine starvation. J. Bacteriol. 100:260–268.

Donachie, W. D., Begg, K. J., Vicente, M. (1976). Cell length, cell growth and cell division. Nature 264:328–333.

Donachie, W. D., Hobbs, D. G., Masters, M. (1968). Chromosome replication and cell division in *Escherichia coli* 15T after growth in the absence of DNA synthesis. Nature 219:1079–1080.

Doolittle, W. F. (1969). The cyanobacterial genome, its expression, and the control of that expression. Adv. Microb. Physiol. 20:1–102.

Downard, J. S., Kupfer, D., Zusman, D. R. (1984) Gene expression during development of *Myxococcus xanthus:* Analysis of the genes for protein S. J. Mol. Biol. to be published.

Dudman, W. F., (1977). The role of surface polysaccharides in natural environments. In: Surface carbohydrates of the prokaryotic cell, edited by I. Sutherland. London, New York, San Francisco: Academic Press, pp. 354–414.

Dunny, G., Brown, B., Clewell, D. B. (1978). Induced cell aggregation and mating in *Streptococcus faecalis*. Evidence for a bacterial sex pheromone. Proc. Natl. Acad.

Sci. USA 75:3479–3483.

Dunny, G., Craig, R., Carron, R., Clewell, D. B., (1979). Plasmid transfer in *Streptococcus faecalis*. Production of multiple sex pheromones by recipients Plasmid 2:454–465.

Duxbury, T., Humphrey, B. A., Marshall, K. C. (1980). Continuous observations of bacterial gliding motility in a dialysis microchamber: The effects of inhibitors. Arch. Microbiol. 124:169–175.

Dworkin, M. (1962). Nutritional requirements for vegetative growth of *Myxococcus xanthus*. J. Bacteriol. 84:250–257.

Dworkin, M. (1963). Nutritional regulation of morphogenesis in *Myxococcus xanthus*. J. Bacteriol. 86:67–72.

Dworkin, M. (1966) Biology of the myxobacteria. Annu. Rev. Microbiol. 20:75–106.

Dworkin, M. (1972). The myxobacteria: New directions in studies of procaryotic development. CRC Crit. Rev. Microbiol. 435–452.

Dworkin, M. (1973). Cell-cell interactions in the myxobacteria. In: Microbial Differentiation, Soc. Gen. Microbiol. Symp., edited by J. M. Ashworth, J. E. Smith. Cambridge: Cambridge University Press, Vol. 23, pp. 125–142.

Dworkin, M. (1983). Tactic behavior of *Myxococcus xanthus*. J. Bacteriol. 154: 452–459.

Dworkin, M., Eide, D. (1983). *Myxococcus xanthus* does not respond chemotactically to moderate concentration gradients. J. Bacteriol. 154:437–442.

Dworkin, M., Gibson, S. M. (1964). A system for studying microbial morphogenesis: Rapid formation of microcysts in *Myxococcus xanthus*. Science 146:243–244.

Dworkin, M., Niederpruem, D. J. (1964) Electron transport system in vegetative cells and microcysts of *Myxococcus xanthus*. J. Bacteriol. 87:316–322.

Dworkin, M., Sadler, W. (1966). Induction of cellular morphogenesis in *Myxococcus xanthus*. 1. General description. J. Bacteriol. 91:1516–1519.

Dworkin, M., Voelz, H. (1962). The formation and germination of microcysts in *Myxococcus xanthus*. J. Gen. Microbiol. 28:81–85.

Dworkin, M., Keller, K. H. Weisberg, D. (1983). Experimental observations consistent with a surface tension model of gliding motility of *Myxococcus xanthus*. J. Bacteriol. 155:1367–1371.

Eberhard, A. (1972). Inhibition and activation of bacterial luciferase synthesis. J. Bacteriol. 109:1101–1105.

Edwards, O. E., Gillespie, D. C., (1966). N-dioxides, a new class of amine oxides. Tetrahedron Lett. 4867–4870.

Elander, R. P., Mabe, J. A., Hamill, R. H. and Gorman, M. (1968). Metabolism of tryptophans by *Pseudomonas aureofaciens*. Appl. Microbiol. 16:753–758.

Elton, R. A. (1973). The relationship of DNA base composition and individual protein composition in micro-organisms. J. Mol. Evol. 2:263–276.

Ensign, J. C. Wolfe, R. S. (1965). Lysis of bacterial cell walls by an enzyme isolated from a myxobacter. J. Bacteriol. 90:395–402.

Ensign, J. C., Wolfe, R. S. (1966). Characterization of a small proteolytic enzyme which lyses bacterial cell walls. J. Bacteriol. 91:525–534.

Evans, J. R., Napier, E. J., Fletton, R. A. (1978). G1499-2, a new quinoline compound isolated from the fermentation broth of *Cytophaga johnsonii*. J. Antibiol. 31:952–958.

Fattum, A., Rosenberg, E., (1977). Growth of *Myxococcus xanthus* FB and other bacteria on casein as a function of cell density. M.Sc. Thesis, Tel Aviv University.

Fautz, E., Grotjahn, L., Reichenbach, H. (1981). 2 and 3-hydroxy fatty acids as valuable chemosystematic markers in gliding bacteria. In: The *Flavobacterium-Cytophaga* Group, edited by H. Reichenbach, O. B. Weeks, Weinheim: Verlag Chemie, pp. 127–133.

Fautz, E., Rosenfelder, G., Grotjahn, L. (1979). Iso-branched 2- and 3-hydroxy fatty acids as characteristic lipid constituents of some gliding bacteria. J. Bacteriol. 140:852–858.

Filer, D., Kindler, S. H., Rosenberg, E. (1977). Myxospore coat synthesis in *Myxococcus xanthus:* Enzymes associated with uridine 5′diphosphate-N-acetylgalactosamine formation during myxospore development. J. Bacteriol. 131:745–750.

Filer, D., Rosenberg, E., Kindler, S. H. (1973). Aspartokinase of *Myxococcus xanthus:* "Feedback stimulation" by required amino acids. J. Bacteriol. 115:23–28.

Finck, G. (1950). Biologische und stoffwechselphysiologische Studien an Myxococcaceen. Arch. Mikrobiol. 15:358–388.

Fletcher, M., Floodgate, G. D. (1973). An electron-microscopic demonstration of an acidic polysaccharide involved in the adhesion of a marine bacterium to solid surfaces. J. Gen Microbiol. 74:325–334.

Fluegel, W. (1963a). Simple method for demonstrating myxobacterial slime. J. Bacteriol. 85:1173–1174.

Fluegel, W. (1963b). Fruiting chemotaxis in *Myxococcus fulvus* (Myxobacteria). Proc. Minn. Acad. Sci. 32:120–123.

Fluegel, W. (1964). Induced fruiting in myxobacteria. Proc. Minn. Acad. Sci. 31:114–115.

Forsberg, C. W., Rayman, M. K., Costerton, J. W., MacLeod, R. A. (1972). Isolation, characterization, and ultrastructure of the peptidoglycan layer of a marine pseudomonad. J. Bacteriol. 109:895–905.

Foster, H. A., Parish, J. H. (1973). Synthesis of RNA during myxospore induction in *Myxococcus xanthus.* J. Gen. Microbiol. 75:401–407.

Freese, E. (1981). Initiation of bacterial sporulation. In: Sporulation and Germination, edited by H. S. Levinson, Sonenshein, A. L., D. J. Tipper. Washington, D. C: American Society Microbiology, pp. 1–12.

Fyrberg, E. A., Kindle, K. L. Davidson, N., Sodja, A., (1980). The actin genes of *Drosophila:* A dispersed multigene family. Cell 19:365–378.

Fytlovitch, S., Nathan, P. D., Zafriri, D., Rosenberg, E. (1983). Amino acid precursors of *Myxococcus xanthus* antibiotic TA. J. Antibiot. 36:1525–1530.

Gallant, J. A. (1979). Stringent control in *Escherichia coli.* Annu. Rev. Genet. 13:393–415.

Geiger, L. W., Morris, D. R. (1980). Stimulation of deoxyribonucleic acid replication fork movement by spermidine analogs in polyamine-deficient *Escherichia coli.* J. Bacteriol. 141:1192–1198.

Geisselsoder, J., Campos, J. M., Zusman, D. R. (1978). Physical characterization of bacteriophage Mx4, a generalized transducing phage for *Myxococcus xanthus.* J. Mol. Biol. 119:179–189.

Geitler, L. (1925). Über *Polyangium parasiticum* n. sp., eine submerse, parasitische Myxobacteriacee. Arch. Protistenkunde 50:67–88.

Gerth, K. (1975). Untersuchungen über den molekularen Mechanismus der Induktion von Myxosporen bei *Stigmatella aurantiaca* Sg al (Myxobacterales). Doctor's Thesis, University of Freiburg, Federal Republic of Germany.

Gerth, K., Reichenbach, H. (1978). Induction of myxospore formation in *Stigmatella aurantiaca* (Myxobacterales) I. General characterization of the system. Arch. Microbiol. 117:173–182.

Gerth, K., Irschik, H., Reichenbach, H., Trowitzsch, W. (1980). Myxothiazol, an antibiotic from *Myxococcus fulvus* (Myxobacterales) I. Cultivation, isolation, physico-chemical and biological properties. J. Antibiot. (Tokyo) 33:1474–1479.

Gerth, K., Irschik, H., Reichenbach, H., Trowitzsch, W. (1982). The myxovirescins, a family of antibiotics from *Myxococcus virescens* (Myxobacterales) J. Antibiot. (Tokyo) 1454–1459.

Gerth, K., Trowitzsch W., Wray, V., Höfle, G., Irschik, H., Reichenbach, H. (1982). Pyrrolnitrin from *Myxococcus fulvus* (Myxobacterales) J. Antibiot. (Tokyo) 35:1101–1103.

Gerth, K., Jansen, R., Reifenstahl, G., Höfle, G., Irschik, H., Kunze, B., Reichenbach, H., Thierbach, G. (1983). The myxalamids, new antibiotics from *Myxococcus xanthus* (Myxobacterales) I. Production, physico-chemical and biological properties, and mechanism of action. J. Antibiot. (Tokyo) 36:1150–1156.

Ghuysen, J. M., Shockman, G. D. (1973). Biosynthesis of Peptidoglycan in Bacterial Membrane and Walls, edited by L. Leive, New York: Marcel Dekker, pp. 37–130.

Gilbert, M., and A. B. Fulton, (1982). Pro-actin: A procaryotic actin-like protein of 55kD MW. J. Cell Biol. 95:283a.

Gillespie, D. C., Cook, F. D. (1965). Extracellular enzymes from strains of Sorangium. Can. J. Microbiol. 11:109–118.

Glauert, A. M., Thornley, M. J. (1969). The topography of the bacterial cell wall, In: Annual Review of Microbiology, edited by C. E. Clifton, S. Raffel, M. P. Starr. Palo Alto, California: Annual Reviews Inc., vol. 23, pp. 159–198.

Gnosspelius, G. (1978a). Myxobacterial slime and proteolytic activity. Arch. Microbiol. 116:51–59.

Gnosspelius, G. (1978b). Purification and properties of an extracellular protease from *Myxococcus viresens*. J. Bacteriol. 133:17–25.

Gogol, E. M. Rosenberg, E. (1964). Observations on the regulation of DNA biosynthesis. Biochm Biophys. Res. Commun. 14:565–570.

Goldman, R. C., White, D., Orskov, F., Orskov, I., Rick, P. D., Lewis, M. S., Bhattacharjee, A. K., Leive, L. (1982). A surface polysaccharide of *Escherichia coli* 0111 contains O-antigen and inhibits agglutination of cells by O-antiserum. J. Bacteriol. 151:1210–1221.

Gräf, W. (1975). Myxobakterien der Gattung *Myxococcus* als indirekte Fäkalstoffindikatoren in Oberflächengewässern. 1. Mitteilung. Zentralblatt für Bakteriologie, Parasitenkunde, Infektionskrankheiten und Hygiene I. Abteilung, Originale B 160:28–39.

Grand, R. J. A. Nairn, A. C., Perry, S. V. (1980). The preparation of calmodulins from barley (*Hordeum* sp.) and basidiamycete fungi. Biochem. J. 185:755–760.

Greenberg, E. P., Hastings, J. W., Ulitzur, S. (1979). Induction of luciferase synthesis in *Beneckea harveyi* by other marine bacteria. Arch. Microbiol. 120:87–91.

Greene, J. M., Leadbetter, E. R. (1962). Pigments of some myxococci. Bacteriol. Proc. p. 34.

Grilione, P. L., Pangborn, J. (1975). Scanning electron microscopy of fruiting body formation by myxobacteria. J. Bacteriol. 124:1558–1565.

Grimm, K. (1967). Klonkulturen von *Archangium violaceum*. Arch. Mikrobiol. 57:283–284.

Grimm, K. (1978). Comparison of spontaneous, UV-induced, and nitrosoguandine-induced mutability to drug resistance in myxobacteria. J. Bacteriol. 135:748–753.

Grimm, K., Herdrich, K. (1980). Host-cell reactivation and Weigle reactivation of ultraviolet-irradiated *Myxococcus xanthus* phages. Photochem. Photobiol. 31:229–233.

Grimm, K., Kühlwein, H. (1973a). Untersuchungen an spontanen Mutanten von *Archangium violaceum* (Myxobacterales) I. Bewegliche und unbewegliche Zellen von *A. violaceum.* Arch. Mikrobiol. 89:105–119.

Grimm, K., Kühlwein, H. (1973b) Untersuchungen an spontanen Mutanten von *Archangium violaceum* (Myxobacterales) II. Über den Einfluβ des Schleims auf die Bewegung der Zellen und die Entstehung stabiler Suspensionskulturen. Arch. Mikrobiol. 89:121–132.

Grimm, K., Galle, H. K., Heunert, H. H. (1971). *Archangium violaceum* (Myxobacteriales)—Bewegungsaktivität und Kolonieform. Encyclopaedia Cinematographica E 1588, film of the Institut für den wissenschaftlichen Film, Göttingen, Federal Republic of Germany.

Grossman, N., Ron, E. Z. (1980). Initiation of deoxyribonucleic acid replication in *Escherichia coli* B. Uncoupling from mass/ deoxyribonucleic acid ratio. J. Bacteriol. 143:100–104.

Grunberg, E., Berger, J., Beskid, G., Cleeland, R., Prince, H. H., Titswort, E. (1967). Studies on in vitro and in vivo chemotherapeutic properties of antibiotic myxin. Chemotherapy 12:272.

Hagen, D. C., Bretscher, A. P., Kaiser, D. (1978). Synergism between morphogenetic mutants of *Myxococcus xanthus.* Dev. Biol. 64:284–296.

Hamilton, W. D., (1964). The genetical theory of social behaviour. J. Theor. Biol. 7:1–16, 17–32.

Hanawalt, P. C., Cooper, P. K., Ganesan, A. K., Smith, C. A. (1979). DNA repair in bacteria and mammalian cells. Annu. Rev. Biochem. 48:783–836.

Hanson, C. W., Andreoli, A. J. (1973). Some aspects of amino acid metabolism in the fruiting myxobacterium, *Myxococcus xanthus.* Arch. Mikrobiol. 92:1–10.

Hanson, C. W., M. Dworkin, (1974). Intracellular and extracellular nucleotides and related compounds during the development of *Myxococcus xanthus.* J. Bacteriol. 118:486–496.

Harcke, E., Hüttermann, A., Kühlwein, H. (1971). Studies on lytic activities of *Chondrococcus coralloides* (Myxobacteriales) (sic). I. Purification and some properties of the bacteriolytic and proteolytic activity. Arch. Mikrobiol. 77:86–95.

Harcke, E., Hütterman, A., Kühlwein, H. (1972). Studies on lytic activities of *Chondrococcus coralloides* (Myxobacterales). II. Identification of the bacteriolytic enzyme as a muramidase. Arch. Mikrobiol. 85:6–12.

Hart, B. A., Zahler, S. A. (1966). Lytic enzymes produced by *Myxococcus xanthus.* J. Bacteriol. 92:1632–1637.

Hartwell, L. H., Culotti, J., Pringle, J. R., Reid, B. J. (1974). Genetic control of the cell division cycle in yeast. Science 183:46–51.

Haseltine, W. A., Block, R. (1973). Synthesis of guanosine tetra- and pentaphosphate requires the presence of a codon-specific, uncharged transfer ribonucleic acid in the acceptor site of ribosomes. Proc. Natl. Acad. Sci. USA 70:1564–1568.

Haskå, G. (1974) Extracellular lytic enzymes of Myxococcus *virescens.* IV. Purification and characterization of a D-alanyl-ϵ-N-lysine endopeptidase. Physiol. Plant 31:252–256.

Haskå, G., Stahl, S. (1971). Variants of *Myxococcus virescens* exhibiting dispersed growth. Growth and production of extracellular enzymes and slime. Physiol. Plant 24:136–142.

Haskå, G., Norén, B., Odham, G. (1972). Effect of fatty acids on the activity of bacteriolytic enzymes. Physiol. Plant 27:187–194.

Hastings, J. W., Nealson, K. H. (1977). Bacterial bioluminescence. Annu. Rev. Microbiol. 31:549–595.

Hayes, W. (1968). The Genetics of Bacteria and Their Viruses. New York: John Wiley.

Hedges, R. W., Jacob, A. E., Barth, P. T., Grinter, N. J. (1975). Compatibility properties of P1 and $\phi$ Amp prophages. Mol. Gen. Genet. 141:263–267.

Heilmann, H. D. (1972). On the peptidoglycan of the cell walls of *Pseudomonas aeruginosa*. Eur. J. Biochem. 31:456–463.

Helmstetter, C. E. (1967). Rate of DNA synthesis during the division cycle of *Escherichia coli* B/r. J. Mol. Biol. 24:417–427.

Helmstetter, C. E. (1974). Initiation of chromosome replication in *Escherichia coli*. II. Analysis of the control mechanism. J. Mol. Biol. 84:21–36.

Helmstetter, C. E., Pierucci, O. (1976). DNA synthesis during the division cycle of three substrains of *Escherichia coli* B/r. J. Mol. Biol. 102:477–486.

Helmstetter, C. E., Cooper, S., Pierucci, O., Revelas, E. (1968). On the bacterial life sequence. Cold Spring Harbor Symp. Quant. Biol. 33:809–822.

Helmstetter, C. E., Pierucci, O., Weinberger, M., Holmes, M., Tang, M.-S. (1979). Control of cell division in *Escherichia coli*. In: The Bacteria, edited by L. N. Ornston, J. R. Sokatch, New York: Academic Press, vol. 7, pp. 517–579.

Hemphill, H. E. Zahler, S. A. (1968a). Nutrition of *Myxococcus xanthus* FBa and some of its auxotrophic mutants. J. Bacteriol. 95:1011–1017.

Hemphill, H. E. Zahler, S. A. (1968b). Nutritional induction and suppression of fruiting in *Myxococcus xanthus* FBa. J. Bacteriol. 95:1018–1023.

Henning, U., Rehn, K., Braun, V., Hohn, B. (1972). Cell envelope and shape of *Escherichia coli* K12. Properties of a temperature-sensitive *rod* mutant. Eur. J. Biochem. 26:570–586.

Henrichsen, J. (1972). Bacterial surface translocation: a survey and a classification. Bacteriol. Rev. 36:478–503.

Hess, B., Borteux, A. (1971). Oscillatory phenomena in biochemistry. Annu. Rev. Biochem. 40:237–258.

Heumann, H. G., Kühlwein, H. (1979). Effects of drugs that influence eucaryotic motile processes on motility of *Cystobacter fuscus* (Myxobacterales). Z. Allg. Mikrobiol. 19:547–552.

Hirosawa, R., Wolk, C. P. (1979a). Factors controlling the formation of akinetes adjacent to heterocysts in the cyanobacterium *Cylindrospermum licheniforme* Kutz. J. Gen. Microbiol. 114:423–432.

Hirosawa, R., Wolk, C. P. (1979b). Isolation and characterization of a substance which stimulates the formation of akinetes in the cyanobacterium *Cylindrospermum licheniforme* Kutz. J. Gen. Microbiol. 114:433–441.

Hirsch, H. J. (1977). Bacteriocins from *Myxococcus fulvus* (Myxobacterales). Arch. Mikrobiol. 115:45–49.

Hirsch, H. J., Tsai, H. Geffers, I. (1978). Purification and effects of fulvocin C, a bacteriocin from *Myxococcus fulvus* Mx f16. Arch. Microbiol. 119:279–286.

Ho, J., McCurdy, H. D. (1979). Demonstration of positive chemotaxis to cyclic GMP and 5′AMP in *Myxococcus xanthus* by means of a simple apparatus for generating practically stable concentration gradients. Can. J. Microbiol. 25:1214–1218.

Hodgkin, J., Kaiser, D. (1977). Cell-to-cell stimulation of movement in nonmotile mutants of *Myxococcus*. Proc. Natl. Acad. Sci. USA 74:2938–2942.

Hodgkin, J., Kaiser, D. (1979a). Genetics of gliding motility in *Myxococcus xanthus* (Myxobacterales): genes controlling movements of single cells. Mol. Gen. Genet. 171:167–176.

Hodgkin, J., Kaiser, D. (1979b). Genetics of gliding motility in *Myxococcus xanthus* (Myxobacterales): two gene systems control movement. Mol. Gen. Genet. 171:177–191.

Holliday, R., Pugh, J. E. (1975). DNA modification mechanisms and gene activity during development. Science 187:226–232.

Hook, L. A. (1977). Distribution of myxobacters in aquatic habitats in an alkaline bog. Applied and Environmental Microbiology 34:333–335.

Humphrey, B. A., Dickson, M. R., Marshall, K. C. (1979). Physicochemical and in situ observations on the adhesion of gliding bacteria to surfaces. Arch. Microbiol. 120:231–238.

Hüttermann, A. (1969). Studies on a bacteriolytic enzyme of *Archangium violaceum* (Myxobacterales). II. Partial purification and properties of the enzyme. Arch. Mikrobiol. 67:306–317.

Ikeda, H., Tomizawa, J. (1965). Transducing fragments in generalized transduction by phage P1. I. Molecular origin of the fragments. J. Mol. Biol. 14:85–109.

Imschenezki, A. A. (1959). Mikrobiologie der Cellulose. [Translated from the Russian edition of 1953.] Berlin: Akademie Verlag.

Inouye, M., Franceschini, T., Inouye, M. (1983). Structural similarities between the development specific protein S from a gram negative bacterium, *Myxococcus xanthus,* and calmodulin. Proc. Natl. Acad. Sci. USA 80:6829–6833.

Inouye, M., Inouye, S., Zusman, D. R. (1979a). Gene expression during development of *Myxococcus xanthus:* Pattern of protein synthesis. Dev. Biol. 68:579–591.

Inouye, M., Inouye, S., Zusman, D. R. (1979b). Biosynthesis and self-assembly of protein S, a development specific protein of *Myxococcus xanthus.* Proc. Natl. Acad. Sci. USA 76:209–213.

Inouye, S., White, D., Inouye, M. (1980a). Development of *Stigmatella aurantiaca:* Effects of light and gene expression. J. Bacteriol. 141:1360–1365.

Inouye, S., Harada, W., Zusman, D., Inouye, M. (1981). Development specific protein S of *Myxococcus xanthus:* Purification and characterization. J. Bacteriol. 148:678–683.

Inouye, S., Ike, Y., Inouye, M. (1983). Tandem repeat of the genes for protein S, a development specific protein of *Myxococcus xanthus.* J. Biol. Chem. 258:38–40.

Inouye, S., Inouye, M., McKeever, B., Sarma, R. (1980b). Preliminary crystallographic data for protein S, a development-specific protein of *Myxococcus xanthus.* J. Biol. Chem. 255:3713–3713

Irschik, Gerth, K., Kemmer, T., Steinmetz, H., Reichenbach, H. (1983). The myxovalargins, new peptide antibiotics from *Myxococcus fulvus* (Myxobacterales). J. Antibiotics 36:6–12.

Isler, O., edit. (1971). Carotenoids. Basel: Birkhäuser Verlag.

Itwasa, Y., Yonemitsu, K., Matsui, K. (1981). Calmodulin-like activity in the soluble fraction of *Escherichai coli.* Biochem. Biophys. Res. Commun. 98:565–660.

Jacob, F., Brenner, S., Cusin, F. (1963). On the regulation of DNA replication in bacteria. Cold Spring Harbor Symp. Quant. Biol. 28:329–348.

Jackson, R. L., Wolfe, R. S. (1968). Composition, properties and substrate specificities of myxobacter AL-1 protease. J. Biol. Chem. 243:870–888.

Jahn, E. (1924). Beiträge zur botanischen Protistologie. I. Die Polyangiden. Leipzig: Gebrüder Borntraeger.

Janssen, G., (1983). Cell–cell interaction during developmental lysis of *Myxococcus xanthus*. Ph.D. dissertation. University of Minnesota, Minneapolis.

Janssen, G. E., Dworkin, M. (1980). Abstract Annual Meeting of ASM 1:113.

Jarvick, J., Botstein, D. (1973). A genetic method for determining the order of events in a biological pathway. Proc. Natl. Acad. Sci. USA 70:2046–2050.

Jeffers, E. E. (1964). Myxobacters of a freshwater lake and its environs. Int. Bull. Bacterial Nomenclature Taxonomy 14:115–136.

Jennings, J. (1961). Association of a steroid and a pigment with a diffusible fruiting factor in *Myxococcus virescens*. Nature 190:190.

Johnson, J. L., Ordal, E. J. (1968). Deoxyribonucleic acid homology in bacterial taxonomy: Effect of incubation temperature on reaction specificity. J. Bacteriol. 95:893–900.

Johnson, J. L., Ordal, E. J. (1969). Deoxyribonucleic acid homology among the fruiting myxobacteria. J. Bacteriol. 98:319–320.

Johnson, R. J., White, D. (1972). Myxospore formation in *Myxococcus xanthus:* Chemical changes in the cell wall during cellular morphogenesis. J. Bacteriol. 112:849–855.

Jones, G. W. Rutter, J. M. (1974). The association of K88 antigen with haemagglutinating activity in porcine strains of *Escherichia coli*. J. Gen Microbiol. 84:135–144.

Jones, M. V. (1978). Inhibition of fruiting body formation in *Myxococcus xanthus* by cytokinins. FEMS Lett. 3:147–149.

Jones, N. C., Donachie, W. D. (1973). Chromosome replication, transcription and control of cell division in *Escherichia coli*. Nature New Biol. 243:100–103.

Juengst, F. W., Dworkin, M. (1973). Ribonucleic acid and protein synthesis during germination of *Myxococcus xanthus* myxospores. J. Bacteriol. 113:786–797.

Jurasek, L. Whitaker, D. R. (1965). Lytic enzymes of *Sorangium* sp. A comparison of some physical properties of the a and b lytic proteases. Can. J. Microbiol. 43:1955–1960.

Kaiser, D. (1978). Genetics of cell interactions in myxobacteria. Birth Defects: Original Article Series 14:391–399.

Kaiser, D. (1979). Social gliding is correlated with the presence of pili in *Myxococcus xanthus*. Proc. Natl. Acad. Sci. USA 76:5952–5956.

Kaiser, D., Crosby, C. (1983). Cell movement and its coordination in swarms of *Myxococcus xanthus*. Cell Motility 3: 227–245.

Kaiser, D., Dworkin, M. (1975). Gene transfer to a myxobacterium by *Escherichia coli* phage P1. Science 187:653–654.

Kaiser, D., Manoil, C., Dworkin, M. (1979). Myxobacteria: Cell interactions, genetics and development. Annu. Rev. Microbiol. 33:595–639.

Katz, W., Strominger, J. L. (1967). Structure of the cell wall of *M. lysodeikticus*. II. Study of the structure of the peptides produced after lysis with the Myxobacterium enzyme. Biochemistry 6:930–937.

Katznelson, H. K., Gillespie, D. C., Cook, F. D. (1964). Studies on the relationships between nematodes and other soil microorganisms. III. Lytic action of soil myxobacters on certain species of nematodes. Can. J. Microbiol. 10:699–704.

Keller, K. H. Grady, M., Dworkin, M. (1983). Surface tension gradients: Feasible model for gliding motility in *Myxococcus xanthus*. J. Bacteriol. 155:1358–1366.
Kharchenko, E. P. (1975). Evolution of nucleotide composition of DNA: First hypotheses and new facts. J. Evolutionary Biochem. Physiol. 11:30–34.
Kimchi, A., Rosenberg, E. (1976). Linkages between deoxyribonucleic acid synthesis and cell division in *Myxococcus xanthus*. J. Bacteriol. 128:69–79.
Kleckner, N., Roth, J., Botstein, D. (1977). Genetic Engineering in Vivo using translocatable drug-resistance elements. J. Mol. Biol. 116:125–159.
Kleinig, H. (1972). Membranes from *Myxococcus fulvus* (Myxobacterales) containing carotenoid glucosides I. Isolation and composition. Biochimica et Biophysica Acta 274:489–498.
Kleinig, H. (1974). Inhibition of carotenoid synthesis in *Myxococcus fulvus* (Myxobacterales). Arch. Microbiol. 97:217–226.
Kleinig, H. (1975). On the utilization in vivo of lycopene and phytoene as precursors for the formation of carotenoid glucoside ester and on the regulation of carotenoid biosynthesis in *Myxococcus fulvus*. Eur. J. Biochem. 57:301–308.
Kleinig, H., Reichenbach, H. (1969). Carotenoid pigments of *Stigmatella aurantiaca* (Myxobacterales) I. The minor carotenoids. Arch. Mikrobiol. 68:210–217.
Kleinig, H., Reichenbach, H. (1973a). Biosynthesis of carotenoid glucoside esters in *Myxococcus fulvus* (Myxobacterales): Inhibition by nicotine and carotenoid turnover. Biochim. Biophy. Acta 306:249–256.
Kleinig, H., Reichenbach, H. (1973b). A new carotenoid glucoside ester from *Chondromyces apiculatus*. Phytochemistry 12:2483–2485.
Kleinig, H., Reichenbach, H., Achenbach, H. (1970). Carotenoid pigments of *Stigmatella aurantiaca* (Myxobacterales) II. Acylated carotenoid glycosides. Arch. Mikrobiol. 74:223–234.
Kleinig, H., Englert, G., Meister, A., Reichenbach, H. (1985). The pigments of *Nannocystis exedens* (Myxobacterales): novel aromatic carotenoids. Helv. Chim. Acta to be published.
Kleinig, H., Reichenbach, H., Achenbach, H. Stadler, J. (1971) Carotenoid pigments of *Sorangium compositum* (Myxobacterales) including two new carotenoid glycoside esters and two new carotenoid rhamnosides. Arch. Mikrobiol. 78:224–233.
Kleinig, H., Reichenbach, H., Theobald, N., Achenbach, H. (1974). *Flexibacter elegans* und *Myxococcus fulvus:* Aerobic Gram-negative bacteria containing menaquinones as the only isoprenoid quinones. Arch. Microbiol. 101:91–93.
Kletter, B., Henis, Y. (1963). Comparative studies on the growth of myxobacteria on bacterial suspensions. Can. J. Microbiol. 9:577–584.
Koch, A. L. (1977). Does the initiation of chromosome replication regulate cell division? Adv. Microb. Physiol. 16:49–97.
Koch, A. L. (1983). The surface stress theory of microbial morphogenesis. Adv. Microb. Physiol. to be published.
Kohl, W., Gloe, A., Reichenbach, H. (1983). Steroids from the myxobacterium *Nannocystis exedens*. J. Gen. Microbiol. 129:1629–1635.
Komano, T., Inouye, S., Inouye, M. (1980). Patterns of protein production in *Myxococcus xanthus* during spore formation induced by glycerol, dimethyl sulfoxide, and phenethyl alcohol. J. Bacteriol. 144:1076–1082.
Kornberg, A. (1980). DNA Replication. San Francisco: W. H. Freeman.
Kornberg, A. (1982). 1982 Supplement to DNA Replication. San Francisco: W. H. Freeman.

Kottel, R., White, D. (1974). Autolytic activity associated with myxospore formation in *M. xanthus*. Arch. Microbiol. 95:91–95.

Kottel, R. H., Bacon, K., Clutter, D., White, D. (1975). Coats from Myxococcus xanthus: Characterization and synthesis during myxospore differentiation. J. Bacteriol. 124:550–557.

Kottel, R. H., Orlowski, M., White, D., Grutsch, J. (1974). Presence of amino acid dehydrogenases and transaminases in *Myxococcus xanthus* during vegetative growth and myxospore formation. J. Bacteriol. 119:650–651.

Krassilnikov, N. A. (1959). Diagnostik der Bakterien und Actinomyceten. [Translated from the Russian edition of 1949]. Jena: Gustav Fischer.

Krestinger, R. H., Nockohds, C. E. (1973) Carp muscle calcium-binding protein. J. Biol. Chem. 248:3313–3326.

Kriegstein, H. J., Hogness, D. S. (1974). Mechanism of DNA replication in Drosophila chromosomes: structure of replication forks and evidence for bidirectionality. Proc. Natl. Acad. Sci. USA 71:125–139.

Krzemieniewska, H., Krzemieniewski, S. (1926). Die Myxobakterien von Polen. Acta Soc. Bot. Pol. 4:1–54.

Krzemieniewska, H., Krzemieniewski, S. (1927a). Über die Verbreitung der Myxobakterien im Boden. Acta Soc. Bot. Pol. 5:102–139.

Krzemieniewska, H., Krzemieniewski, S. (1927b). Die Myxobakterien von Polen, Anhang. Acta Soc. Bot. Pol. 5:79–98.

Krzemieniewska, H., Krzemieniewski, S. (1928). Zur Morphologie der Myxobakterienzelle Acta Soc. Bot. Pol. 5:(46)–(90).

Krzemieniewska, H., Krzemieniewski, S. (1930). Die Myxobakterien von Polen. III. Teil Acta Soc. Bot. Pol. 7:250–273.

Krzemieniewska, H., Krzemieniewski, S. (1937a). Die zellulosezersetzenden Myxobakterien. Bulletin de l'Académie Polonaise des Sciences et des Lettres. Classe des Sciences Mathématiques et Naturelles. Série (I):11–31.

Krzemieniewska, H., Krzemieniewski, S., (1937b). Über die Zersetzung der Zellulose durch Myxobakterien. Bulletin de l'Académie Polonaise des Sciences et des Lettres. Classe des Sciences Mathématiques et Naturelles. Série B (I):33–59.

Kühlwein, H. (1950). Beiträge zur Biologie und Entwicklungsgeschichte der Myxobakterien. Arch. Mikrob. 14:678–704.

Kühlwein, H. (1953). Weitere Untersuchungen an Myxobakterien. Arch. Mikrobiol. 19:365–371.

Kühlwein, H. (1957). Untersuchungen über die Bewegung der Myxobakterien. Ber. Dtsch. Bot. Ges. 70:227–232.

Kühlwein, H. (1960). Zur Systematik und Verbreitung der Myxobakterien. Zentralbl. Bakteriol. Parasit., Infek und Hyg. II Abt. 113:480–490.

Kühlwein, H. (1963). Beiträge zur Kenntnis der Myxobakterienzelle. I. Licht-und electronenmikroskopische Untersuchungen von Zellwänden. Zentralbl. Bakteriol. Parasit., Infek. und Hyg. I-Orig. 189:468–481.

Kühlwein, H., Finck, G. (1950). Zur Stuffwechselphysiologie der Myxobakterien. Naturwissenschaften 37:43–44.

Kühlwein, H., Gallwitz, E. (1958). *Polyangium violaceum* nov. spec. Ein Beitrag zur Kenntnis der Myxobakterien. Arch. Mikrobiol. 31:139–145.

Kühlwein, H., Reichenbach, H. (1964). Ein Neuer Vertreter der Myxobakteriengattung *Archangium* Jahn. Arch. Mikrobiol. 48:179–184.

Kühlwein, H., Reichenbach, H. (1965). Anreicherung und Isolierung von Myxobakterien. Zentralblatt für Bakteriologie, Parasitenkunde, Infektionskrankheiten und Hygiene.I. Abt. Orig., Suppl. 1:57–80.

Kühlwein, H., Reichenbach, H., (1968). Schwarmentwicklung und Morphogenese bei Myxobakterien/*Archangium–Myxococcus–Chondrococcus–Chondromyces.* Text to film C893/1965. Göttingen: Inst. Wiss. Film.

Kühlwein, H., Schlicke, B., Galle, H. K., Heunert, H. H. (1971a). *Polyangium fuscum* (Myxobacteriales)—Cystenkeimung und Schwarmentwicklung. Encyclopaedia Cinematographica E 1582, film of the Institut für den wissenschaftlichen Film, Göttingen, Federal Republic of Germany.

Kühlwein, H., Schlicke, B., Galle, H. K., Heunert, H. H. (1971b). *Polyangium fuscum* (Myxobacteriales)—Morphogenese. Encyclopaedia Cinematographica E1583, film of the Institut für den wissenschaftlichen Film, Göttingen, Federal Republic of Germany.

Kuner, J. M., Kaiser, D. (1981). Introduction of transposon Tn*5* into *Myxococcus* for analysis of developmental and other nonselectable mutants. Proc. Natl. Acad. Sci. USA 78:425–429.

Kuner, J. M., Kaiser, D. (1982). Fruiting body morphogenesis in submerged cultures of *Myxococcus xanthus.* J. Bacteriol. 151:458–461.

Kuner, J. M., Avery, L., Berg, D., Kaiser, D. (1981). Uses of transposon Tn*5* in the genetic analysis of *Myxococcus xanthus.* In: Microbiology 1981, edited by D. Schlessinger. ASM, Washington, D.C.

Kunze, B., Reichenbach, H., Augustiniak, H., Höfle, G. (1982). Isolation and identification of althiomycin from *Cystobacter fuscus* (Myxobacterales). J. Antibiot. (Tokyo) 35:635–636.

Lampky, J. R. Brockman, E. R. (1977). Fluorescence of *Myxococcus stipitatus.* Int. J. Systematic Bacteriol. 27:161.

Lapidus, I. R., Berg, H. C (1982). Gliding motility of *Cytophaga* sp. strain U67. J. Bacteriol. 151:384–398.

Lark, K. G. (1972). Evidence for the direct involvement of RNA in the initiation of DNA replication in *Escherichia coli* 15T. J. Mol. Biol. 64:47–60.

Lark, K. G., Lark, C. (1965). Regulation of chromosome replication in *Escherichia coli:* Alternate replication of two chromosomes at slow growth rates. J. Mol. Biol. 13:105–126.

LaRossa, R., Kuner, J., Hagen, D., Manoil, C., Kaiser, D. (1983). Developmental cell interactions of *Myxococcus xanthus:* Analysis of mutants. J. Bacteriol. 153:1394–1404.

Larpent, J. P. (1970). De la cellule à l'organisme (acrasiales, myxomycètes, myxobactériales). Paris: Masson.

Lechevalier, H. A. (1975). Production of the same antibiotics by members of different genera of microorganisms. Adv. Appl. Microbiol. 19:25–45.

Lesk, A. M. (1973). On hypothesized selective pressure by u.v. on DNA base compositions. J. Theor. Biol. 40:201–202.

Lesley, S. M., Behki, R. M. (1967). Mode of action of myxin on *Escherichia coli.* J. Bacteriol. 94:1837–1845.

Lesley, S. M. Behki, R. M. (1971). Recovery of metabolic activity in *E. coli* following limited exposure to myxin. Can. J. Microbiol. 17:1327–1333.

Lev, M. (1954). Demonstration of a diffusible fruiting factor in myxobacteria. Nature 173:501.

Lewin, B. (1980). Gene Expression v. 2. New York: John Wiley, 2nd ed.

Liaaen-Jensen, S. (1973). Structural elucidation of carotenoids—a progress report.

Pure Applied Chem. 35:81–112.

Liaaen-Jensen, S., Andrewes, A. G. (1972). Microbial carotenoids. Annu. Rev. Microbiol. 26:225–248.

Lindberg, A. (1973). Bacteriophage receptors. Annu. Rev. Microbiol. 27:205–241.

Lippincott, J. A., Lippincott, B. B. (1978). Cell walls of crown-gall tumors and embryonic plant tissues lack *Agrobacterium* adherence sites. Science 199:1075–1077.

Lloyd, D., Poole, R. K, Edwards, S. W. (1982). The Cell Division Cycle: Temporal Organization and Control of Cellular Growth and Reproduction. New York: Academic Press.

Loebeck, M. E., Klein, H. P. (1956). Substrates for *Myxococcus virescens* with special reference to eubacterial fractions. J. Gen. Microbiol. 14:281–289.

Lonski, J., Heromin, R., Ingraham, D., (1977). A quantitative assay to study cell movement in the myxobacteria. J. Cell Sci. 25:173–178.

Lopez, J. M., Dromerick, A., Freese, E. (1981). Response of guanosine 5′-triphosphate concentration to nutritional changes and its signifcance for *Bacillus subtilis* sporulation. J. Bacteriol. 146:605–613.

Ludwig, W., Schleifer, K. H., Reichenbach, H., Stackebrandt, E., (1983). A phylogenetic analysis of the myxobacteria *Myxococcus fulvus, Stigmatella aurantiaca, Cystobacter fuscus, Sorangium cellulosum* and *Nannocystis exedens*. Arch. Microbiol. 135:58–62.

MacNab, R. M., Koshland, D. E., Jr. (1972). The gradient-sensing mechanism in bacterial chemotaxis. Proc. Nat. Acad. Sci. USA 69:2509–2512.

MacRae, T. H., McCurdy, H. D. (1975). Ultrastructural studies of *Chondromyces crocatus* vegetative cells. Can. J. Microbiol. 21:1815–1826.

MacRae, T. H., McCurdy, H. D. (1976). The isolation and characterization of gliding motility mutants in *Myxococcus xanthus*. Can. J. Microbiol. 22:1282–1292.

MacRae, T. H., Dobson, W. J., McCurdy H. W. (1977). Fimbriation in gliding bacteria. Can. J. Microbiol. 23:1096–1108.

Mandel, J. L., Chambone, P. (1979). DNA methylation: Organ specific variations in the methylation pattern within and around ovalbumin and other chicken genes. Nucleic Acids Res. 7:1081–1203.

Mandel, M., Leadbetter, E. R. (1965). Deoxyribonucleic acid base composition of myxobacteria. J. Bacteriol. 90:1795–796.

Maniatis, T., Fritsch, E. F., Sambrook, J. (1982). Molecular Cloning. New York: Cold Spring Harbor Lab.

Manoil, C., Kaiser, D., (1980a). Accumulation of guanosine tetraphosphate and guanosine pentaphosphate in *Myxococcus xanthus* during starvation and myxospore formation. J. Bacteriol. 141:297–304.

Manoil, C., Kaiser, D. (1980b). Guanosine pentaphosphate and guanosine tetraphosphate accumulation and induction of *Myxococcus xanthus* fruiting body development. J. Bacteriol. 141:305–315.

Manoil, C., Kaiser, D. (1980c). Purine-containing compounds, including cyclic adenosine 3′5′-monophosphate, induce fruiting of *Myxococcus xanthus* by nutritional imbalance. J. Bacteriol. 141:374–377.

Margalith, P. (1962). Bacteriolytic principles *M. fulvus*. Nature 196:1335–1336.

Marinus, M. G., Morris, N. R. (1975). Pleiotropic effects of a DNA adenine methylation mutation (*dam*-3) in *Escherichia coli* K12. Mutat. Res. 28:15–26.

Marshall, K. C., Bitton, G. (1980). Microbial adhesion in perspective. In: Adsorption of Microorganisms to Surfaces, edited by G. Bitton, and K. C. Marshall, New York: John Wiley, pp. 1–5.

Martin, S., Sodergren, E., Masuda, T., Kaiser, D. (1978). Systematic isolation of transducing phages for *Myxococcus xanthus*. Virology 88:44–53.

Marvin, D. A. (1968). Control of DNA replication by membrane. Nature 219:485–486.

Mason, O. J., Powelson, D. (1958a). Lysis of *Myxococcus xanthus*. J. Gen. Microbiol. 19:65–70.

Mason, D. J., Powelson, D. (1958b). The cell wall of *Myxococcus xanthus*. Biochem. Biophys. Acta 29:1–7.

Mathew, S., Dudani, A. (1955). Lysis of human pathogenic bacteria by myxobacteria. Nature 175:15–126.

Mayer, D. (1967). Ernährungsphysiologische Untersuchungen an *Archangium violaceum*. Arch. Mikrobiol. 58:186–200.

Mayer, H., Reichenbach, H. (1978). Restriction endonucleases: General survey procedure and survey of gliding bacteria. J. Bacteriol. 136:708–713.

Maynard-Smith, J. (1976). Group Selection. Q. Rev. Biol. 51:277–283.

McCann, K., Clarke, C. H. (1981). Plasmid-mediated UV-protection in *Myxococcus xanthus*. Mol. Gen. Genet. 182:137–142.

McCurdy, H. D. (1963). A method for the isolation of myxobacteria in pure culture. Can. J. Microbiol. 9:282–285.

McCurdy, H. D. (1969a). Studies on the taxonomy of the Myxobacterales. I. Record of Canadian isolates and survey of methods. Can. J. Microbiol. 5:1453–1461.

McCurdy, H. D. (1969b). Light and electron microscope studies on the fruiting bodies of *Chondromyces crocatus*. Arch. Mikrobiol. 65:380–390.

McCurdy, H. D. (1971). Studies on the taxonomy of the Myxobacterales III. Chondromyces and Stigmatella. Int. J. Systematic Bacteriol. 21:40–49.

McCurdy, H. D. (1974). Myxobacterales. In: Bergey's Manual of Determinative Bacteriology (8th edition), edited by R. Buchanan, N. Gibbons. Baltimore: The Williams and Wilkins Corp., pp. 76.-98.

McCurdy, H. D. MacRae, T. H. (1974). Xanthacin. A bacteriocin of *Myxococcus xanthus* fb. Can. J. Microbiol. 20:131–135.

McCurdy, H. D., Wolf, S. (1967). Deoxyribonucleic acid base composition of fruiting Myxobacterales. Can. J. Microbiol. 13:1707–1708.

McCurdy, H. D., Ho, J., Dobson, W. J. (1978). Cyclic nucleotides, cyclic nucleotide phosphodiesterase, and development in *Myxococcus xanthus*. Can. J. Microbiol. 24:1475–1481.

McGhee, J. D., Ginder, G. D. (1979). Specific DNA methylation sites in the vicinity of the chicken $\beta$-globin genes. Nature 280:419–420.

McVittie, A., Zahler, S. A. (1962). Chemotaxis in *Myxococcus*. Nature 194:1299–1300.

McVittie, A., Messik, F., Zahler, S. A. (1962). Developmental biology of *Myxococcus*. J. Bacteriol. 84:546–551.

Meijer, M., Beck, E., Hansen, F. G., Bergmans, H. E. N., Messer, W., von Meyenburg, K., Schaller, H. (1979). Nucleotide sequence of the origin of replication of the *Escherichia coli* K-12 chromosome. Proc. Natl. Acad. Sci. USA 584.

Mendelson, N. H. (1982). Bacterial growth and division: Genes, structures, forces, and clocks. Microbiol. Rev. 46:341–375.

Meng, K. E., Pfister, R. M. (1980). Intracellular structures of *Mycoplasma pneumoniae* revealed after membrane removal. J. Bacteriol. 144:390–399.

Mennigmann, H. D. (1979). Biological effects in bacteria of long-wave UV light. Acta Biol. Med. Germ. 38:1247–1257.

Meyer-Pietschmann, K. (1951). Lebendbeobachtungen an *Myxococcus rubescens.* Arch. Mikrobiol. 16:163–176.

Mitchell, P. (1956). Hypothetical thermokinetic and electrokinetic mechanisms of locomotion in micro-organisms. Proc. R. Phys. Soc. 25:32–34.

Mizrahi, A., Arnan, J., Miller, G., Liron, Z., Mana, M., Batus, Y., Rosenberg, E. (1976). Scaling-up of process for production of antibiotic of *Myxococcus xanthus* TA. J. Appl. Chem. Biotechnol. 26:160–166.

Moon, H. W., Kohler, E. M., Schneider, R. A., Whipp, S. C. (1980). Prevalence of pilus antigens, enterotoxin types, and enteropathogenicity among K88-negative enterotoxigenic *Escherichia coli* from neonatal pigs. Infect. Immun. 27:222–230.

Moore, D. B., Sowa, A., Ippen-Ihler, K. (1981). Location of an F-pilin pool in the inner membrane. J. Bacteriol. 146:251–259.

Morris, D. W. Parish, J. M. (1976). Restriction in *Myxococcus virescens.* Arch. Mikrobiol. 108:227–230.

Morrison, C. E., Zusman, D. R. (1979). *Myxococcus xanthus* mutants with temperature-sensitive, stage-specific defects: Evidence for independent pathways in development. J. Bacteriol. 140:1036–1042.

Mosteller, R. D. (1978). Evidence that glucose starvation-sensitive mutants are altered in *relB* locus. J. Bacteriol. 133:1034–1037.

Murooka, Y., Harada, T. (1979). Expansion of the host range of coliphage P1 and gene transfer from enteric bacteria to other gram-negative bacteria. Appl. Environ. Microbiol. 38:754–757.

Murray, M. J. Shilo, B.-Z., Shih, C., Cowing, D., Hsu, H. W., Weinberg, R. A. (1981). Three different human tumor cell lines contain different oncogenes. Cell 25:355–361.

Murray, R. G. E., Steed, P., Elson, H. E. (1965). The location of the mucopeptide in sections of the cell wall of *Escherichia coli* and other gram-negative bacteria. Can. J. Microbiol. 11:547–559.

Nakamura, K., Inouye, M. (1979). DNa sequence of the gene for the outer membrane lipoprotein of *E. coli:* An extremely AT-rich promoter. Cell 18:1109–1117.

Nanninga, N., Koppes, L. J. H., de Vries-Tijssen, F. C. (1979). The cell cycle of *Bacillus subtilis* as studied by electron microscopy. Arch. Microbiol. 123:173–181.

Nasmyth, K., Tatchell, K., Hall, B. D., Astell, C., Smith, M. (1981). A position effect in the control of transcription at yeast mating type loci. Nature 289:244–250.

Nealson, K. H., Hastings, J. W., (1979). Bacterial bioluminescence: its control and ecological significance. Microbiol. Rev. 43:496–518.

Nealson, K. H., Platt, T., Hastings, J. W. (1970). Cellular control of the synthesis and activity of the bacterial luminescent system. J. Bacteriol. 104:313–322.

Nelson, D. R. Zusman, D. R. (1983a). Evidence for long-lived mRNA during fruiting body formation in *Myxococcus xanthus.* Proc. Natl. Acad. Sci. USA 80:1467–1471.

Nelson, D. R. Zusman, D. R. (1983b). Transport and localization of protein S, a spore coat protein during fruiting body formation by *Myxococcus xanthus.* J. Bacteriol. 154:547–553.

Nelson, D. R., Cumsky, M., Zusman, D. R. (1981a). Development-specific localization of myxobacterial hemagglutinin during fruiting body formation of *Myxococcus xanthus.* In: Sporulation and Germination, edited by H. S. Levinson, A. L. Sonenshein, D. J. Tipper, American Society for Microbiology, pp. 276–280. Washington, D. C.

Nelson, D. R., Cumsky, M. G., Zusman, D. R. (1981b). Localization of myxobacterial hemagglutinin in the periplasmic space and on the cell surface of *Myxococcus xanthus* during developmental aggregation. J. Biol. Chem. 256:12589–12595.

Nicoladis, A. A., Holland, I. B. (1978). Evidence for specific association of the chromosomal origin with outer membrane fractions isolated from *Escherichia coli*. J. Bacteriol. 135:178–189.

Nikaido, H. (1973). Biosynthesis and assembly of lipopolysaccharide and the outer membrane layer of gram-negative cell wall. In: Bacterial membranes and walls, edited by L. Leive (ed.), New York: Marcel Dekker, pp. 131–208.

Nikaido, H. (1979). Permeability of the outer membrane of bacteria. Angew. Chem. [Engl] 18:337–420.

Nishida, M., Matsubara, T., Watanabe. (1965). Pyrrolnitrin, a new antifungal antibiotic. J. Antibiot (Tokyo) 18:211–219.

Norén, B. (1952). Further notes on the distribution of myxobacteria in Swedish soils. Sv. Bot. Tidskr.46:446–453.

Norén, B. (1953). On the production of antibiotics by myxobacteria. Sv. Bot. Tidskr. 47:402–410.

Norén, B. (1955a). Studies on myxobacteria III. Organic factors in nutrition. Botaniska Notiser 108:81–134.

Norén, B. (1955b). Studies on myxobacteria VI. Lytic activity on different eubacteria. Sv. Bot. Tidskr. 49:282–294.

Norén, B. (1960A). Notes on the bacteriolytic activity of *Myxococcus virescens*. Sv. Bot. Tidskr. 54:550–560.

Norén, B. (1960b). Lytic activity on autoclaved and on intact eubacterial cells by a preparation U2D, obtained from a metabolic solution of *Myxococcus cirescens*. Bataniska Notiser 113:320–336.

Norén, B. and Raper, K. B. (1962). Antibiotic activity of Myxobacteria in relation to their bacteriolytic capacity. J. Bacteriol. 84:157–162.

Normore, W. M. (1976). Guanine-plus-cytosine (GC) composition of the DNA of bacteria, fungi, algae, and protozoa. In: CRC Handbook of Biochemistry and Molecular Biology, edited by G. D. Fasman. Cleveland, Ohio: CRC Press, 3rd ed. Vol. II. pp. 65–235.

O'Connor, K. A., Zusman, D. R. (1983) Coliphage P1 mediated transduction of cloned DNA from *Escherichia coli* to *Myxococcus xanthus:* Use for complementation and recombinational analyses. J. Bacteriol. 155:317–329.

Oetker, H. (1953). Untersuchungen über die Ernährung einiger Myxobakterien. Arch. Mikrobiol. 19:206–246.

Okano, P., Bacon, K., Rosenberg, E. (1970). Ribonucleic acid synthesis during microcyst formation in *Myxococcus xanthus:* Characterization by deoxyribonucleic acid-ribonucleic acid hybridization. J. Bacteriol. 104:275–282.

Orlowski, M. (1980). Cyclic adenosine 3′,5′ -monophosphate binding protein in developing myxospores of *Myxococcus xanthus*. Can. J. Microbiol. 26:905.-911.

Orlowski, M. White, D. (1974). Intracellular proteolytic activity in developing myxospores of *Myxococcus xanthus*. Arch. Microbiol. 97:347–357.

Orlowski, M., Martin, P., White, D., Wong, M. C. (1972). Changes in activity of glyoxylate cycle enzymes during myxospore development in *Myxococcus xanthus*. J. Bacteriol. 111:784–790.

Orndorff, P. E., Dworkin, M. (1980). Separation and properties of the cytoplasmic and outer membranes of vegetative cells of *Myxococcus xanthus*. J. Bacteriol. 141:914–927.

Orndorff, P. E., Dworkin, M. (1982). Synthesis of several membrane proteins during developmental aggregation in *Myxococcus xanthus*. J. Bacteriol. 149:29–39.

Orndorff, P. E., Stellwag, E., Starich, T., Dworkin, M. Zissler, J. (1983). Genetic and physical characterization of lysogeny by bacteriophage MX8 in *Myxococcus xanthus*. J. Bacteriol. 154:772–779.

Osborn, M. J., Wu, H. C. P. (1980). Proteins of the outer membrane of gram-negative bacteria. In: Annual Review of Microbiology, edited by M. P. Starr, J. L. Ingraham, A. Balows. Palo Alto, California: Annual Reviews, Inc., pp. 369–422.

Oshima, T., Imahori, K. (1974). Physicochemical properties of deoxyribonucleic acid from an extreme thermophile. J. Biochem. 75:179–183.

Ottow, J. C. G. (1975). Ecology, physiology and genetics of fimbriae and pili. In: Annual Review of Micribiology, edited by M. P. Starr, J. L. Ingraham, S. Rafel. Palo Alto, California: Annual Reviews, Inc., vol. 29. pp. 79–108.

Oxford, A. E. (1947). Observations concerning the growth and metabolic activities of myxococci in a simple protein-free medium. J. Bacteriol. 53:129–138.

Oyaizu, H., Komgata, K. (1981). Chemotaxonomic and phenotypic characterization of the strains of species in the *Flavobacterium-Cytophaga*-complex. J. Gen. Appl. Micribiol. 27:57–107.

Ozeki, H. Ikeda, H. (1968). Transduction mechanisms. Annu. Rev. Genet. 2:245–278.

Pachler, P. F., Koch, A. L., Schaechter, M. (1965). Continuity of DNA synthesis in *Escherichia coli*. J. Mol. Biol. 11:650–653.

Parish, J. H. (1975). Transfer of drug resistance to *Myxococcus* from bacteria carrying drug-resistance factors. J. Gen Microbiol. 87:198–210.

Parish, J. H. (1979). Myxobacteria. In: Developmental Biology of Prokaryotes, edited by J. H. Parish. Berkeley, California, University of California Press: pp. 227–253.

Parker, J., Watson, R. J., Freson, J. D. (1976). A relaxed mutant with an altered ribosomal protein L11. Molec. Gen. Genet. 144:111–114.

Pate, J. L. Chang, L. Y. E. (1979). Evidence that gliding motility in prokaryotic cells is driven by rotary assemblies in the cell envelopes. Curr. Microbiol. 2:59–64.

Pate, J. L., Petzold, S. J., Chang, L. Y. E. (1979). Phages for the gliding bacterium *Cytophaga johnsonae* that infect only motile cells. Curr. Microbiol. 2:257–262.

Paulton, R. J. L. (1971). Nuclear and cell division in filamentous bacteria. Nature New Biol. 231:271–274.

Pestka, S., Brot, N. (1971). Studies on the formation of transfer ribonucleic acid-ribosome complexes. IV. Effect of antibiotics on steps of bacterial protein synthesis: some new ribosomal inhibitors of translocation. J. Biol. Chem. 246:7715–7122.

Peterson, E. A., Gillespie, D. C., Cook, F. D. (1966). A wide-spectrum antibiotic produced by a species of *Sorangium*. Can. J. Microbiol. 12:221–230.

Peterson, J. E. (1959). New species of myxobacteria from the bark of living trees. Mycologia 51:163–172.

Peterson, J. E. (1969). Isolation, cultivation and maintenance of the myxobacteria. In: Methods in microbiology, edited by J. R. Norris, D. W. Ribbons. London, New York: Academic Press, vol. 3B, pp. 185–210.

Planque, K. Kijne, J. W. (1977). Binding of pea lectins to a glycan type polysaccharide in the cell walls of *Rhizobium leguminosarum*. FEBS Lett. 73:64–66.

Pritchard, R. H. (1974). On the growth and form of a bacterial cell. Philos. Trans. R. Soc. Lond. (Biol.) 267:303–336.

Pritchard, R. H., and Zaritsky, A. (1971). Effect of thymine concentration on the replication velocity of DNA in a thymineless mutant of *Escherichia coli.* 1971. Nature 226:126–131.

Pritchard, R. H., Barth, P. T., Collins, T. (1969). Control of DNA synthesis in bacteria. Symp. Soc. Gen. Microbiol. 19:263–297.

Pronina, N. I. (1962). A description of new species and varieties of cellulose decomposing myxobacteria. Microbiology [English translation of Mikrobiologiya] 31:384–390.

Qualls, G. T., Stephens, K., White, D. (1978a). Morphogenetic movements and multicellular development in the fruiting myxobacterium, *Stigmatella aurantiaca.* Dev. Biol. 66:270–274.

Qualls, G. T., Stephens, K., White, D. (1978b). Light-stimulated morphogenesis in the fruiting myxobacterium, *Stigmatella aurantiaca.* Science 201:444–445.

Ramsey, W. S. Dworkin, M. (1968). Microcyst germination in *Myxococcus xanthus.* J. Bacteriol. 95:2249–2257.

Ramsey, W. S. Dworkin, M. (1970). Stable messenger ribonucleic acid and germination of *Myxococcus xanthus* microcysts. J. Bacteriol. 101:531–540.

Raverdy, J. (1973). Sur l'isolement et l'activite bacteriolytique de quelques myxobacteries isolee de l'eau. Water Res. 7:687–693.

Razin, A., Riggs, A. D. (1980). DNA methylation and gene function. Science 210:604–6100.

Rees, D. A. (1972). Shapely polysaccharides. Biochem. J. 126:257–273.

Reeve, J. (1979). Use of minicells for bacteriophage-directed polypeptide synthesis. Methods in Enzymol. 68:493–503.

Reichenbach, H. (1962). Über verschiedene Arten von Cystenmustern bei *Chondrococcus coralloides* (Myxobacteriales). Berichte der deutschen Botanischen Gesellschaft 75:85–90.

Reichenbach, H. (1965a). Rhythmische Vorgänge bei der Schwarmentfaltung von Myxobakterien. Berichte der deutschen Botanischen Gesellschaft 78:(102)–(105).

Reichenbach, H. (1965b). Untersuchungen an *Archangium violaceum.* Ein Beitrag zur Kenntnis der Myxobakterien. Arch. Mikrobiol. 52:376–403.

Reichenbach, H. (1966). *Myxococcus* spp. (Myxobacteriales). Schwarmentwicklung und Bildung von Protocysten. Publikationen zu wissenschaftlichen Filmen, Göttingen 1A:557–578.

Reichenbach, H. (1968). *Archangium violaceum* (Myxobacteriales)—Schwarmentwicklung und Bildung von Protocysten. Encycl. Publikationen zu wissenschaftlichen Filmen, Göttingen 2A:317–333.

Reichenbach, H. (1970). *Nannocystis exedens* gen. nov., spec. nov., a new myxobacterium of the family Sorangiaceae. Arch. Mikrobiol. 70:119–138.

Reichenbach, H. (1974a). Die Biologie der Myxobakterien. Biologie in unserer Zeit 4:33–45 (1974).

Reichenbach, H. (1974b) *Chondromyces apiculatus* (Myxobacteriales)—Schwarmentwiklung und Morphogenese. Publikationen zu wissenschaftlichen Filmen, Göttingen, Sektion Biologie 7(3):245–263; sowie Sektion Medizin 2 (4):397–415.

Reichenbach, H. (1981). Taxonomy of the gliding bacteria. Annu. Rev. Micribiol. 35:339–364.

Reichenbach, H. (1983). A simple method for the purification of myxobacteria. J. Microbiol. Meth. 1:77–79.

Reichenbach, H. Dworkin, M. (1969). Studies on *Stigmatella aurantiaca* (Myxobac-

terales). J. Gen. Microbiol. 58:3–14.
Reichenbach, H. Dworkin, M. (1970). Induction of myxospore formation in *Stigmatella aurantiaca* (Myxobacterales) by monovalent cations. J. Bacteriol. 101:325–326.
Reichenbach, H. Dworkin, M. (1981a). Introduction to the gliding bacteria. In: The Prokaryotes, edited by M. P. Starr, H. Stolp, H. G. Trüper, A. Balows, H. G. Schlegel. Berlin: Springer-Verlag, vol. 1, chap. 19, pp. 315–327.
Reichenbach, H., Dworkin, M. (1981b). The order Myxobacterales. In: The Prokaryotes: A Handbook on Habitats, Isolation, and Identification of Bacteria, edited by M. P. Starr, H. Stolp, H. G. Truper, A. Balows, H. G. Schlegel. Berlin: Springer-Verlag, pp. 328–355.
Reichenbach, H., Kleinig, H. (1971). The carotenoids of *Myxococcus fulvus* (Myxobacterales). Arch. Mikrobiol. 76:364–380.
Reichenbach, H., Kleinig, H. (1972). Die Carotinoide der Myxobakterien. Zentralblatt für Bakteriologie und Hygiene, I. Abteilung Originale A 220:458–463.
Reichenbach, H., Galle, H. K., Heunert, H. H. (1975–1976). *Stigmatella aurantiaca* (Myxobacterales). Schwarmentwicklung und Fruchtkorperbildung. Encyclopaedia Cinematographica E2421, film of the Institut für den wissenschaftlichen Film, Göttingen, Germany.
Reichenbach, H., Heunert, H. H. Kuczka, H. (1965a). Schwarmentwicklung und Morphogenese bei Myxobakterien—*Archangium, Myxococcus, Chondrococcus, Chondromyces*. Film C893. Inst. f. Wissen. Film. Göttingen, Germany.
Reichenbach, H., Heunert, H. H. and Kuczka, H. (1965b). *Chondromyces apiculatus* (Myxobacteriales)—Schwarmentwicklung und Morphogenese. Film E779. Inst. f. Wissen. Film. Göttingen, Germany.
Reichenbach, H., Heunert, H. H. and Kuczka, H. (1965c). *Myxococcus* spp. (Myxobacteriales)—Schwarmentwicklung und Bildung von Protocysten. Film E778. Inst. f. Wissen. Film. Göttingen, Germany.
Reichenbach, H., Heunert, H. H. and Kuczka, H. (1965d). *Archangium violaceum* (Myxobacteriales)—Schwarmentwicklung und Bildung von Protocysten. Film E777. Inst. f. Wissen. Film. Göttingen, Germany.
Reichenbach, H., Kohl, W., Achenbach, H. (1981). The flexirubin-type pigments. In: The *Flavobacterium-Cytophaga* Group, edited by H. Reichenbach, O. B. Weeks. Weinheim: Verlag Chemie, pp. 101–108.
Reichenbach, H., Voelz, H., Dworkin, M. (1969). Structural changes in *Stigmatella aurantiaca* during myxospore induction. J. Bacteriol. 97:905–911.
Renwick, G. A. Peterson, J. E. (1969). Carotenoids of a cellulolytic, fruiting myxobacterium. *Sorangium cellulosum*. Bacteriological Proceedings 42.
Rhaese, H. J., Groscurth, R., Vetter, R., Gilbert, H. (1979). Regulation of sporulation by highly phosphorylated nucleotides in *Bacillus subtilis*. In: Regulation of Molecular Synthesis by Low Molecular Weight Mediators, edited by G. Koch, D. Richter. New York: Academic Press,pp. 145–159.
Rickenberg, H. V. (1974). Cyclic AMP in prokaryotes. Annu. Rev. Microbiol. 18:353–389.
Ridgway, H. F. (1977). Source of energy for gliding motility in *Flexibacter polymorphus:* effects of metabolic and respiratory inhibitors on gliding movement. J. Bacteriol. 131:544–556.
Riggs, A. D. (1975). X inactivation, differentiation, and DNA methylation. Cytogenet. Cell. Genet. 14:9–25.

Ringel, S. M., Greenough, R. C., Roemer, S., Connor, D., von Strandtmann, M. (1977). Ambruticin (W 7783), a new antifungal antibiotic. J. Antibiot. (Tokyo) 30:371–375.

Rodriques, F. K. Virrankoski-Castrodeza, V., Parish, J. H. and Grimm, K. (1980). Isolation and characterization of new bacteriophages for *Myxococcus xanthus*. Arch. Microbiol. 126:175–180.

Rogers, M. J., Forsberg, C. W. (1971). Role of autolysins in the killing of bacteria by some bactericidal antibiotics. J. Bacteriol. 108:1235–1243.

Roper, M. M. Marshall, K. C. (1977). Lysis of *Escherichia coli* by a marine myxobacter. Microbial Ecology. 3:167–171.

Roper, M. M., Marshall, K. C. 1978). Biological control agents of sewage bacteria in marine habitats. Aust. J. Mar. Freshwater Res. 29:335–343.

Rosen, S. D., Kafka, J. A., Simpson, D. L., Barondes, S. H. (1973). Developmentally regulated carbohydrate-binding protein in *Dictyostelium discoideum*. Proc. Natl. Acad. Sci. USA 70:2554–2557.

Rosen, S. D., Simpson, D. L., Rose, J. E., Barondes, S. H. (1974). Carohydrate-binding protein from *Polysphondylium pallidum* implicated in intercellular adhesion. Nature 252:149–151.

Rosenberg, B. H., Cavalieri, L. F., Ungers, G. (1969). The negative control mechanism for *E. coli*. DNA rplication. Proc. Natl. Acad. Sci. USA 63:1410–1417.

Rosenberg, E., Katarski, M., Gottlieb, P. (1967). Deoxyribonucleic acid synthesis during exponential growth and microcyst formation in *Myxococcus xanthus*. J. Bacteriol. 93:1402–1408.

Rosenberg, E., Keller, K. H., Dworkin, M. (1977). Cell density-dependent growth of *Myxococcus xanthus* on casein. J. Bacteriol. 129:770–777.

Rosenberg, E., Vaks, B., Zuckerberg, A. (1973). Bactericidal action of an antibiotic produced by *Myxococcus xanthus*. Antimicrob. Agents Chemother. 4:507–513.

Rosenberg, E., Filer, D., Zafriti, D., Kindler, S. H. (1973). Aspartokinase activity and the developmental cycle of *Myxococcus xanthus*. J. Bacteriol. 115:29–34.

Rosenberg, E., Fytlovitch. S., Carmeli, S. Kashman, Y. (1982). Chemical properties of *Myxococcus xanthus* antibiotic TA. J. Antibiot. (Tokyo) 35:788–793.

Rosenfelder, G., Luderitz, O., Westphal. O. (1974). Composition of lipopolysaccharides from *Myxococcus fulvus* and other fruiting and non-fruiting myxobacteria. Eur. J. Biochem. 44:411–420.

Rossier, C., Eitle, E., Driel, R. V., Gerisch, G. (1980). Biochemical regulation of cell development and aggregation in *Dictyostelium discoideum*. Symp. Soc. Gen. Microbiol. 30:405–425.

Rothstein, S. J., Jorgensen, R. A., Postle, K., Reznikoff, W. S. (1980). The inverted repeats of Tn*5* are functionally different. Cell 19:795–805.

Rothstein, S. J., Jorgensen, R. A., Yin, J. C. P., Young-di, Z., Johnson, R. C., Reznikoff, W. S. (1981). Genetic organization of Tn*5*. Cold Spring Harbor Symp. Quant. Biol. 45:99–105.

Rückert, G. (1973). Biogeographische Untersuchungen über die Verbreitung Fruchtkörper-bildender Myxobakterien in Island. Acta Botanica Islandica 2:56–63.

Rückert, G. (1978a). Myxobakterien in Substraten aus ariden Gebieten unter besonderer Berücksichtigung Südwestafrikas. Namib und Meer 8:53–66.

Rückert, G. (1978b). Förderung der Furchtkörper-Bildung von *Myxococcus virescens* Thaxter (*Myxobacterales*) in Rohkulturen durch Salzzusatz. Z. Allg. Mikrobiol. 18:69–71.

Rückert, G. (1979). Myxobakterien-Artenspektren von Böden in Abhängigkeit von bodenbildenden Faktoren unter besonderer Berücksichtigung der Bodenreaktion. Z. Pflanzenernährung Bodenkunde 142:330–343.

Rückert, G. (1981). Myxobakterien (*Myxobacterales*) auf Blattoberflächen. Z. Allg. Mikrobiol. 21:761–763.

Rudd, K., Zusman, D. R. (1979). Rifampin resistant mutants of *Myxococcus xanthus* defective in development. J. Bacteriol. 137:295–300.

Rudd, K., Zusman, D. R. (1982). RNA polymerase of *Myxococcus xanthus:* Purification and selective transcription in vitro with bacteriophage templates. J. Bacteriol. 151:89–105.

Sabados-Sarić, A. (1957). Istrazivanja o miksobacterijama u thina Jugoslavije (Myxobacteria in different types of soils in Yugoslavia). [In Serbo-Croatian, with English summary.] Jugos. Akademija Znanosti Umjetnosti. Zagreb 312:5–29.

Sadler, W., Dworkin, M. (1966). Induction of cellular morphogenesis in *Myxococcus xanthus.* II. Macromolecular synthesis and mechanism of inducer action. J. Bacteriol. 91:1520–1525.

Sakakibara, H., Naganawa, H., Ohno, M., Maeda, K. Umezawa, H. (1974). The structure of althiomycin. J. Antibiot. (Tokyo) 27:897–899.

Sancar, A., Hack, A. M., Rupp, W. D. (1979). Simple method for identification of plasmid-coded proteins. J. Bacteriol. 137:692–693.

Sanger, F., Coulson, A. R., Hong, G. F., Hill, D. F. Peterson, G. B. (1982). Nucleotide sequence of bacteriophage λ DNA. J. Mol. Biol. 162:729–773.

Schmidt-Lorenz, H. (1969). The fine structure of the swarm cells of myxobacteria. J. Appl. Bacteriol. 32:22–23.

Schmidt-Lorenz, W., Kühlwein, H. (1968). Intracelluläre Bewegungsorganellen der Myxobakterien. Arch. Mikrobiol. 60:95–98.

Schröder, J., Reichenbach, H. (1970). The fatty acid composition of vegetative cells and myxospores of *Stigmatella aurantiaca* (Myxobacterales). Arch. Mikrobiol 71:384–390.

Schweizer, M., Henning, U. (1977). Action of a major outer cell envelope membrane protein in conjugation of *Escherichia coli* K12. J. Bacteriol. 129:1651–1652.

Shabtai, Y., Rosenberg, E., Kindler, S. H. (1975). Aspartokinase isoenzymes of the fruiting myxobacterium *Myxococcus xanthus*. Biochim. Biophys. Acta 403:345–354.

Shakhova, I. K. (1978). Ultraviolet irradiation and repair as possible factors determining the evolution of DNA composition. J. Evolutionary Biochem. Physiol. 14:176–180.

Shenk, T. E., Rhodes, C., Rigby, P. W. J., Berg, P. (1975). Biochemical method for mapping mutational alterations in DNA with S1 nuclease: The localization of deletions and temperature-sensitive mutations in Simian Virus 40. Proc. Natl. Acad. Sci. USA 72:989–993.

Shetlar, M. D. (1980). Cross-linking of proteins to nucleic acids by ultraviolet light. Photochem. Photobiol. 5:105–197.

Shigeura, H. T., Gordon, C. N. (1962a). Hadacidin, a new inhibitor of purine biosynthesis. J. Biol. Chem. 237:1932–1936.

Shigeura, H. T., Gordon, C. N. (1962b). The mechanism of action of hadacidin. J. Biol. Chem. 237:1937–1940.

Shilo, M. (1970). Lysis of blue-green algae by Myxobacter. J. Bacteriol. 104:453–461.

Shimkets, L. J. (1980). The role of purines in the development of *Myxococcus xanthus*. Ph.D. Dissertation. University of Minnesota, Minneapolis.

Shimkets, L. J. Dworkin, M. (1981). Excreted adenosine is a cell density signal for the initiation of fruiting body formation in *Myxococcus xanthus*. Dev. Biol. 84:51–60.

Shimkets. L. J., Kaiser, D. (1982a). Murein components rescue developmental sporulation of *Myxococcus xanthus*. J. Bacteriol. 152:462–470.

Shimkets, L. J. Kaiser, D. (1982b). Induction of coordinated cell movement in *Myxococcus xanthus*. J. Bacteriol. 152:451–461.

Shimkets, L. J., Dworkin, M., Keller, K. H. (1979). A method for establishing stable concentration gradients in agar suitable for studying chemotaxis on a solid surface. Can. J. Microbiol. 25:1460–1467.

Shimkets, L. J., Gill, R. E., Kaiser, D. (1983). Developmental cell interactions in *Myxococcus xanthus* and the spoC locus. Proc. Natl. Acad. Sci. USA 80:1406–1410.

Sigg, H. P., Toth, A. (1967). Uber diestruktur der Phenazin-N-Oxide. Helvetica Chimica Acta 50:716–719.

Simon, M., Zieg, J., Silverman, M., Mandel, G., Doolittle, R. (1980). Phase variation: Evolution of a controlling element. Science 209:1370–1374.

Singer, C. E., Ames, B. N. (1970). Sunlight ultraviolet and bacterial DNA base ratios. Science 170:822–826.

Singer, C. E., Ames, B. N. (1972). Reply to Bak *et al.* (1972). Science 175:1393.

Singh, B. N. (1947). Myxobacteria in soils and composts; their distribution, number and lytic action on bacteria. J. Gen. Microbiol. 1:1–10.

Singh, B. N., Sing, N. B. (1971). Distribution of fruiting myxobacteria in Indian soils, bark of trees and dung of herbivorous animals. Indian J. Microbiol. 11:47–92.

Singh, B. N. Yadava, J. N. S. (1976). Fructification and antagonistic effect of myxobacteria on eubacteria: Lytic effect and fruiting body formation of *Myxococcus, Chondrococcus* and *Angiococcus* spp. Indian J. Exp. Biol. 14:68–70.

Siu, C. H., Lerner, R. A., Ma, G., Firtel, R. A., Loomis, W. F. (1976). Developmentally regulated proteins of the plasma membrane of *Dictyostelium discoideum.* The carbohydrate-binding protein. J. Mol. Biol. 100:157–178.

Smith, B. A., Dworkin. M. (1980). Adenylate energy charge during fruiting body formation by *Myxococcus xanthus*. J. Bacteriol. 142:1007–1009.

Smith, B. A., Dworkin, M. (1981). *Myxococcus xanthus* synthesizes a stabilized messenger RNA during fruiting body formation. Curr. Microbiol. 6:95–100.

Smith, J. A., Martin, L. (1973). Do cells cycle? Proc. Natl. Acad. Sci. USA 70:1263–1267.

Snieszko, S. F., McAllister, J., Hitchnu. (1942). Further studies on the biology of certain myxobacteria. J. Bacteriol. 43:28–29.

Sodergren, E., Kaiser, D. (1983). Insertions of Tn*5* near genes that govern stimulatable cell motility in *Myxococcus*. J. Mol. Biol. 167:295–310.

Sodergren, E., Cheng, Y., Avery, L., Kaiser, D. (1983). Recombination in the vicinity of insertions of transposon Tn*5* in Myxococcus xanthus. Genetics 105:281–291.

Sojka, W. J. (1971). Enteric diseases in new-born piglets, calves, and lambs due to *Escherichia coli* infection. Vet. Bull. 41:509–522.

Somerville, C. R., Ahmed, A. (1979). Mutants of *Escherichia coli* defective in the degradation of guanosine 5′-triphosphate, 3′-disphosphate (pppGpp). Mol. Gen. Genet. 169:315–323.

Sompayrac, L., Maaløe, O. (1973). Autorepressor model for control of DNA replication. Nature New Biol. 241:133–135.

Sonntag, I., Schwarz, H., Hirota, Y., Henning, U. (1978). Cell envelope and shape of *Escherichia coli:* Multiple mutants missing the outer membrane lipoprotein and other major outer membrane proteins. J. Bacteriol. 136:280–285.

Sorhaug, T., (1974). Glycerol ester hydrolase, lipase, of *Myxococcus xanthus* Fb. Can. J. Microbiol. 20:611–615.

Soriano, S. (1973). Flexibacteria. Annu. Rev. Microbiol. 27:155–170.

Soriano, S., Lewin, R. A. (1965). Gliding microbes: some taxonomic reconsiderations. Antonie van Leeuwenhoek 31:66–80.

Southern, E. M. (1975). Detection of specific sequences among DNA fragments separated by gel electrophoresis. J. Mol. Biol. 98:503–517.

Spratt, B. G. (1975). Distinct penicillin binding proteins involved in the division, elongation, and shape of *Escherichia coli* K-12. Proc. Natl. Acad. Sci. USA 72:2999–3003.

Spratt, B. G. Pardee, A. B. (1975). Penicillin-binding proteins and cell shape in *E. coli.* Nature 254:516–517.

Stackebrandt, E., Woese, C. R. (1981). The evolution of prokaryotes. Symp. Soc. Gen. Microbiol. 32:1–31.

Stanier, R. Y. (1942). A note on elasticotaxis in myxobacteria. J. Bacteriol. 44:405–412.

Stein, R., Gruenbaum, Y., Pollack, Y., Razin, A., Cedar, H. (1982). Clonal inheritance of the pattern of DNA methylation in mouse cells. Proc. Natl. Acad. Sci. USA 79:61–65.

Stephens, K., White, D. (1980a). Scanning electron micrographs of fruiting bodies of the myxobacterium *Stigmatella aurantiaca* lacking a coat and revealing a cellular stalk. FEMS Microbiol. Lett. 9:189–192.

Stephens, K., White, D. (1980b). Morphogenetic effects of light and guanine derivatives on the fruiting myxobacterium *Stigmatella aurantiaca.* J. Bacteriol. 144:322–326.

Stephens, K., Hegeman, G. D., White, D. (1982). Pheromone produced by the myxobacterium *Stigmatella aurantiaca.* J. Bacteriol. 149:739–747.

Sternberg, N., Powers, M. Yarmolinsky, M., Austin, S. (1981). Group Y incompatibility and copy control of P1 prophage. Plasmid 5:138–149.

Stirm, S. F., Orskov, I. Orskov, Birch-Anderson, A., (1967). Episome-carried surface antigen K88 of *Escherichia coli.* III. Morphology. J. Bacteriol. 93:740–748.

Streisinger, G., Enrich, J., Stahl, M. M. (1967). Chromosome structure in phage T4, III. Terminal redundancy and length determination. Proc. Natl. Acad. Sci. USA 57:292–295.

Sudo, S. Z., Dworkin, M. (1969). Resistance of vegetative cells and microcysts of *Myxococcus xanthus.* J. Bacteriol. 98:883–887.

Sudo, S. Z., Dworkin, M. (1972). Bacteriolytic enzymes produced by *Myxococcus xanthus.* J. Bacteriol. 110:236–245.

Sudo, S. Z., Dworkin, M. (1973). Comparative biology of prokaryotic resting cells. Adv. Microb. Physiol. 9:153–224.

Sugimoto, K., Oka, A., Sugisaki, H., Takanami, M., Nishimura, A., Yesuda, Y., Hirota, Y., (1979). Nucleotide sequence of *Escherichia coli* K-12 replication origin. Proc. Natl. Acad. Sci. USA 76:575–579.

Susskind, M. M., Botstein, D. (1978). Molecular genetics of bacteriophage P22. Microbiol. Rev. 42:385–413.

Sutherland, I. W. (1979). Polysaccharides produced by *Cystobacter, Archangium, Sorangium* and *Stigmatella* species. J. Gen. Microbiol. 111:211–216.

Sutherland, I. W., Smith, M. L. (1973). The lipopolysaccharides of fruiting and non-fruiting myxobacteria. J. Gen. Microbiol. 74:259–266.

Sutherland, I. W., Thomson, S. (1975). Comparison of polysaccharides produced by *Myxococcus* strains. J. Gen. Microbiol. 89:124–132.

Swenson, P. A., Setlow, R. B. (1966). Effects of ultraviolet radiation on macromolecular synthesis in *Escherichia coli.* J. Mol. Biol. 15:201–219.

Sy, J. (1977). In vitro degradation of guanosine 5′-diphosphate, 3′-diphosphate. Proc. Natl. Acad. Sci. USA 74:5529–5533.

Teta, L. A., Hanna, M. H. (1981). Relationship between light and an aggregation stimulating factor in the cellular slime mold *Polysphondelium violaceum.* Abstracts of Annual Meeting of A.S.M. I 149.

Thaxter, R. (1892). On the Myxobacteriaceae, a new order of Schizomycetes. Botan. Gazette 17:389–406.

Thierbach, G., Reichenbach, H. (1981a). Myxothiazol, a new antibiotic interfering with respiration. Antimicrob. Agents Chemother. 19:504–507.

Thierbach, G., Reichenbach, H. (1981b). Myxothiazol, a new inhibitor of the cytochrome b-$c_1$ segment of the respiratory chain. Biochim. Biophys. Acta 638:282–289.

Thierbach, G., Reichenbach, H. (1983). The effect of the new antibiotic myxothiazol on the respiration of *Paracoccus denitrificans.* Arch. Mikrobiol. 134:104–107.

Tipper, D. J., Strominger, J. L., Ensign, J. C. (1967). Structure of the cell wall of *S. aureus,* strain Copenhagen VII. Mode of action of the bacteriolytic peptidase from Myxobacter and the isolation of intact cell wall polysaccharides. Biochemistry 6:906–920.

Tonegawa, S., Sakano, H., Maki, R., Traunecker, G. H., Roeder, W., Kurosawa, Y. (1980). Somatic reorganization of immunoglobulin genes during lymphocyte differentiation. Cold Spring Harbor Symp. Quant. Biol. 45:839–858.

Tormo, A., Martinez-Salas, E., Vicente, M. (1980). Involvement of the *ftsA* gene product in late stages of the *Escherichia coli* cell cycle. J. Bacteriol. 141:806–813.

Torti, S., Zusman, D. (1981). Genetic characterization of aggregation-defective developmental mutants of *Myxococcus xanthus.* J. Bacteriol. 147:768–775.

Trowitzsch, W., Reifenstahl, G., Wray, V., Gerth, K. (1980). Myxothiazol, an antibiotic from *Myxococcus fulvus* (Myxobacterales). II. Structure elucidation. J. Antibiot. (Tokyo) 33:1480–1490.

Trowitzsch, W., Wray, V., Gerth, K., Höfle, G. (1982). Structure of myxovirescin A., a new macroyclic antibiotic from gliding bacteria. J. Chem. Soc. Chem. Commun. 1340–1432.

Tsai, C. S., Whitaker, D. R., Jurasek, L., Gillespie, D. C. (1965). Lytic enzymes of *Sorangium* sp. Action of the $\alpha$ and $\beta$-lytic proteases on two bacterial mucopeptides. Can. J. Biochem. 43:1971–1983.

Tsai, H., Hirsch, H. J. (1981). The primary structure of fulvocin C from *Myxococcus fulvus.* Biochim. Biophys. Acta. 667:213–217.

Tsai, W. C., Westby, C. A. (1978). Synthesis and salvage of purines during cellular morphogenesis of *Myxococcus xanthus.* J. Bacteriol. 136:582–587.

Tyler, B. (1978). Regulation of the assimilation of nitrogen compounds. Annu. Rev. Biochem. 47:1127–1161.

Ulitzur, S., Hastings, J. W. (1979). Autoinduction in a luminous bacterium: a confirmation of the hypothesis. Curr. Microbiol. 2:345–348.

Umbarger, H. E. (1978). Amino acid biosynthesis and its regulation. Annu. Rev. Biochem. 47:533–606.

Vacheron, M. J., Arpin, N., Michel, G. (1970). Isolement d'esters de phleixanthophylle de *Nocardia kirovani*. Comptes Rendus Hebdomadaires des Séances de l'Académie des Sciences, Paris, Serie C 271:881–884.

Vaks, B., Zuckerberg, A., Rosenberg, E. (1974). Purification and partial characterization of an antibiotic produced by *Myxococcus xanthus*. Can. J. Microbiol. 20:155–161.

Van de Putte, P., Cramer, S., Giphart-Gassler, M. (1980). Invertible DNA determines host specificity of bacteriophage Mu. Nature 286:218–222.

Van der Ploeg, L. H. T., Flavell, R. A. (1980). DNA methylation in the human $\gamma\delta\beta$-globin locus in erythroid and nonerythroid tissues. Cell 19:947–958.

Verma, J. P., Martin, H. H. (1967). Uber die Oberflächenstruktur von Myxobakterien. I. Chemie und Morphologie der Zellwande von *Cytohaga hutchinsonii* und *Sporcytophaga myxococcoides*. Arch. Mikrobiol. 59:355–380.

Voelz, H. G., Dworkin, M. (1962). Fine structure of *Myxococcus xanthus* during morphogenesis. J. Bacteriol. 84:943–952.

Voelz, H. G., Reichenbach, H. (1969). Fine structure of fruiting bodies of *Stigmatella aurantiaca* (Myxobacterales). J. Bacteriol. 99:856–866.

Waalwijk, C., Flavell, R. A. (1978). DNA methylation at a CCGG sequence in the large intron of the rabbit $\beta$-globin gene: Tissue-specific variations. Nucleic Acids RES. 4631–4641.

Walker, R. W. (1969). Cis-11-hexadecanoic acid from *Cytophaga hutchinsonii* lipids. Lipids 4:15–18.

Wallace, D. C., Morowitz, H. J. (1973). Genome size and evolution. Chromosoma 40:121:126.

Wang, J. H., Waisman, D. M. (1979). Current topics in cellular regulation. 15:47–107.

Ward, C. B., Glaser, D. A. (1969). Analysis of the chloramphenicol-sensitive and chloramphenicol-resistant steps in the initiation of DNA synthesis in *E. coli* B/r. Proc. Natl. Acad. Sci. USA 64:905–912.

Ward, C. B., Glaser, D. A. (1970). Control of initiation of DNA synthesis in *E. coli* B/r. Proc. Natl. Acad. Sci. USA 67:255–262.

Ware, J. C., Dworkin, M. (1973). Fatty acids of *Myxococcus xanthus*. J. Bacteriol. 115:253–261.

Watson, B. F., Dworkin, M. (1968). Comparative intermediary metabolism of vegetative cells and microcysts of *Myxococcus xanthus*. J. Bacteriol. 96:1456–1473.

Watterson, D. M., Sharief, F., Vanaman, T. C. (1980). The complete amino acid sequence of the $Ca^{++}$-dependent modulator protein (calmodulin) of bovine brain. J. Biol. Chem. 255:962–975.

Weckesser, J., Rosenfelder, G., Mayer, H., Lüderitz, O. (1971). The identification of 3-0-methyl-D-xylose and 3-0-methyl-L-xylose as constituents of the lipopolysaccharides of *Myxococcus fulvus* and *Rhodopseudomonas viridis*, respectively. Eur. J. Biochem. 24:112–115.

Weidel, W., Pelzer, H. (1965). Bagshaped macromolecules—a new outlook on bacterial cell walls. In: Advances in Enzymology, edited by F. F. Nord., New York: John Wiley.

Weidel, W., Frank, H., Martin, H. H. (1960). The rigid layer of the cell wall of *Escherichia coli* strain. B. J. Gen. Microbiol. 22:158–166.

Weigele, M., Leimgruber, W. (1967). The structure of myxin. Tetrahedron Lett. 715–718.

Weiss, P. (1961). Guiding principles in cell locomotion and cell aggregation. Exptl. Cell Res. Suppl. 8:260–281.

Westby, C. A., Tsai, W. C. (1974). De novo purine synthesis in vegetative cells and myxospores of *Myxococcus xanthus*. J. Bacteriol. 117:1099–1107.

Whatley, M. H., Bodwin, J. S., Lippincott, B. B., Lippincott, J. A. (1976). Role for *Agrobacterium* cell envelope lipopolysaccharide in infection site attachment. Infect. Immun. 13:1080–1083.

Whitaker, D. R. (1965). Lytic enzymes of *Sorangium* sp. Isolation and enzymatic properties of the a and β-lytic proteases. Can. J. Biochem. 43:1935–1954.

White, D. (1975). Myxospores of *Myxococcus xanthus*. In: Spores VI., edited by P. Gerhardt, H. L. Sadoff, R. N. Costilow. Washington, D.C: American Society for Microbiology, pp. 44–51.

White, D. (1981). Cell interactions and the control of development in myxobacteria populations. Intl. Rev. Cytol. 71:203–227.

White, D., Dworkin, M., Tipper, D. J. (1968). Peptidoglycan of *Myxococcus xanthus:* Structure and relation to morphogenesis. J. Bacteriol. 95:2186–2197.

White, D., Johnson, J. A., Stephens, K. (1980a). Effects of specific cations on aggregation and fruiting body morphology in the Myxobacterium *Stigmatella aurantiaca*. J. Bacteriol. 144:400–405.

White, D., Shropshire, W. Jr., Stephens, K . (1980b). Photocontrol of development by *Stigmatella aurantiaca*. J. Bacteriol. 142:1023–1024.

Wilson, E. O., (1975). Sociobiology, the new synthesis. Cambridge: Massachusetts. Belknap Press.

Wilson, V. L. Jones, P. A. (1983). Inhibition of DNA methylation by chemical carcinogens *in vitro*. Cell 32:239–246.

Wingard, M. Matsueda, G. Wolfe, R. S. (1972). Myxobacter AL-1 protease II: specific peptide bond cleavage on the amino side of lysine. J. Bacteriol. 112:940–949.

Wireman, J. W. (1979). Developmental induction of *Myxococcus xanthus* myxospores. J. Bacteriol. 140:147–153.

Wireman, J. W., Dworkin, M. (1975). Morphogenesis and developmental interactions in myxobacteria. Science 189:516–522.

Wireman, J. W., Dworkin, M. (1977). Developmentally induced autolysis during fruiting body formation by *Myxococcus xanthus*. J. Bacteriol. 129:796–802.

Witkin, S., Rosenberg, E. (1970). Induction of morphogenesis by methionine starvation in *Myxococcus xanthus:* Polyamine control. J. Bacteriol. 103:641–469.

Woese, C. R., Bleyman, M. A. (1972). Genetic code limit organisms—do they exist? J. Molec. Evolution 1:223–229.

Wolff, J. Cook, G. H. Goldhammer, A. R., Berkowitz, S. A. (1980). Calmodulin activates prokaryotic adenylate cyclase. Proc. Natl. Acad. Sci. USA 77:3841–3844.

Wolpert, J. S., Albersheim, P. (1976). Host-subunit interactions. I. The lectins of legumes interact with the O-antigen-containing lipopolysaccharides of their symbiont *Rhizobia*. Biochim. Biophys. Res. Commun. 70:729–737.

Wu, T. T. (1966). A model for three-point analysis of random general transduction. Genetics 54:405–410.

Wynne-Edwards, V. C., (1962). Animal dispersion in relation to social behaviour. Edinburg: Oliver & Boyd Ltr.

Yagi, Y., Kessler, R., Brown, B., Lopatin, D., Clewell, D., (1981). Pheromone-induced aggregation substance in *Streptococcus faecalis*. In: Molecular Biology, Pathogenicity, and Ecology of Bacterial Plasmids. edited by S. B. Levy, R. C. Clowes, E. L. Koenig New York: Plenum Press, pp. 191–205.

Yajko, D. M., Zusman, D. R. (1978). Changes in cAMP levels during development of *Myxococcus xanthus*. J. Bacteriol. 133:1540–1542.

Yazawa, M., Yagi, K., Toda, H., Kondo, K., Narita, K., Yamazaki, R., Sobui, K., Kakiuchi, S., Nagao, S., Nazawa, Y. (1981). The amino acid sequence of the Tetrahymena calmodulin which specifically interacts with quanylate cyclase. Biochem. Biophys. Res. Commun. 99:1051–1057.

Yee, T., Inouye, M. (1981). Reexamination of the genome size of myxobacteria, including the use of a new method for genome size analysis. J. Bacteriol. 145:1257–1265.

Yee, T., Inouye, M. (1982). Two-dimensional DNA electrophoresis applied to the study of DNA methylation and the analysis of genome size in *Myxococcus xanthus*. J. Mol. Biol. 154:181–196.

Yee, T., Inouye, M. (1984). Two-dimensional S1 nuclease heteroduplex mapping: Detection of rearrangements in bacterial genomes. Proc. Natl. Acad. Sci. USA, In press.

Zafriri, D., Rosenberg, E., Mirelman, D. (1981). Mode of action of *Myxococcus xanthus* antibiotic TA. Antimicrob. Agents Chemother. 19:349–351.

Zahavi, A., (1977). Reliability in communication systems and evolution of altruism. In: Evolutionary Ecology, edited by B. Stonehouse. C. Perrins, London: Macmillan Press, 253–259.

Zahavi, A., (1981). Some comments on sociobiology. Auk 98:412–415.

Zannis-Hadjopoulos, M., Persico, M., Martin, R. G. (1981). The remarkable instability of replication loops provides a general method for the isolation of origins of DNA replication. Cell 27:155–163.

Zaritsky, A. (1975). Rate stimulation of deoxyribonucleic acid synthesis after inhibition. J. Bacteriol. 122:841–846.

Zusman, D. R. (1980). Genetic approaches to the study of development in the myxobacteria. In: The Molecular Genetics of Development, edited by T. Leighton, W. F. Loomis. New York: Academic Press, pp. 41–78.

Zusman, D. R. (1982). "Frizzy" mutants: a new class of aggregation-defective developmental mutants of *Myxococcus xanthus*. J. Bacteriol. 150:1430–1437.

Zusman, D., Rosenberg, E. (1968). Deoxyribonucleic acid synthesis during microcyst germination in Myxococcus xanthus. J. Bacteriol. 96:981–986.

Zusman, D., Rosenberg, E. (1970). DNA cycle of *Myxococcus xanthus*. J. Mol. Biol. 49:609–619.

Zusman, D., Rosenberg, E. (1971). Division cycle of *Myxococcus xanthus:* II. Kinetics of stable and unstable ribonucleic acid synthesis. J. Bacteriol. 105:801–810.

Zusman, D., Gottlieb, P., Rosenberg, E., (1971). Division cycle of *Myxococcus xanthus*. III. Kinetics of cell growth and protein synthesis. J. Bacteriol. 105:811–819.

Zusman, D. R., Krotoski, D. M., Cumsky, M. (1978). Chromosome replication in *Myxococcus xanthus*. J. Bacteriol. 133:122–129.

Zyskind, J. W., Dean, L. T., Smith, D. W. (1977). Temporal sequence of events during the initiation process in *Escherichia coli* deoxyribonucleic acid replication: Roles of the *dnaA* and *dnaC* gene products and ribonucleic acid polymerase. J. Bacteriol. 129:1466–1475.

# Index

MIX
Papier aus verantwortungsvollen Quellen
Paper from responsible sources
FSC® C105338
www.fsc.org

If you have any concerns about our products,
you can contact us on
**ProductSafety@springernature.com**

In case Publisher is established outside the EU,
the EU authorized representative is:
**Springer Nature Customer Service Center GmbH**
**Europaplatz 3, 69115 Heidelberg, Germany**

Printed by Libri Plureos GmbH
in Hamburg, Germany